Big Data in Oncology: Impact, Challenges, and Risk Assessment

RIVER PUBLISHERS SERIES IN BIOMEDICAL ENGINEERING

Series Editors:

DINESH KANT KUMAR
RMIT University, Australia

The "River Publishers Series in Biomedical Engineering" is a series of comprehensive academic and professional books which focus on the engineering and mathematics in medicine and biology. The series presents innovative experimental science and technological development in the biomedical field as well as clinical application of new developments.

Books published in the series include research monographs, edited volumes, handbooks and textbooks. The books provide professionals, researchers, educators, and advanced students in the field with an invaluable insight into the latest research and developments.

Topics covered in the series include, but are by no means restricted to the following:

- Biomedical engineering
- Biomedical physics and applied biophysics
- Bio-informatics
- Bio-metrics
- Bio-signals
- Medical Imaging

For a list of other books in this series, visit www.riverpublishers.com

Big Data in Oncology: Impact, Challenges, and Risk Assessment

Editors

Neeraj Kumar Fuloria
AIMST University, Malaysia

Rishabha Malviya
Galgotias University, Greater Noida, India

Swati Verma
Galgotias University, Greater Noida, India

Balamurugan Balusamy
Galgotias University, Greater Noida, India

River Publishers

Routledge
Taylor & Francis Group
NEW YORK AND LONDON

Published 2023 by River Publishers
River Publishers
Alsbjergvej 10, 9260 Gistrup, Denmark
www.riverpublishers.com

Distributed exclusively by Routledge
605 Third Avenue, New York, NY 10017, USA
4 Park Square, Milton Park, Abingdon, Oxon OX14 4RN

Big Data in Oncology: Impact, Challenges, and Risk Assessment / by Neeraj Kumar Fuloria, Rishabha Malviya, Swati Verma, Balamurugan Balusamy.

Routledge is an imprint of the Taylor & Francis Group, an informa business

ISBN 978-87-7022-813-8 (hardback)
ISBN 978-87-7022-999-9 (paperback)
ISBN 978-10-0096-523-0 (online)
ISBN 978-10-0344-263-9 (master ebook)

OUR SINCERE THANKS TO

(Prof)Dr. Preeti Bajaj
Vice-Chancellor
Galgotias University
Without her encouragement
and support this task would not have been possible

Contents

Preface

In the biomedical sector, access to large-scale real-world data to promote fundamental and computational science in scientific research and growth is both a benefit and an obstacle. New transformative technologies, both laboratories based and computational, are making it possible to collect high-resolution clinical and biological data, extract meaningful insights, and then offer patients just this sort of personalized medical advice. And that means cancer patients can hope for earlier diagnosis and therefore better outcomes. Big data help to solve a variety of limitations, including the difficulties of monitoring intensive care patients, predicting the efficacy of drugs, and determining the spread of epidemics. Big data aid in the development and reshaping of cancer prevention strategies. Big data is also used to analyze existing preventive efforts and uncover new insights that can be used to enhance them. Big data has shown to be useful in oncology, and its potential in this area cannot be overstated. Therefore, any technology or system that is tailored towards improving the cancer healthcare system is most likely to enjoy favorable public perception. This book discusses emerging technology such as machine learning, artificial intelligence, deep learning, etc., used for cancer management.

This book covers basic to advanced-level topics related to applications of big data in cancer management. It can be offered as an elective course for graduate and postgraduate students as this is to greatly influence and impact young minds.

This book gives compiled information about targeted treatment in oncology, diagnostic imaging, medical imaging, and personalized therapy using big data analytics.

This book contains 13 chapters subdivided into various section that is written by profound researchers from many parts of the world. The book is profusely referenced and copiously illustrated. This should be noted that all chapters were deliberately reviewed and all were suitably revised once or two times. So, the content presented in this book is of the greatest value and meets the highest standard of publication.

This book should be of immense interest and usefulness to researchers and industrialists working in clinical research, disease management, pharmacist, formulation scientist working in R&D, remote healthcare management, health analysts, and researchers from the pharmaceutical industry.

Finally comes the best part thank you to everyone who helped to make this book possible. First and foremost, we express our heartfelt gratitude to the authors for their contribution, dedication, participation, and willingness to share their significant research experience in the form of written testimonials, which would not have been possible without them. lastly, we are feeling fortunate to express our gratitude to River Publishers for its unwavering support.

Editors:

Dr. Neeraj Kumar Fuloriais presently working as a Senior Associate Professor at the Faculty of Pharmacy, AIMST University, Malaysia; with an extensive experience of 20 years in academics, research & industry. He completed his B.Pharm in 1998 (Gulbarga University, India), M.Pharm in 2003 (Rajiv Gandhi University Health Sciences, India), MBA in 2004 (Madurai Kamaraj University, India), and Ph.D. in Pharmacy in 2010 (Gautam Buddha Technical University, India). So far, he supervised 6 Postgraduate and 24 undergraduate research scholars: and currently, he is supervising 6 Ph.D. scholars. He published 96 Research and review articles, 4 books, 3 MOOC, and 2 patents (Australia). For his research work, Dr. Fuloria received 8 national and international grants. He is also a member of various professional bodies like the Indian society of analytical Scientists, and the NMR society of India. Apart from it his work in academics and research Dr. Fuloria received various awards such as the Inspirational Scientist Award 2020 (at Trichy, India), the Young achiever award 2019 (at 10th Indo-Malaysian Conference, India), and the Appreciation Award 2017 (at Teachers Award Ceremony of Kedah State Malaysia, at AIMST University, Malaysia).

Dr. Rishabha Malviya completed B. Pharmacy from Uttar Pradesh Technical University and M. Pharmacy (Pharmaceutics) from Gautam Buddha Technical University, Lucknow Uttar Pradesh. His PhD (Pharmacy) work was in the area of Novel formulation development techniques. He has 12 years of research experience and presently working as Associate Professor in the Department of Pharmacy, School of Medical and Allied Sciences, Galgotias University since past 8 years. His area of interest includes formulation optimization, nanoformulation, targeted drug delivery, localized drug delivery and characterization of natural polymers as pharmaceutical excipients. He has authored more than 150 research/review papers for national/international journals of repute. He has 58 patents (19 grants, 38 published, 1 filed) and publications in reputed National and International journals with total of 191 cumulative impact factor. He has also received an Outstanding Reviewer award from Elsevier. He has authored/edited/editing 42 books (Wily, CRC Press/Taylor and Francis, Springer, River Publisher, IOP publishing and OMICS publication) and authored 26 book chapters. His name has included in word's top 2% scientist list for the year 2020 and 2021 by Elsevier BV and Stanford University. He is Reviewer/Editor/Editorial board member of more than 50 national and international journals of repute. He has invited as author for "Atlas of Science" and pharma magazine dealing with industry (B2B) "Ingredient south Asia Magazines".

Swati Verma has completed B.Pharm from KIET (AKTU), Ghaziabad and M.Pharm (pharmaceutical chemistry) from Banasthali Vidyapith, Tonk, Rajasthan. She has joined BBDNIIT as Assistant Professor and currently, she is working at Galgotias University, Greater Noida. Her area of interest is computer-aided drug design (CADD), peptide chemistry, analytical chemistry, medicinal chemistry, artificial intelligence, neurodegeneration, and gene therapy. She has attended and well organized more than 15 national and international seminars/conferences/workshops.

Prof. Balamurugan Balusamy has served up to the position of Associate Professor in his stint of 14 years of experience with VIT University, Vellore. He had completed his Bachelor, Master, and Ph.D. degrees from top premier institutions in India. His passion is teaching and adapting different design thinking principles while delivering his lectures. He has published 30+ books on various technologies and visited 15+ countries for his technical course. He has several top-notch conferences in his resume and has published over 150 quality journal, conference, and book chapters combined. He serves in the advisory committee for several startups and forums and does consultancy work for the industry on Industrial IoT. He has given over 175 talks at various events and symposiums. He is currently working as a professor at Galgotias University and teaches students and does research on Blockchain and IoT.

Acknowledgement

Having an idea and turning it into a book is as hard as it sounds. The experience is both internally challenging and rewarding. At the very outset, we fail to find adequate words, with a limited vocabulary to our command, to express our emotion to the almighty, whose eternal blessing, divine presence, and masterly guidelines help us to fulfill all our goals.

When emotions are profound, words sometimes are not sufficient to express our thanks and gratitude. We especially want to thank the individuals that helped make this happen. Without the experiences and support from my peers and team, this book would not exist.

No words can describe the immense contribution of our parents and friends, without whose support this work would have not been possible.

Last but not least, we would like to thank, our publisher for their support, innovative suggestions, and guidance in bringing out this edition.

List of Figures

List of Tables

List of Reviewers

S. No.	Name	Affiliation	Email ID
1	Dr. Vetriselvan Subramaniyan	Faculty of Medicine, Bioscience and Nursing, MAHSA University, Selangor, Malaysia	drvetriselvan@mahsa.edu.my
2	Dr. Neeraj Kumar Fuloria	Faculty of Pharmacy & Centre of Excellence for Biomaterials Engineering, AIMST University, Bedong, Kedah, Malaysia	nfuloria@gmail.com
3	Dr. Rajendra Awasthi	Amity Institute of Pharmacy, Amity University, Noida, India	awasthi02@gmail.com
4	Akansha Sharma	Department of Pharmacy, Monad University, Hapur, India	akankshasona012@gmail.com
5	Dr. Md. Aftab Alam	Department of Pharmacy, Galgotias University, Greater Noida.	aftab.alam@galgotiasuniversity.edu.in

List of Contributors

Ahamad, Javed, *Department of Pharmacognosy, Faculty of Pharmacy, Tishk International University, Iraq*

Ahuja, Alka, *Department of Pharmaceutics College of Pharmacy, National University of Science and Technology, Oman*

Alam, Md. Aftab, *Department of Pharmacy, School of Medical and Allied Science, Galgotias University, India*

Bairagee, Deepika, *Oriental College of Pharmacy and Research, Oriental University, Sanwer Road, India*

Chaubey, Ratnesh, *Department of Pharmacy, Dr MC Saxena College of Pharmacy, India*

Chauhan, Akash, *Department of Pharmacy, School of Medical and Allied Sciences, Galgotias University, India*

Dahiya, Rajiv, *Department of Pharmacy, Faculty of Medical Sciences, The University of the West Indies, Trinidad & Tobago*

Dahiya, Sunita, *Department of Pharmaceutical Sciences, School of Pharmacy, University of Pureto Rico, Medical Sciences Campus, USA*

Dua, Kamal, *Department of Pharmacy, Graduate School of Health, University of Technology Sydney, Australia*

Dubey, Ayush, *Department of Pharmacy, School of Medical and Allied Sciences, Galgotias University, India*

Dwivedi, Sumeet, *Oriental College of Pharmacy and Research, Oriental University, India*

Francis, Arul Prakash, *Centre for Molecular Medicine and Diagnostics (COMManD), Saveetha Dental College and Hospitals, Saveetha University, India*

Fuloria, Neeraj Kumar, *Faculty of Pharmacy, AIMST University, Malaysia*

Fuloria, Shivkanya, *Faculty of Pharmacy, AIMST University, Malaysia*

Gupta, Saurabh Kumar, *Department of Pharmaceutical Sciences and Technology, Madan Mohan Malviya University of Technology, India*

Jain, Neelam, *Oriental College of Pharmacy and Research, Oriental University, India*

Jain, Neetesh Kumar, *Oriental College of Pharmacy and Research, Oriental University, India*

Jain, Surendra, *School of Pharmacy, University of Mississippi, USA*

Khan, Neelam, *Oriental College of Pharmacy and Research, Oriental University, India*

Khan, Shah Alam, *Department of Pharmaceutical Chemistry College of Pharmacy, National University of Science and Technology, Oman*

Kishore, Narra, *Department of Pharmaceutics, Vignan Pharmacy College, India*

Kumar, Neeraj, *Department of Pharmacy, Dr MC Saxena College of Pharmacy, India*

Kurra, Pallavi, *Department of Pharmaceutics, Vignan Pharmacy College, India*

Malviya, Rishabha, *Department of Pharmacy, School of Medical and Allied Science, Galgotias University, India*

Meenakshi, Dhanalekshmi Unnikrishnan, *Department of Pharmacology and Biological Sciences, College of Pharmacy, National University of Science and Technology, Oman*

Mishra, Ayush Chandra, *Department of Pharmacy, Dr MC Saxena College of Pharmacy, India*

Mishra, Sudhanshu, *Department of Pharmacy, Birla Institute of Technology and Science, Pilani-Rajasthan, India*

Naim, Mohammad Javed, *Department of pharmaceutical Chemistry, Tishk International University, Iraq*

Nandakumar, Selvasudha, *Department of Biotechnology, Pondicherry University, India*

Ojha, Smriti, *Department of Pharmaceutical Science & Technology, Madan Mohan Malaviya University of Technology, India*

Pandey, Akanksha, *Department of Pharmacy, School of Medical and Allied Sciences, Galgotias University*

Sekar, Mahendran, *Department of Pharmaceutical Chemistry, Faculty of Pharmacy and Health Sciences, Royal College of Medicine Perak, University Kuala Lumpur, Malaysia*

Sharma, Urvashi, *Medi-caps University, India*

Shukla, Amrita, *Department of Pharmacy, Dr MC Saxena College of Pharmacy, India*

Singh, Arun Kumar, *Department of Pharmacy, Galgotias University, India*

Singh, Nitu, *Oriental College of Pharmacy and Research, Oriental University, India*

Srivastava, Shobhit Prakash, *Department of Pharmacy, Dr MC Saxena College of Pharmacy, India*

Subramaniyan, Vetriselvan, *Faculty of Medicine, Bioscience and Nursing, MAHSA University, India*

Sundram, Sonali, *Department of Pharmacy, School of Medical and Allied Science, Galgotias University, India*

Sweety, Pushpa, *Department of Biotechnology, Anna University, India*

Thakur, Arun, *Department of Pharmacy, School of Medical and Allied Sciences, Galgotias University, India*

Uthirapathy, Subasini, *Department of Pharmacognosy, Tishak International University, Iraq*

Verma, Swati, *Department of Pharmacy, School of Medical and Allied Science, Galgotias University, India*

Yadav, Shalini, *Department of Pharmaceutical Sciences and Technology, Madan Mohan Malviya University of Technology, India*

List of Abbreviations

ADE	Adverse drug events
ADR	Adverse drug reaction
AE	Autoencoder
AI	Artificial intelligence
ANN	Artificial neural networks
APOLLO	Adaptive patient-oriented longitudinal learning and optimization
ASA	Aspirin
ASCO	American society of clinical oncology
AUC	Area under the curve
AWS	Amazon web services
BAM	Binary alignment map
BC	Breast cancer
BD	Big data
BDI	Big data institute
BIC	Biomedical intelligence cloud
BNs	Bayesian networks
BRAF	Human gene that encodes a protein called B-Raf
CADD	Computer-aided drug design
CCCs	Comprehensive cancer centres
CCLE	Cancer cell line encyclopedia
CDSL	Cancer data science laboratory
CGHub	Cancer genomics hub
CGWB	Cancer genome work bench
CNN	Convolutional neural networks
CNV	Copy number variations
CoC	Commission on cancer
COD	Crystallography open database
COSMIC	Catalogue of somatic mutations in cancer
CpG	Cytosine phosphate guanine

CPRG	Cancer program resource gateway
CSD	Cambridge structural database
CSIR	Council of scientific and industrial research
CT	Computed tomography
DAC	Data access committee
DBN	Deep belief network
DDNN	Deep deconvolution neural network
DFS	Disease free survival
DHNA	Dutch head and neck audit
DICA	Dutch institute for clinical auditing
DICOM	Digital imaging and communication in medicine
DLCA	Dutch lung cancer audit
DLNN	Deep learning neural networks
DMTR	Dutch melanoma treatment registry
DNA	Deoxyribonucleic acid
DTI	Drug-target interactions
DTs	Decisional trees
DWAS	Data-wide association study
EB	Exabytes
EBI	European bioinformatics institute
EEG	Electro encephalogram
EGA	European genome archive
EHR	Electronic health records
EMR	Electronic medical record
EOSC	European open science
EPW	European program of work
ER	Early returns
ETL	Extract, transform, load
EU	European union
EU	Europe
FAIR	Findable, accessible, interoperable, and reusable
FDA	Food and drug administration
FHIR	Fast healthcare interoperability resources
GAN	Generative adversarial networks
GB	Gigabytes
GDAC	Genome data analysis center
GDPR	General data protection regulation
GDSC	Genomics of drug sensitivity in cancer

GEO	Gene expression omnibus
GEPIA	Gene expression profiling interactive analysis
GPCRs	G protein-coupled receptors
GPW	General program of work
GRNN	General regression neural network
GWAS	Genome-wide association study
HMF	Hartwig medical foundation
HNC	Head and neck cancer
HPV	Human papillomavirus
HRD	Health and retirement study
HTPS	Hypertext transfer protocol secure
HTS	High throughput screening
IARC	International agency for research on cancer
IBM	International business machines
ICGC	International cancer genome consortium
ID	Identity document
IDC	International data corporation
IoT	Internet of things
IP	Intellectual property
IT	Information technology
KEGG	Kyoto encyclopedia of genes and genomes
LHS	Learning health system
MASTER	Molecularly aided stratification for tumor eradication research
MDR	Multidrug resistance
MIMIC	Medical information mart for intensive care
ML	Machine learning
MMN	Mismatch negativity
MPM	Medical practise management
MRF	Magnetic resonance fingerprinting
MRI	Magnetic resonance imagining
NB	Network-based
NCDB	National cancer databases
NCI	National cancer institute
NCRP	National cancer registry program
NFC	Near field communication
NGS	Next-generation sequencing
NHGRI	National human genome research institute
NIH	National institutes of health

NLP	Natural language processing
NSCLC	Non-small cell lung cancer
OS	Overall survival
PACS	Picture archiving and communication system
PALGA	Pathologisch anatomisch landelijk gegevens archief
PB	Petabytes
PC	Personal computers
PCGP	Pediatric cancer genome project
PDB	Protein data bank
PET	Positron emission tomography
PHI	Protected health information
PHR	Personal health records
PM	Precision medicine
PNN	Probabilistic neural network
PPIs	Protein–protein interactions
PREMs	Patient related experience measurements
PROMs	Programmable read-only memory
PSI	Percent spliced in
QSAR	Quantitative structure-activity relationship
RBF	Radial basis function
RCT	randomized control trials
RFID	Radio frequency identification
RNA	Ribonucleic acid
RPPA	Reverse phase protein array
RWE	Real-world evidence
SEER	Surveillance, epidemiology, and end results
SNPs	Single nucleotide polymorphisms
SQL	Structured query language
SRS	Spontaneous reporting system
SVM	support vector machines
TARGET	Therapeutically applicable research to generate effective treatments
TB	Terabyte
TCGA	The cancer genome atlas
TCPA	The cancer proteome atlas
THPA	Human protein atlas
TL	Transfer learning
TPS	Treatment planning system

UCSC	University of california, santa cruz
UPST	US preventive services task
VAE	Variational autoencoder
VS	Virtual screening
WHO	World health organization
ZB	Zettabytes

1

Big Data Analysis – A New Revolution in Cancer Management

Saurabh Kumar Gupta[1], Sudhanshu Mishra[2], Shalini Yadav[3], Smriti Ojha[4*], Shivkanya Fuloria[5], and Sonali Sundram[6]

[1]Rameshwaram Institute of Technology and Management, India
[2]Department of Pharmacy, Birla Institute of Technology and Science, Pilani-Rajasthan, India
[3]Dr MC Saxena College of Pharmacy, India
[4]Department of Pharmaceutical Science & Technology, Madan Mohan Malaviya University of Technology, India
[5]Faculty of Pharmacy, AIMST University, Malaysia
[6]Department of Pharmacy, School of Medical and Allied Sciences, Galgotias University, Greater Noida, India
*Corresponding Author: Department of Pharmaceutical Science & Technology, Madan Mohan Malaviya University of Technology, Gorakhpur, Uttar Pradesh, India, EMail: smritiojha23@gmail.com.

Abstract

"Big data" refers to huge datasets that can be utilized as problem solver practice in various carcinomas. It has provoked people's imagination over the past 20 years because of the enormous potential it has. Big data is created, stored, and analyzed by people and private organizations in order to improve the services they provide. Hospital records, patient medical records, medical examination findings, and internet of things devices are just a few examples of big data sources in the healthcare system. To get useful information from this data, it must be properly managed and analyzed. Otherwise, finding a solution through studying vast data is akin to looking for a needle in a haystack. Each phase of processing large data comes with its own set of obstacles that can only be overcome by adopting high-end computing solutions for big data analysis. As a result, healthcare providers must be fully equipped with sufficient infrastructure to create and access big data in order to deliver possible solutions for maintaining public health. Modern healthcare organizations may be able to restructure medical therapies and personalized medicine through a good correlation of biomedical and healthcare data. Big data analytics, especially in data-rich sectors like cancer, have enormous promise to enhance risk classification.

Keywords: Big Data Analytics, Healthcare, Personalized Medicine, Internet of Things, Cancer.

1.1 Introduction

Think of a world in which each piece of information about an individual or organization, each agreement signed, as well as every aspect which can be documented is discard as soon as it is used. As a result, organizations would lose their capacity to retrieve critical data and knowledge, conduct in-depth analyses, and generate new opportunities and advantages. Data on users' personal details, items available, transactions made, and employees hired big data, as the name implies, refer to huge volumes of are now essential for the functioning of day-to-day life [1]. Data is the backbone of any business's success. The key to a greater organization and new advances has been informed. We can better organize ourselves to give the best results if we have more data. That is why the collection of data is such a crucial aspect of any business. We may also utilize this information to forecast present trends in specific metrics as well as future events [2]. As we become more conscious of this, we have begun to produce and collect more data on nearly everything, using technical advancements in this domain. Today, we are bombarded with massive amounts of data from every aspect of our lives, including social activities, science, work, health, and so on.

In some ways, the current situation resembles a data flood. Technological advancements have aided us in generating ever-increasing amounts of data, to the point where it is now unmanageable with currently available technologies. As a result, the term "big data" was coined to characterize data that is enormous and unmanageable [3]. We need to find innovative techniques to organize this data and derive relevant information in order to meet our current and future social needs. Healthcare is one such specific social necessity, represented in Figure 1.1. Healthcare firms, like every other industry, generate a significant amount of data, which brings both opportunities and challenges.

Big data, as the name implies, refers to huge volumes of data that are unmanageable with standard software or web-based platforms. It outperforms

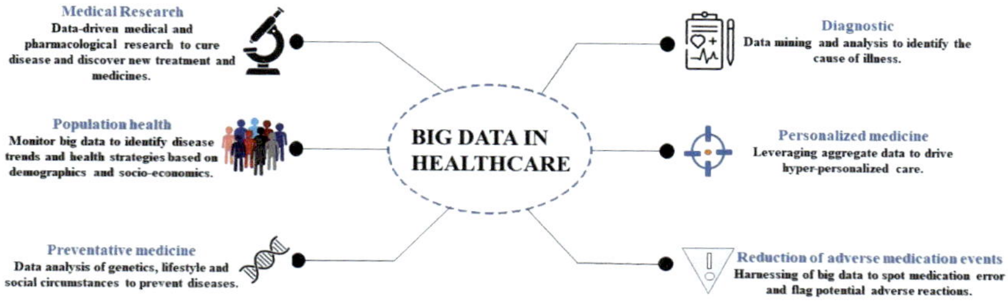

Figure 1.1 Big data analysis in different fields of healthcare.

standard storage, processing, and analytical capabilities. Even though there are other definitions for big data, Douglas Laney's definition is the most prominent and widely recognized. Laney noticed that (big) data was expanding in three dimensions: volume, velocity, and diversity [4].

Data analytics solutions use algorithms generated from historical patient data to anticipate future health outcomes for individuals or communities [5]. As the volume of EHR, radiological, genetic, and other data in cancer has grown, there have been a number of use cases that are potentially able to generalize. Various observables approximating thousands to millions are frequently created and stored in oncology databases, despite the fact that average patient cohort sizes are very small. In rarer forms of cancer, like neck and head cancer, the disparity between the depth of information for each patient and the population size is much more pronounced. However, if there are instances to learn over, current methodological developments in machine learning and neural networks are especially useful. Object recognition in photos, for example, works well, but optimizing such systems requires thousands to millions of samples. As a result, if we want to use this to improve tailored therapies, we will need more data depth in terms of sample size [6]. This necessitates strong data management, data sharing, data standardization, data provenance, and data exchange procedures in cancer, especially in the area of neck and head oncology. This paper examines the existing research on use cases and obstacles in using data analytics to increase oncology risk classification.

1.2 Big Data for Personalized Medicine

Personalized medicine relies entirely on big data to convert existing information and data into meaningful intelligence that can be utilized to enhance treatment outcomes [7]. The (EHR) electronic health record of the patient is a resource for a huge volume of information that consists of data on socio-demographics, health problems, heredity, and therapies; however, the ability of the human mind to assess this data is restricted without appropriate decision. The goal in healthcare is to create a system that is predictive, preventative, and participatory by providing a constantly learning infrastructure with real-time knowledge production. Models of computers are needed to assist doctors in the organization of data, recognize sequences, evaluate outcomes, and determine action thresholds in order to attain this goal [8]. There are already examples of big data analytics being used to generate new information, improve healthcare, and streamline surveillance of public health. For example, the EHR has been effectively accessed for enhanced pharmacovigilance and post-marketing drug surveillance.

The amount of data available to the biomedical community grows at an exponential rate, especially when new technologies, such as sequencing and imaging, generate terabytes of data. The majority of data comes from computed automatic analysis of the data, like radiomics and digital image assessment, rather than direct patient-related records present in ordinary healthcare practices. Because of their complicated architecture and variability, neck and head malignancies provide a unique set of therapeutic and diagnostic hurdles. Radiomics can overcome these obstacles [9]. Radiomics is a less costly and non-invasive method of extracting and mining a vast number of medical imaging properties. In precision oncology and cancer care, radiomics enables predictive and efficient machine learning techniques for categorization (or individualization), that is observing dissimilarities in (predicted/expected) survival of (clusters of) patient populations and therapeutic result estimation to assist in choosing the best treatment option for patients with head and neck cancer [10]. This could make it easier for health and radiation oncologists to decrease the rate of therapy and doses of irradiation in certain population groups.

Big data for health is projected to have a significant impact on pharmacogenetics and stratified healthcare. When given the same chemotherapeutics, patients with the same cancer subtype frequently respond differently. For instance, a polymorphic gene linked to Tamoxifen response [17] in non-small cell lung cancer, BRAF mutations (Y472C) which have been found to be linked with Dasatinib response [18], and several gene signatures that have recently been linked to rectal cancer, differ in their response to chemoradiotherapy [19]. The observed variability of medication response is thought to be due to genomic instability. The complicated interplay of genomes, resistance, chemotherapeutic sensitivity, and toxicity has been the focus of recent investigations [20–22]. The Pan-Cancer project was launched by the Cancer Genome Atlas research network to explore a range of tumor types and molecular irregularities in different cancers, as well as to assist researchers in discovering novel anomalies. Similarly, organizations like the Genomics of Drug Sensitivity in Cancer and the Cancer Cell Line Encyclopedia are compiling massive genomic information to look into the connection between genetic biomarkers and drug sensitivity in a number of tumor cell lines.

1.3 Big Data Analytics in Oncology: Challenges

1.3.1 Data acquisition

The development of solid risk classification models based on the experience of a sufficient number of patients is likely to improve the total expenditure of therapy as well as its results, but there is a key limitation: a dearth of high-quality data. The most significant challenge for risk-based frameworks, especially in cancer therapy, is the limited availability of key components of patient information. In claims-based datasets, ER sessions and hospital stays are usually not collected and compiled into a huge databank of information that is accessible in a time-wise manner. Accurate death dates, for example, frequently necessitate searching many data sources, making mortality prediction challenging. Furthermore, almost no data is obtained for patients at home, where they spend the majority of their time.

Unique approaches to gathering data of real-time on cancer patients regularly could help to avoid unnecessary hospital stays by uncovering trends that occur in patients at the onset of sickness. EHR-based data could be collected in real-time using real-world data sources [11, 12]. Algorithmic prediction on the basis of real-world data may be more appropriate than clinical trial-based algorithms, which often ignore relevant sectors of the population. Real-world information, such as those from ASCO Cancer Lin-Q and Flatiron Health, might be useful; however, they have significant drawbacks due to manual curation and the heterogeneity of the user interlinked with the pathogenic and medicinal record [13].

1.3.2 Ensuring representativeness and mitigating bias

If previous data can be used to train analytic models, predictions may aggravate existing inequalities in clinical care. Algorithms relying on subjective clinical data or healthcare access may systematically discriminate against particular patient groups [14]. Consider the case of a cancer-specific forecasting algorithm based on tumor genetic data. Patients from certain ethnic minorities may be underrepresented in the datasets used to train the algorithm. This could lead to erroneous tumor genetic variation classification in minority groups [15]. In contrast, a lack of data from under-represented populations may make it impossible to uncover predictive genetic variations, putting the prediction model's generalizability in jeopardy. It will be critical to guarantee that all interested people in a training set are represented, as well as to implement audit methods when the data tool is developed, to ensure that represented populations do not experience systematic variation in predicted output [16].

1.3.3 Sources of big data

Big data sources abound and come in a variety of shapes and sizes. Patient-derived data is the most evident in oncology. This data is frequently stored in computerized patient files for therapeutic purposes and includes a variety of data points/subjects. These files contain clinical information about patients, tumors, treatment, and outcomes, as well as demographic information like race, age, and gender, confirming various disease symptoms, family genetics, comorbidity information, radiology-based results (like MRI, CT scan, ultrasonography, positron emission tomography), and various crystal tissues-based analysis including diagnostic features of histopathology, immune-chemistry, DNA and RNA genetic material sequence and related experiments, hemolytic analysis, and whole-genome BAM standard alignment format files [17]. However, data obtained from *in vitro* trials is valuable and can be valuable information. The analytical computation of these data is also the main source of big data. Secondary and computational analytic data, such as radiology-based and digital analyses, as well as gene expression and mutational analyses, are among the processed sources of data [18]. Machine learning is a growing source of processed data, which typically consists of big computer data files including structured data. However, data from *in vitro* trials is valuable and can be a valuable source. The various sources of big data in cancer care hospitals are summarized below.

1.3.3.1 Media

Social media and interactive platforms such as Facebook, Google, YouTube, Twitter, and Instagram, as well as generic media such as videos, photographs, audios, and podcasts, provide qualitative and quantitative insights on all aspects of user involvement.

1.3.3.2 Cloud

Cloud computing refers to the processing and computing of any set of data. *In vitro* data, on the other hand, is valuable and can be a valuable source.

Computational analysis of big data is a second source. Radiological and digital imaging-based analysis, along with gene expression and mutation assessments, are examples of processed data. Machine learning is becoming a more common source of processed data – it constructs and processes the data to make it easier for consumers; cloud computing suppliers frequently use a "software programming as a service provider" paradigm. A solace is usually offered to receive customized commands and arguments, but every task can also be completed through the website's user interface. Systems for database management, identity management, machine learning capabilities and cloud-based virtual machines and containers, and other items are typically included in this bundle [19].

Large, network-based systems, in turn, frequently generate big data. It can take the shape of a standard or non-standard document. If the dataset is received in a nonstandard format, the Cloud Computing provider's artificial intelligence, along with machine learning, may be utilized to standardize the data, which typically consists of big computer data files including structured data [20].

1.3.3.3 Web

A large quantity of data is publicly indulged and freely accessible on the common public webpage. On the World Wide Web (or "internet"), both individuals and corporations have access to this data. Additionally, digital services like Wikipedia provide free and easy access to informative data for anybody [21]. Because of the breadth of the internet, it is widely used, which is especially beneficial for start-ups and small enterprises because they do not need to establish their big databases and data storehouses before exploiting big data.

1.3.3.4 Internet of Things (IoT)

Machine-created information or data derived using the Internet of Things is an important source of big data. This information is typically obtained through sensors attached to electronic gadgets. The source capacities are determined by the sensing potential of different sensors to give precise data. The Internet of Things is gaining traction, and it now encompasses big data created not only by computers and smartphones but by any devices that can generate and analyze data. Data can now be gathered from health-related equipment, armored procedures, video games, meters, cameras, and other household goods thanks to the Internet of Things [22].

1.4 Databases

1.4.1 National Cancer Databases

The National Cancer Database (NCDB) is a statewide carcinoma analytical database for more than 1500 commissioned and accredited cancer programs in Puerto Rico and the United States. This is a collaborative venture of the American College of Surgeons and Commission on Cancer (CoC) of the American Cancer Society. Around 70% of all newly diagnosed cancer cases in the US are registered with records and reports and are included with NCDB at the institutional level. Since its inception in 1989, the NCDB has grown to contain roughly 34 million data points from registries of cancer hospitals around the country. All forms of cancer data are collected and examined. The data is utilized to investigate cancer care trends, to establish regional and state standards for the hospital as participants, and serves as the foundation for research. The NCDB keeps with security a variety of web-based data apps that were created to make NCDB data more accessible [23]. These informative data apps are intended for use by CoC-accredited cancer programs to review and compare cancer care, from the treatment provided to patients diagnosed and/or treated at their facility, to cancer care provided in the state, regional, and national cancer facilities. Nationally standardized data and information, along with defining coding and decoding, as stated in the CoC' s Facility Cancer Reports Data Standards, and nationally standardized data transmission specific layout in coordination with the North American Association of Central Cancer Registries, are used to collect and submit data elements to the NCDB from the CoC-accredited cancer program My Registries. Patient characteristics, cancer stage, tumor histological characteristics, first-course treatment type, and outcomes data are all included [24].

1.4.2 Cancer genomics databases

Cancerous phenotypes demand changes at several stages of information with proper flow patterns, from genomics to proteomics. The cancer phenotype emerges as a result of molecular changes at various information processing levels. Interrogating tumor cells using several types and stages of information with the correct flowchart depicted by proper omics – such as genome sequencing, epigenomic studies and related data, transcriptomics, and protein structures with 3D shapes, is essential to recognize the fundamental action that drives the accession of cancer identification [25]. The benefit of integrating multi-tasking omics data is always associated with a cost: an additional layer of complexity stemming from the congenitally various types of omics database sets, which can make it difficult to integrate the data in a biologically meaningful way. If a variety of carcinoma-specific online omics-data sources can be connected efficiently and systematically, new human biology-based discoveries for cancer research could be facilitated. In this chapter, we present a detailed sketch of the online single- and multi-omics resources dedicated to cancer. Among the online omics-data, resources we catalogue are the Cancer Genome Atlas (TCGA) and its related data channels and equipment for multi-omics investigative process and visualization, the International Cancer Genome Consortium (ICGC), The Pathology Atlas, Catalogue of Somatic Mutations in Cancer (COSMIC), Gene Expression Omnibus (GEO), and Proteomics identification. By analyzing the strong and weak points of the various websites, we intend to highlight present biological and technical barriers, as well as possible solutions. Using integrated molecular signatures of cancer, we examine the many ways for integrating multi-omics dimensions for cancer patient classification and biomarker prediction. Finally, we explore multi-omics-driven systems biology methodologies for fulfilling the promise of precision medication as the future of cancer research. We expect that this prospective study will enable researchers and practitioners throughout the world to use online tools to examine and integrate the current omics datasets, which might lead to new biological discoveries and improved cancer research [26]. Role of big data in health care system is shown in the Figure 1.2.

1.4.3 Commercial–private databases

Using the integration of various molecular databases of cancer, we examine the many ways for integrating multi-omics proportions for cancer patient classification and predicting various stages with the aid of biomarkers. Finally, we propose multi-omics-driven systems-biology methodologies for realizing precision on-promise medicines as the future of cancer research. It is believed that these data sources will motivate researchers and clinicians around the world to use online resources to examine and integrate the current omics information, which could lead to new biological discoveries and improved cancer research [27].

1.4.4The practical applicability of various sources of big data in cancer care

- To answer crucial practice and policy concerns, administrative and various clinical trial data can be combined.

Figure 1.2 Role of big data in healthcare.

- Each data resource has potential and limits that must be addressed during its creation, analysis, and interpretation.
- Analytical techniques should provide the proper quality, reproducibility, and integrity of the available data, which are varied both within and between sources.
- Data aggregation and analysis can benefit from the use of similar data items across diverse data source categories.
- Machine learning optimization may be able to alleviate some of the present limits of large data analysis.

1.5 Inferring Clinical Information with Big Data

The large volume of administrative data and more advanced analytics tools have prompted efforts for the excavation of such data to conquer some of the shortcomings, most notably the lack of rich clinical data. The availability of machine-learning technologies to forecast tumor stages from administrative data is one example. As previously reported, researchers utilized data of SEER-Medicare to test several approaches, including machine-learning algorithms, for predicting stages of metastasis in lung cancer patients undergoing treatment. In development and validation groups, the presentation from clinical, algorithm-based, random forest-based, and analytical data was examined. Random forests were the best in performance, as identified while developing the cohort, but their accomplishment was not reproducible in validation analysis. In both the developing and validating cohorts, the performance of logistic regression was consistent. In both the development and validation groups, the 14-variable logistic regression indicated greater accuracy (77% vs. 71%), as well as improved sensitivity for early-stage disease (77% vs. 73%), when compared to the clinical algorithm. As a result, while machine-learning techniques like random forest have the capability to enhance categories of stages of lung cancer, they are susceptible to overfitting [28].

Overfitting can be avoided by using ensembles (meta-algorithms that include various approaches of machine learning), external validation, and cross-validation. However, these observations are encouraging, and the decreased accuracy in validation that is compared to the development group implies that machine-learning algorithms should be used with caution in research and service delivery.

Administrative data is becoming more widely available, and it has a lot of potential for answering concerns regarding cancer treatment delivery, especially when combined with other clinical data sources. Such information can reveal patterns in healthcare utilization, quality, and spending, as well as cancer-care inequities among different patient populations. [29] Apart from this potential, the data has significant limitations, which must be acknowledged and evaluated with methodologic rigor to guarantee that meaningful findings are reached.

1.5.1 Utilization of administrative records for cancer care

Administrative data is increasingly being utilized to better understand how healthcare is delivered to a population of patients, as well as the effectiveness of care and the impact of certain policies or population-level initiatives. Billing claims that are given to a health insurer or payer to remunerate (e.g., private insurance claims or Medicare) are used to measure healthcare expenditure, and form the most prevalent source of administrative records (e.g., Medicare Advantage data). Office appointments, emergency department visits, hospitalizations, diagnostic and therapeutic tests, hospital equipment, hospital care and home care, and, in certain cases, prescription medication are all commonly found in these records. All service has one or more codes for diagnosis connected with it, which can offer information about health issues. These data help researchers, clinicians, and policymakers better understand how care is delivered and can reveal information on healthcare utilization, quality, and costs [30].

1.5.2 Big data and cancer research

Data is everywhere in healthcare, the sources ranging from various healthcare centers, clinical laboratories, and research organizations to monitoring systems; data management is everywhere. In fact, there are many different sorts of data in biological and cancer research (as shown in Figure 1.3), including data gathered and organized through various phases of clinical trials, data obtained by genome sequencing, and data generated by computational drug modeling. The integration of big data and analytics in cancer research is very beneficial. Cancer screening programs generate a large amount of image sourcing and data from the clinical lab, which necessitates deep analysis and repeat testing in order to extract actual value. Researchers can develop good and promising medications, comprehend their properties *in vivo*, and create novel active pharmaceutical ingredients by repeated testing and data analysis [31].

Carcinoma is amongst the most complicated health issues affecting our civilization, and the condition is considered to be worse as the world's population grows. Carcinoma is acknowledged as one of the biggest reasons behind early mortality in the EU, according to the State of Health in the EU reports. It also has an economic impact because it reduces labor market participation and productivity [32]. Cancer researchers now have strong new approaches to extract value from a variety of data sources thanks to advances in big data analytics. For one patient, these various sources include a massive quantity of data. The collection of different types of omics data would provide a distinct molecular profile for every patient, immensely aiding oncologists in their efforts to develop personalized therapy approaches. Cancer is a molecularly serious disease with significant intra- and intertumoral heterogeneity among cancer types and even patients.

1.5.3 FAIR data

The principles of FAIR (Findable, Accessible, Interoperable, and Reusable) must be followed to ensure that data can be utilized in further studies. The principles of FAIR data were initially articulated in 2014. Since then, the G20 (2016) and G7 (2017) have recognized and accepted the principles, and the EU has placed FAIR data at the heart of the EOSC (European Open Science Cloud). Findability (F) necessitates a permanent identifier, accessibility (A)

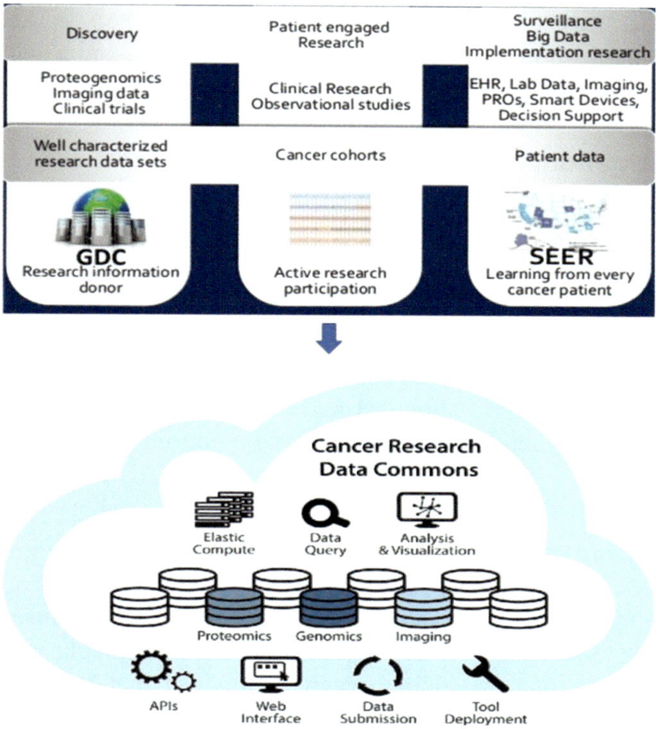

Figure 1.3 Integration of big data for cancer research.

necessitates clearly specified access restrictions (data privacy requirements are incorporated in the specification), and interoperability (I) necessitates the usage of a community-recognized ontology to describe the data. Finally, the provenance of the data, as well as the accuracy and completeness of the metadata, are crucial to its reuse (R) [33].

It is no secret that big data is seen as an infallible way to decipher the complexities of cancer. Companies are investigating data visualization/analysis technologies in order to discover novel methods to cure cancer. *In silico* Medicine, 10X Genomics, Benevolent AI, and NuMedii have all reached their initial watershed in capturing, assembling, putting together, and scrutinizing data from cancer cells. Actually, 10X Genomics has gone beyond to come up with whole-exome sequencing, genome sequencing and single-cell transcriptome analysis services, all of which clearly identify cancer-prone sequences of the gene in the mRNA, DNA, and polypeptide chains. Some researchers are testing cancer medications in a variety of cellular contexts using a wide set of data, unique selection methods, and high-definition data filtering algorithms.

1.5.4 Cancer genome sequencing of humans

Each cell of the human body has an equal set of chromosomes in number and contains approximately a similar amount of genetic material (DNA). However, cells of tumors have different chromosomal content and growth abnormalities that, when visualized *in silico*, can be exploited to extract data down to the level of DNA. The DNA sequence data research could aid cell biologists, bioinformaticians, molecular biologists, and nanobiotechnologists in developing more effective methods for removing chromosomal aberrations, potentially leading to new therapeutic options [34].

1.5.5 High-throughput sequence analysis

All of us are living in an era when personalized therapy is gaining more popularity in oncology therapy and is the most promising area for advancement. The emphasis on computational biology has shifted because of this push toward customized medicine. Professor Olivier Elementa, a Cornell University Computational Medicine expert, points out that because tumor cells are constantly altering, progressing, and redesigning to normal settings, faster technologies of the new generation are more important than ever in determining a tumor's genetic build-up. And it does not stop there; mutation sequences must be recognized, subdivided, and refined about the gene in question [35].

1.5.6 Sequencing genomes of other organisms

The genome of Escherichia coli, a single-celled organism, was the first to be sequenced. Plant genomes, such as Arabidopsis thaliana, and non-vertebrate genomes, such as those of worms, reptiles, and rodents, were then sequenced. Genome sequencing became more important as these organisms became more complicated in terms of knowing a cell's ability to govern, sensitize, or stave off malignant cells. It gave relevant mechanisms about cancer-triggering theories. Many researchers are currently analyzing real-time data by using hamster/chicken/ mouse cell lines of ovarian cancer in order to supplement or enhance carcinoma identification techniques, as well as to increase the accuracy of currently available screening assays [36].

1.5.7 Transcriptome analysis for better cancer monitoring

Throughout, carcinoma-related investigations during the last ten years have big databases of screening and experimental data have been created. Maintaining marker genes, which are currently the primary instruments for observation of oncogenes, research for biocompatibility, and drug discovery have gained critical relevance. Furthermore, several businesses extrapolate cancer genomic data in order to study protein synthesis and transcriptomes. This will be important for locating misplaced gene segments and their products, which aids in the tracking of conservative and non-conservative mutations [37].

1.5.8 Diagnostic modeling and machine learning algorithms

Healthcare systems hold massive volumes of data, which may now be accessed more easily thanks to contemporary technology. High-throughput machine learning-based algorithms that can scan and interact with data and assure the utmost correctness in the amalgamation of massive databases are being used by biotech/interdisciplinary researchers to conduct extensive assessments of those databases. To acquire a better image of tumors, researchers are using machine learning algorithms and high-tech data modeling systems to combine tumor-related data from various resources. New approaches to evaluating cancer cell development and the death of healthy cells are being enabled by genetic data visualization tools, which are making waves in cancer detection [38]. Variation in the information system of genetic and clinical research, which are well-organized, open-source research data management platforms for visualizing HTPS data, have been efficiently put to use.

1.5.9 Disease prognosis

The cancer-related data visualization software applications that have been developed, such as Cancer Lin-Q, allow physicians, surgeons, and researchers to view good quality patient healthcare data. This is significant because it aids in the understanding of prior cancer cases, disease development, and treatment regimens. Doctors are employing screening tools to refer to, to protect the medical data of patients, and are using them to promote clinical research, offer individualized protocols of treatment, and more effectively decide the option for management of cancer. Hospitals with a high rate of cancer admission have begun to use Tumor ID cards, which allow their data to be centralized for clinical investigation [39].

1.6 Correlation of Clinical Data with Cancer Relapses

Many healthcare professionals are using data analytics technologies to figure out why some patients have recurrent cancers and others do not. Healthcare professionals are analyzing a vast number of case reports to check the health risks of the patient in a much broader context than previously. However, though medical patient data have been around for long, their approachability and utilization have just recently increased. This means that laboratory data is evaluated after being compared to other internationally reported instances, rather than being subjected to traditional identification techniques. As a result, data is the most important necessity for tailored treatment. Our treatments are changing at a faster rate than cancer. As a result, if you want to beat, control, or avoid it, you need to focus your efforts on superior target detection. Big data is a critical tool that cancer researchers should use to increase the quality of their research and get the best outcomes [40].

1.7 Conclusion

We investigated the novel topic of big data in this study, which has recently sparked a lot of attention due to its supposed unrivalled prospects and benefits. In the digital age we now live in, massive amounts of high-speed data are produced on a daily basis, containing intrinsic features and patterns of hidden knowledge that should be extracted and utilized. Big data analytics may be used to leverage organizational change and enhance decision-making by applying advanced analytic approaches to large datasets and discovering meaningful insights and relevant information. By providing a constantly learning platform with real-time knowledge generation, the aim in healthcare is to develop a system that is predictive, preventive, and interactive. Big data analysis holds a lot of promise for improving healthcare and transforming people's health. However, realizing this promise will necessitate addressing issues such as data privacy, security, ownership, and governance.

Conflict of Interest

The authors have no conflict of interest.

Funding

The authors have not received any funding for the work.

References

[1] Shaikh, A. R., Butte, A. J., Schully, S. D., Dalton, W. S., Khoury, M. J., & Hesse, B. W. (2014). Collaborative biomedicine in the age of big data: the case of cancer. *Journal of medical Internet research. 16*(4), e2496. doi: 10.2196/jmir.2496.

[2] Yan, X., Xiangxin, L., Xiangquan, Z., Jiankang, C., Weibo, J. (2020) Application of blockchain technology in food safety control:current trends and future prospects. *Critical Reviews in Food Science and Nutrition.* 12, 1-20. DOI: 10.1080/10408398.2020.1858752.

[3] Andreu-Perez, J., Poon, C. C., Merrifield, R. D., Wong, S. T., & Yang, G. Z. (2015). Big data for health. *IEEE journal of biomedical and health informatics. 19*(4), 1193-1208. doi: 10.1109/JBHI.2015.2450362.

[4] Laney, D. (2001). 3D data management: Controlling data volume, velocity and variety. *META group research note. 6*(70), 1.

[5] Elfiky, A. A., Pany, M. J., Parikh, R. B., & Obermeyer, Z. (2018). Development and application of a machine learning approach to assess short-term mortality risk among patients with cancer starting chemotherapy. *JAMA network open. 1*(3), e180926-e180926. doi: 10.1001/jamanetworkopen.2018.0926.

[6] Zhang, C., Bijlard, J., Staiger, C., Scollen, S., van Enckevort, D., Hoogstrate, Y., & Abeln, S. (2017). Systematically linking tranSMART, Galaxy and EGA for reusing human translational research data. *F1000Research.* 16, *6, 1488.* doi: 10.12688/f1000research.12168.1. eCollection 2017.

[7] Govers, T. M., Rovers, M. M., Brands, M. T., Dronkers, E. A., de Jong, R. J. B., Merkx, M. A., & Grutters, J. P. (2018). Integrated prediction and decision models are valuable in informing personalized decision making. *Journal of Clinical Epidemiology. 104*, 73-83. doi: 10.1016/j.jclinepi.2018.08.016.

[8] Hood, L., & Flores, M. (2012). A personal view on systems medicine and the emergence of proactive P4 medicine: predictive, preventive, personalized and participatory. *New biotechnology. 29*(6), 613-624. doi: 10.1016/j.nbt.2012.03.004.

[9] Wong, A. J., Kanwar, A., Mohamed, A. S., & Fuller, C. D. (2016). Radiomics in head and neck cancer: from exploration to application. *Translational cancer research. 5*(4). 371.doi: 10.21037/tcr.2016.07.18.

[10] Willems, S. M., Abeln, S., Feenstra, K. A., De Bree, R., van der Poel, E. F., de Jong, R. J. B., & van den Brekel, M. W. (2019). The potential use of big data in oncology. *Oral Oncology. 98*, 8-12. doi.org/10.1016/j.oraloncology.2019.09.003

[11] Karanis, Y. B., Canta, F. B., Mitrofan, L., Mistry, H., & Anger, C. (2016). 'Research' vs 'real world'patients: the representativeness of clinical trial participants. *Annals of Oncology. 27*, vi542. doi: 10.1016/s0376-8716(98)00161-6.

[12] O'Connor, J. M., Fessele, K. L., Steiner, J., Seidl-Rathkopf, K., Carson, K. R., Nussbaum, N. C., & Gross, C. P. (2018). Speed of adoption of immune checkpoint inhibitors of programmed cell death 1 protein and comparison of patient ages in clinical practice vs pivotal clinical trials. *JAMA oncology. 4*(4), e180798-e180798. doi: 10.1001/jamaoncol.2018.0798.

[13] Parikh, R. B., Gdowski, A., Patt, D. A., Hertler, A., Mermel, C., & Bekelman, J. E. (2019). Using big data and predictive analytics to determine patient risk in oncology. *American Society of Clinical Oncology Educational Book, 39*, e53-e58. DOI: 10.1200/EDBK_238891.

[14] Mullainathan, S., & Obermeyer, Z. (2017). Does machine learning automate moral hazard and error?. *American Economic Review, 107*(5), 476-80. doi: 10.1257/aer.p20171084

[15] Manrai, A. K., Funke, B. H., Rehm, H. L., Olesen, M. S., Maron, B. A., Szolovits, P., & Kohane, I. S. (2016). Genetic misdiagnoses and the potential for health disparities. *New England Journal of Medicine. 375*(7), 655-665. doi: 10.1056/NEJMsa1507092.

[16] Parikh, R. B., Obermeyer, Z., & Navathe, A. S. (2019). Regulation of predictive analytics in medicine. *Science. 363*(6429), 810-812. doi: 10.1126/science.aaw0029.

[17] Goetz, M. P., Knox, S. K., Suman, V. J., Rae, J. M., Safgren, S. L., Ames, M. M., & Ingle, J. N. (2007). The impact of cytochrome P450 2D6 metabolism in women receiving adjuvant tamoxifen. *Breast cancer research and treatment, 101*(1), 113-121. doi: 10.1007/s10549-006-9428-0.

[18] Sen, B., Peng, S., Tang, X., Erickson, H. S., Galindo, H., Mazumdar, T., & Johnson, F. M. (2012). Kinase-impaired BRAF

mutations in lung cancer confer sensitivity to dasatinib. *Science translational medicine*. *4*(136), 136ra70-136ra70. doi: 10.1126/scitranslmed.3003513.

[19] Baik, C. S., Myall, N. J., Wakelee, H. A. (2017). Targeting *BRAF*-Mutant Non-Small Cell Lung Cancer: From Molecular Profiling to Rationally Designed Therapy. *Oncologist.* 22(7), 786-796. doi: 10.1634/theoncologist.2016-0458.

[20] Weinstein, J. N., Collisson, E. A., Mills, G. B., Shaw, K. R., Ozenberger, B. A., Ellrott, K., & Stuart, J. M. (2013). The cancer genome atlas pan-cancer analysis project. *Nature genetics*. *45*(10), 1113-1120. doi.org/10.1038/ng.2764.

[21] Barretina, J., Caponigro, G., Stransky, N., Venkatesan, K., Margolin, A. A., Kim, S., & Garraway, L. A. (2012). The Cancer Cell Line Encyclopedia enables predictive modelling of anticancer drug sensitivity. *Nature*. *483*(7391), 603-607. doi.org/10.1038/nature11003.

[22] Garnett, M. J., Edelman, E. J., Heidorn, S. J., Greenman, C. D., Dastur, A., Lau, K. W., & Benes, C. H. (2012). Systematic identification of genomic markers of drug sensitivity in cancer cells. *Nature*, *483*(7391), 570-575. doi: 10.1038/nature11005.

[23] Das, T., Andrieux, G., Ahmed, M., Chakraborty, S. (2020). Integration of Online Omics-Data Resources for Cancer Research. *Front Genet*. 23, 11, 578345. doi: 10.3389/fgene.2020.578345. PMID: 33193699; PMCID: PMC7645150.

[24] Chakraborty, S., Hosen, M. I., Ahmed, M., Shekhar, H. U. (2018). Onco-multi-OMICS approach: a new frontier in cancer research. *Biomed Res. Int.* 2018:9836256. Doi:10.1155/2018/9836256

[25] Chandrashekar, D. S., Bashel, B., Balasubramanya, S. A. H., Creighton, C. J., Ponce-Rodriguez, I., Chakravarthi, B V. S. K. (2017). UALCAN: a portal for facilitating tumor subgroup gene expression and survival analyses. *Neoplasia*. 19, 649–658. DOI:10.1016/j.neo.2017.05.002.

[26] Furey, T. S. (2012). ChIP-seq and beyond: new and improved methodologies to detect and characterize protein-DNA interactions. *Nat. Rev. Genet.* 13, 840–852. doi:10.1038/nrg3306

[27] Dewdney, S. B., Lachance J.(2016). Electronic records, registries, and the development of "Big Data": Crowd-sourcing quality toward knowledge. *Front Oncol*. 6, 268. doi: 10.3389/fonc.2016.00268

[28] Hilal, T., Sonbol, M. B., Prasad V. (2019). Analysis of control arm quality in randomized clinical trials leading to anticancer drug approval

by the US Food and Drug Administration. *JAMA Oncol.* 5(6),887-892. doi: 10.1001/jamaoncol.2019.0167.

[29] Nazha, B., Mishra, M., Pentz, R., Owonikoko, T. K. (2019). Enrollment of Racial Minorities in Clinical Trials: Old Problem Assumes New Urgency in the Age of Immunotherapy. *Am Soc Clin Oncol Educ Book.* 39, 3-10.doi: 10.1200/EDBK_100021.

[30] Nikhil, I., Khushalani, Thach-Giao, T., John F. T. (2021). Current Challenges in Access to Melanoma Care: A Multidisciplinary Perspective. *American Society of Clinical Oncology Educational Book.* 41, e295-e303. doi.org/10.1200/EDBK_320301.

[31] Hutchins, L. F., Unger, J. M., Crowley, J. J., Coltman, Jr. C. A., Albain, K. S. (1999). Underrepresentation of patients 65 years of age or older in cancer-treatment trials. *N Engl J Med.* 341, 2061-2067. DOI: 10.1056/NEJM199912303412706

[32] Elting, L. S., Cooksley, C., Bekele, B. N. (2006). Generalizability of cancer clinical trial results: Prognostic differences between participants and nonparticipants. *Cancer.* 106, 2452-2458, 2006. doi: 10.1002/cncr.21907.

[33] Surveillance, Epidemiology, and End Results (SEER) Program. 2019. https://seer.cancer.gov/ztml(Accessed 1 May 2019).

[34] National Program of Cancer Registries (NPCR). 2019. https://www.cdc.gov/cancer/npcr/about.htm.

[35] Menck, H. R., Garfinkel, L., Dodd, G. D. (1991). Preliminary report of the National Cancer Data Base. *CA Cancer J Clin.* 41(1), 7-18. doi: 10.3322/canjclin.41.1.7.

[36] Fleshner, N. E., Herr, H. W., Stewart, A. K., Murphy, G. P., Mettlin, C., Menck, H. R. (1996) The National Cancer Data Base report on bladder carcinoma. The American College of Surgeons Commission on Cancer and the American Cancer Society.*Cancer.* 1, 78(7), 1505-13. doi: 10.1002/(sici)1097-0142(19961001)78:7<1505::aid-cncr19>3.0.co;2-3.

[37] National Cancer Database (NCDB). 2019. https://www.facs.org/quality programs/cancer/ncdb(Accessed1May2019).

[38] Siegel, R. L., Miller, K. D., Jemal, A. (2015). Cancer statistics. CA Cancer J Clin. 65, 5–29. doi: 10.3322/caac.21254.

[39] Sledge, Jr. G. W., Miller, R. S., Hauser, R. (2013). CancerLinQ and the future of cancer care. *Am Soc Clin Oncol Educ Book.* 430-434. doi: 10.14694/EdBook_AM.2013.33.430.

[40] Arshad, Z., Smith, J., Roberts, M. Open access could transform drug discovery: a case study of JQ1. *Expert Opin Drug Discov*. 2016;11:321–332. doi: 10.1517/17460441.2016.1144587.

Author Biographies

Saurabh Kumar Gupta

I am Saurabh Kumar Gupta. I received the M. Pharm degree in Pharmaceutical Chemistry form NSCBIP W. B., India in 2011. I am currently pursuing a Ph.D. Degree with AKTU, Lucknow, India. I have more than 10 years of experience in teaching and research, working with Rameswaram Institute of Technology and Management, Lucknow, Uttar Pradesh, India as Assistant Professor. My research interests include CADD as well as synthetic work, especially on Aldose reductase inhibitors.

Sudhanshu Mishra

I am Sudhanshu Mishra. I completed my M. Pharm (Pharmaceutics) from Rajiv Gandhi Proudyogiki Vishwavidyalaya and currently work as a teaching faculty in the Department of Pharmaceutical Science & Technology, Madan Mohan Malaviya University of Technology, Gorakhpur. I have worked on the herbal topical formulation for the treatment of arthritis during my M. Pharm research. Meanwhile, I am working on various literature work like writing review articles for different novel approaches and technologies targeting chronic diseases. I have been participating in various academic activities like international seminars, conferences, workshops, and oral presentations. This book chapter is one of the important contributions to my interest in technology and the future research area.

Shalini Yadav

Dr. M. C. Saxena College of Pharmacy, Lucknow, is where I received my bachelor's degree. I gained interest in writing papers during my graduation and have also presented oral and poster presentations and attended a variety of online and offline conferences, webinars, and other events. I have always prioritized learning new technology and developing innovative approaches to combat chronic conditions. One of the significant influences to my interest in technology has been this book chapter.

Dr. Smriti Ojha

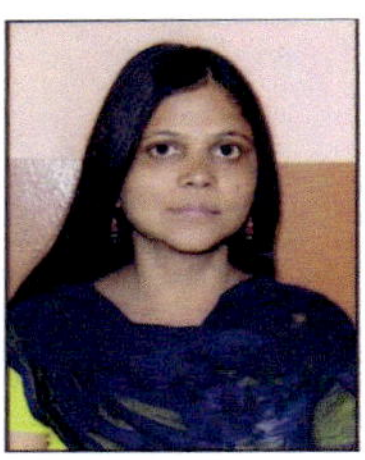

I earned my PhD degree from Dr. A. P. J. Abdul Kalam Technical University, Lucknow, Uttar Pradesh, India and am now working with Madan Mohan University of Technology, Gorakhpur, Uttar Pradesh, India. I have been associated with teaching and research in the field of pharmaceutical sciences for the last 15 years. This book chapter is one of my important contributions towards research and technology.

2

Implications of Big Data in the Targeted Therapy of Cancer

Arun Kumar Singh[1], Rishabha Malviya[1*], and Subasini Uthirapathy[2]

[1]Department of Pharmacy, School of Medical and Allied Sciences, Galgotias University, Greater Noida, India
[2]Faculty of Pharmacy, Tishak International University, Iraq
*Corresponding Author:
Department of Pharmacy, School of Medical and Allied Sciences, Galgotias University, Greater Noida, India, EMail: rishabhamalviya19@gmail.com

Abstract

Big data analysis is becoming more common in biomedical research as a result of advances in information technology. But what exactly is big data, where does it come from, and how can we use it? There are several sources of big data, particularly in cancer, that are discussed in this chapter. In addition, it emphasizes the need for integrating data from diverse sources, including clinical, pathological, and quality-of-life data sources. In the Netherlands, several efforts have been made to connect such databases on a national basis. As a final note, the paper discusses the need to establish the essential infrastructures for using large-scale datasets to better understand head and neck cancer.

Keywords: Big Data, Targeted Therapy, Cancer, Machine Learning.

2.1 Introduction

Datasets that were too large to be analyzed by normal software were dubbed "big data" in the 1990s. High-volume, high-velocity, and diverse information assets necessitated specialist technology and analytic approaches to make effective use of data in 2016 [1]. Quality, validity, and utility, in addition to quantity, velocity, and diversity, are some of the more nuanced aspects of big data, according to some [2, 3]. However, big data's promise for society has been limited by difficulties such as security and privacy concerns and opaque international standards that have made it difficult for big data to be exploited to its full potential during the last many years. This detailed assessment of the present status of research emphasizes big data, the shift in a research paradigm, and the desire for rapid replies. There has been an ever-increasing worldwide volume of data since the 1980s, as the amount of data collected has doubled every 40 months. "Big data" began in 2002, and since then, the quantity of information in the form of statistics, as well as sounds and pictures, has increased massively. Terabytes of data are generated each year by smartphones, computers, wearable technologies, the Internet of Things (IoT), electronic health records (EHRs), insurance websites, and mobile health. Other less visible data sources include clickstream information, device data processing, geolocation information, and video and audio inputs [4]. For example, it is impossible to know how much data is being created. By calculating data and analytics for years, Facebook has been able to improve its product offerings for advertising objectives, whereas Google Pictures used up 13.7 petabytes of storage [5, 6] in its first year. Recent years have witnessed considerable growth in data gathering, storage, processing, and distribution.

Additional requirements include real-time access to the data so that it may be examined and put to use quickly. Software, technology, and human resources are all necessary for managing large amounts of data. Big data may be used to study a wide range of topics, from retail commerce to criminal records to weather trends to disease outbreaks [7]. The UN Statistical Commission established the UN Global Working Group on big data in 2014 to take advantage of the revolutionary potential of big data. Data exchange and economic advantage were envisioned for a worldwide statistical community using advanced analytics linked into the UN global platform [8].

2.1.1 Big data in medicine

According to the report, science and technology are predicted to consume more than one zettabyte (zetta = 10^{21}) of big data every day by 2025. This includes everything from astrophysics to social media to genomics. Big data includes a wide range of physiological, genomic, and electronic health data (Figure 2.1).

Regional, national, and worldwide biobanks are all examples of biological repositories known as Biobanks. They comprise many different institutions,

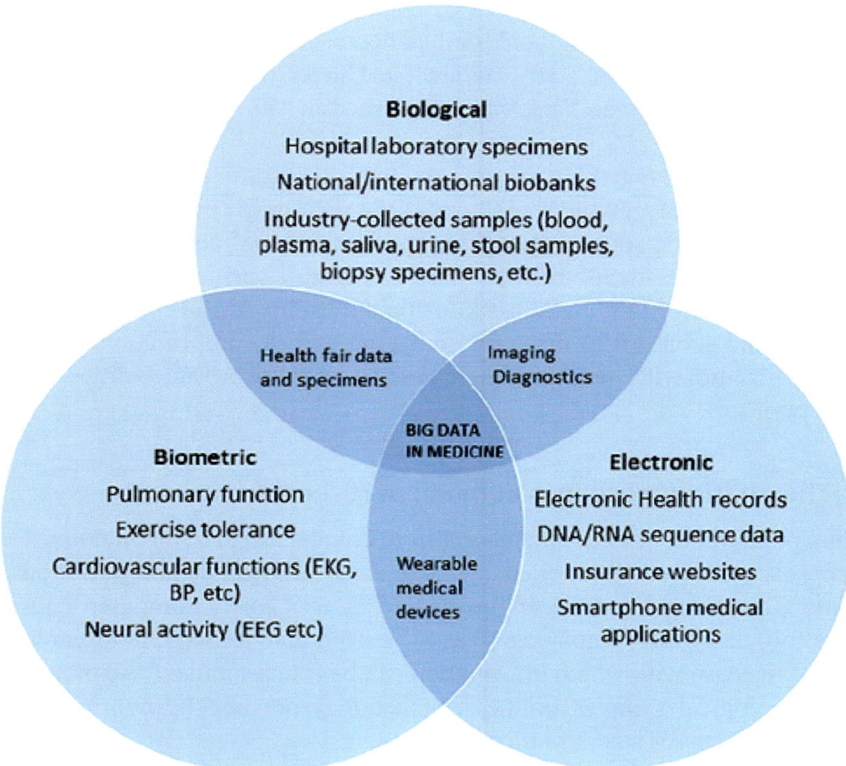

Figure 2.1 Big data in medicine.

including the National Cancer Institute and the China Kadoorie Biobank [9]. Health fairs have non-profit organizations that may give free heart rate, urine, and blood testing for those who need them. Ancestry may be established by saliva testing in commercial biobanks [10].

The processing and preservation of biological materials are required before the digitization of data. Preserving a biospecimen has probably affected you if you have ever touched one. This year's Biospecimen Research Network Symposia will be conducted in conjunction with the annual Biospecimen Research Network Symposia, which was founded by the National Cancer Institute in 2005 [11]. The first quality standard for biobanks was issued in 2009 and has been approved by some biobanks throughout the globe with international backing. As clinical and computational computer technology has progressed, biobanking has become an essential part of the biological sciences field of study. When the University of California San Francisco's AIDS specimen bank was established in 1987, the biologic was 30 years old [12].

All biobanks have one thing in common: they all require a lot of resources to keep track of, analyze, and make use of the data they collect. There are commercial biobanks, such as those run by multinational corporations that gather biological specimens from people for ancestry verification purposes [13]. The payment for DNA analysis kit is paid by subject who collect it i.e., Investigator and submit it to firms for analyzing and storing. In compliance with the legislation, the data can thereafter be sold to private entities for research purposes.

2.1.2 The shifting paradigm in medical research

An aging population has necessitated a shift in clinical research methodology. This approach has been propelled by large-scale biological data collected (biobanks), created, processed, and preserved by less costly computer technologies (big data). Many non-scientists have been able to assist in research because of the ease with which information can be found online. Researchers may buy biological samples online, whether they are healthy controls or relevant specimens for a certain medical problem.

In the past, drug discovery may have been spurred on by chance [15]. After World War II, the therapeutic research technique grew lengthy and costly. To begin, researchers conducted a thorough examination of prospective therapies, which was then followed by a series of tests, including *in vitro* and *in vivo* comparisons to existing standards of care, safety testing, and testing for efficacy. Additionally, new pharmaceuticals were required to undergo FDA clearance, along with randomized control trials (RCTs), and subsequent post-release research before they could enter the general public. The firm could go bankrupt due to uncommon but major side effects once the medicine was put to the market, leaving patients who depended on it without access to appropriate treatment options. Patients with uncommon disorders found it especially tough since repeat research would need a significant commitment of money and effort on the side of the business. Those with short life spans were unable to benefit from effective medicines because of the lengthy procedure. In light of this, the FDA has pushed to accelerate the release of novel therapies, such as HIV meds, during epidemics like the AIDS pandemic [16, 17].

2.2 Changes in Research

2.2.1 Changes in study design

Currently, a more focused and systematic approach is being employed to identify the underlying reason and serve as a springboard for further treatment [18]. The ability to precisely locate mutations has improved since the Human Genome Project's conclusion. Over 3000 genome-wide association studies have been used to study roughly 1800 illnesses in [19] large-scale sweeps of the human genome [20]. Using microarrays and the results of GWAS and QTL studies, we were able to identify potential genes of interest [21]. Large biobanks with patient and control data are analyzed to see whether allelic variations may be linked to illness. Patients with the condition may have a greater frequency of a mutant allele, which may be used as a therapeutic target.

An effort at the treatment that specifically targets a genetic change may be made if an aberrant growth driver mutation is discovered in a tumor [22]. The molecular profile of tumors may be varied, making it difficult to distinguish between driver alterations and bystander or passenger mutations when numerous mutations are present [23, 24]. Precision medicine is based on pharmacogenomics, which is now being used in cancer and is being applied to other disciplines of medicine. It is possible to identify novel biomarkers from large data utilizing molecular pathological epidemiology (MPE) [23, 24] (Table 2.1). It is possible to identify medications that target a specific mutation in an individual's genome, which may lead to safe and effective therapy options.

Biological specimens may be gathered in vast numbers and stored, handled, and studied using big data technologies. For example, machine learning algorithms might create additional data that differs from the original data at the time of analysis. Big data may be turned into knowledge by AI [25, 26] (Table 2.2). With the use of an AI-based computation pathology model in breast cancer specimens, Beck *et al.* revealed earlier unidentified traits to be predictive of unfavorable outcomes [25]. For RLHC models to be useful, they must be compared against verified datasets, which can only be found via the use of artificial intelligence (AI) [29]. Knowledge-driven healthcare may then be put into reality with the help of decision support systems (DSS), which are software programs that can be used in the field. Knowledge-based AI and data-driven AI are two types of AI. AI that relies on human knowledge to answer questions in a specific area is known as knowledge-based AI. Human activity generates a lot of data that may be used to forecast the future. Big data and cheap processing make data-driven AI an attractive business option [30, 31].

Combining artificial intelligence (AI) and data-driven systems (DSS) can improve healthcare delivery. Using AI and DSS, for example, allowed treatment adjustments to be implemented more promptly and without

Table 2.1 New study concepts and experiments based on big data.

Input data	Population	Possibleprediction/conclusion
In molecular pathology, the PIK3CA mutation is used for identification	Colon cancer sufferers	In need of aspirin treatment
Biomarkers are detected by collecting DNA and RNA from the patient as well as any prescription or non-prescription medication, as well as any vitamins or herbs taken by the subject [24]	Alzheimer's disease sufferers' close friends and family.	To identify persons who are at risk for acquiring Alzheimer's disease in its earliest stages
A computational histopathology concept of breast cancer was investigated using AI to discover 6642 quantifiable morphological traits [25]	Patients with breast cancer	Accurately predicted unfavorable outcomes; in addition, stromal morphologic structure, a previously unknown negative prognostic factor, was discovered
A total of 163 social media platforms contributed 99,693 records relating to suicides. Over the course of two years, 2.35 billion posts were analyzed. Other factors, such as overall well-being, were also considered [26]		In their study, researchers found that academic pressure was a major factor in the suicide risk for Korean adolescents
Linagliptin and glimepiride initiators were matched using propensity score (PS) matching to balance more than 120 confounding variables [27]	A combination of Medicare and two private insurance companies' records were utilized to locate people with type 2 diabetes who are at high risk of cardiovascular disease	Comparing linagliptin and glimepiride, the researchers found that neither drug had significantly worse cardiovascular outcomes than the other
The Lung-MAP trial protocol is a phase II/III umbrella study [28]. Genomic screening is used to assign participants to sub-studies	Small-cell lung cancer recurrence or metastasis patients	To select the best treatment for either matched or non-matched therapy, and to do it in the most efficient manner possible

Table 2.2 Big data technology, including examples of how it is being implemented.

	Operational	Analytical
Advantages	Captures and stores data in real time.	Allows complicated analysis of data to be performed quickly to produce results.
System format	NoSQLa has minimal response latency and is good for handling several requests for the same data at the same time [42].	High throughput (measured in outcomes per unit of time) is an important consideration in system design
Data forms	This may not be a standard table-to-row connection. Faster big-volume calculations in the cloud are possible since it is cheaper and faster than traditional relational databases and can leverage the cloud.	Examples include: MPPb – specialized analytic data systems that can collect and analyze large datasets over many nodes, allowing tasks to be performed simultaneously while minimizing computation time and expense [43]. In addition to SQLc and secure analytic platforms such as Hadoop, MapReduce provides another approach for analyzing data [44].
Computer network capability	It may be used in a variety of groupings.	It works with a wide range of clusters.

increased adverse events, than for patients waiting for their next caregiver appointment, in a small trial of 12 people with type 1 diabetes [32].

2.2.2 New study design

The RCT design has to be tweaked in light of recent advancements in illness diagnosis, management, and treatment. Over the recent decade, new clinical trial procedures, platforms, basket/bucket designs, and umbrella designs have emerged [33]. When enrolling patients, the existence of a particular genetic mutation is taken into consideration, rather than their histology or cell type, in a clinical study using a basket design. The data of hundreds of patients must be filtered to find the correct genetic alteration that will allow a small number of patients to participate in a sub-trial. For initial-stage and single-arm investigations, sub-trials are often used. If one or two phases fail, the study can be terminated early on. As part of the research strategy, tumor pathogenic mechanisms are determined and linked to therapeutic theories. A larger confirmatory investigation, such as a screening test, would be required for a responding sub-study. In the US and Europe, "rare malignancies" constitute the fourth most common cancer type despite their low individual incidence [34]. Certain cancers have a lower survival rate than more common malignancies, making them harder to detect and cure. These patients may benefit from a therapeutic study that focuses on tumor genetic instability rather than organ histology. A characteristic driver mutation has been used as a test subject for medications rather than a sickness specific to an organ. Larotrectinib was well tolerated by patients with Tropomyosin Receptors Kinase Fusion-positive malignancies; Larotrectinib is a research medication that has demonstrated considerable sustained anticancer therapy in humans with this molecular characterization of the targets [35, 36]. Instead of a specific disease, the FDA approved this innovative treatment for malignancies that had one unique mutation in their DNA.

Also possible in basket studies is an off-label use of a pharmaceutical in individuals with a hereditary condition for whom the treatment was initially approved, or a repurposed drug [37].

2.2.3 Umbrella design

As a result of this approach, researchers can test some treatments for a particular illness, such as lung cancer, at the same time (Ferrarotto and colleagues; [28] Table 2.2).

2.2.4 Platform trials

The combining of resources is made possible by big data. Patients who have biomarker data may be able to take part in clinical studies [38]. There is one control arm in platform trials that may be compared to several experimental arms and may not need randomization; hence they can be called large screening procedures. An RCT can nevertheless be more cost-effective and time-efficient if it uses artificial intelligence (AI) to compare diverse datasets. The drug testing process could be expedited and the RCT phase could begin sooner as a result.

2.2.5 Adverse drug events (ADE)

For the ADE, reporting is an ongoing process. It is possible that data mining utilizing artificial intelligence (AI) could enhance the precision and accuracy of literature searches for the treatment of Alzheimer's disease (ADE). There may be ADE interactions between medications in big data, which is continually updated to reflect any shifts in these interactions [39].

2.2.6 Real-world evidence (RWE)

Since the introduction of electronic health records (EHRs), the amount of real-world evidence (RWE) has grown. RWE's digital form might benefit greatly from big data. The National Comprehensive Cancer Network (NCCN) has released a clinical practice recommendation based on RWE results. As an additional benefit, the American Society of Clinical Oncology recommends RWE in combination with randomized controlled research [40]. Reduced drug R&D costs may be achieved in large part by more rapid treatment assessments in clinical settings. Because of this legislation, which was put into effect on December 13, 2016, the FDA has developed a framework to examine whether RWE might be used to support approval of a new indication or post-approval research requirements [41]. In the pharmaceutical sector, data from electronic health records (EHRs) is being used as a foundation for prescription approval. This might include observational research studies of high-enough quality to enable the approval of new pharmacological indications using natural language and artificial intelligence. The influence of comorbidity on therapeutic efficacy and subgroups within certain disease entities may also be identified using AI technologies [42]. Researchers could be able to predict future illness risk based on RWE data [43] based on variables such as age, race, family history, and genetics. For novel therapies or comparisons of medications, RWE and RCT commercialization may speed up the FDA clearance process. RWE was used in the Cardiovascular Results Study of Linagliptin vs Glimepiride in Type 2 Diabetes to evaluate the cardiovascular outcomes of different treatments (CAROLINA). Additional data about Patron *et al.* [27] may be found in Table 2.1.

2.3 Big data: technology and security concern

Due to the decreasing cost of computing power, large amounts of big data can now be stored and processed [44]. Operational or analytic are two ways to categorize examples of big data technologies (Table 2.2). For each system, there are distinct benefits as well as data formats and networking capabilities that must be considered (Figure 2.2).

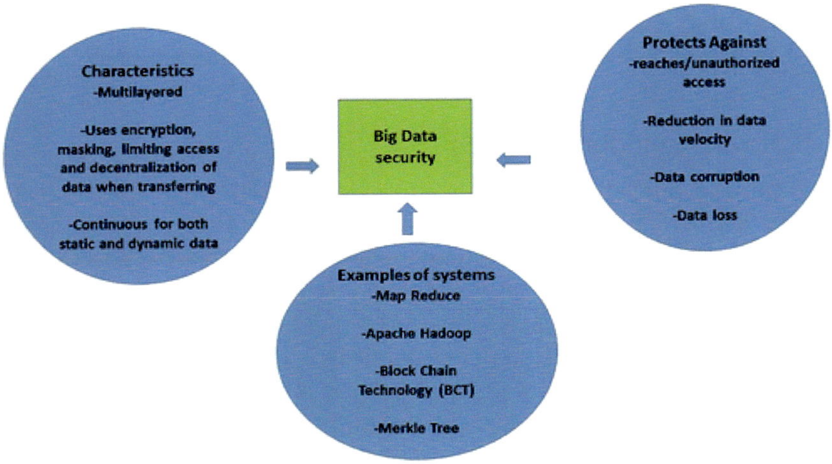

Figure 2.2 Big data security.

Data collection, transport, analysis, storage, and processing are all areas where big data security controls and technologies should be implemented. Enormous parallel processing systems, as well as the security required to safeguard massive volumes of dynamic data, are all part of this. Data might be stolen, lost, or corrupted as a result of human mistakes, poor technology, or criminal intent. When it comes to health-related data, the risk of data breach and the resulting penalties and lawsuits increase as a result. Each access point must have processes in place to avoid data loss and corruption. For example, while collecting data, threats must be interrupted. Precautions include encrypting data at both the inlet and outlet points as well as allowing only partial transfers and analysis, securing cloud storage, blocking access using firewalls, and more [45]. As an example, a blockchain network is a protection mechanism that can authenticate users and trace access to data as well as restrict large data retrieval because of its decentralized nature [46]. As far as standardizing big data security is concerned, there is still more to be done. With a large amount of personal health information, data security was found to be the most important concern [47].

2.3.1 Utility of big data

Biomedical research has yet to grasp the full promise of big data 48–50]. There is little doubt that in the near future, big data will have a favorable influence on clinical diagnosis, healthcare delivery, and biological research [51, 52]. Some instances of current-day programs will be discussed [53, 54].

2.3.2 Daily diagnostics

Big data can be used in clinical practice right now. Patients' histological follow-up can be accessed in near-real-time by Dutch pathologists, for example. For more than 40 years already, the PALGA foundation has been in charge of the Netherlands' whole computerized histology database (www.palga.nl). An enormous biomedical database in the Netherlands, PALGA contains the records of more than 12 million patients and has more than 72 million entries. In the local hospital's information system and the national PALGA database, each histology report that has been validated by a Dutch pathologist and is stored for future use. Participants of PALGA will be capable of following the development of each person in real time, allowing them to make informed decisions about treatment. As an example, pathological documentation of previously significant pathological findings (such as resection boundaries and positive lymph nodes) may have been omitted in situations where pathology was performed elsewhere. Using this information, researchers can look for associations between seemingly unrelated diseases with a low overall prevalence that initially appears to be unconnected [55].

Patients' electronic medical records provide enormous volumes of medical data that can be used in predictive modeling. One of the first predictive models for HNC patients receiving treatment in developed countries can be found online at www.oncologiq.nl [56]. Automatic model updates are now possible because of advancements in statistical prognostication methods [57]. This data could be used to construct clinical choice tools for improved patient counseling and quasi-patient outcome assessments.

2.3.3 Quality of care measurements

To date, the benefits of linking database data with information on patient outcomes, features, and therapy have gone unappreciated [58]. Recent French research emphasized the landscape of molecular analysis for targeted therapy, as well as the information that follows this, for non-small-cell lung cancer (NSCLC) therapy [59, 60]. You will know right away whether the test and therapy are a good match. Failing laboratories may be motivated by the desire to provide the best service possible by changing their procedures and operations. The PALGA database and the national cancer registry in the

Netherlands have been used to highlight the wide range of clinical treatments for head and neck cancer in the Netherlands. Even while laboratories and institutions may fear reputational damage or naming-and-shaming if such data is made public, it is necessary to do so in order to improve treatment [61, 62]. According to previous experience, most hospitals are eager to participate in mirror feedback as long as it is presented discreetly to the public and solely given to individuals. A method for regular automated feedback on pathology and treatment-related information was therefore created by the Dutch Institute for Clinical Auditing (NICA). There may be faults in the underlying chain that may be uncovered by looking at recurrent data from similar hospitals.

2.3.4 Biomedical research

Big data is likely to have the greatest impact on research. "Data-wide association studies" (DWAS) are taking over from "genome-wide association studies" (GWAS) as the main age of "genome-wide association studies." Imaging and molecular analysis, as well as combinations with other data, provide an unmatched opportunity for data scientists and bioinformaticians alike. Biomedical research is in desperate need of big data. Among the existing medical constraints is a paucity of information on the biology of sickness. It will be possible to include more accurate models of cancer behavior which patients would benefit much from, using certain therapies by combining enormous amounts of big data from several sources, such as DNA, RNA, protein, and metabolomics. For example, these multi-omics datasets will help us better understand the molecular pathways that control HNSCC's growth pattern, metastatic spread, and response to targeted treatment, among other things.

2.3.5 Personalized medicine

Big data is essential to personalized medicine if we are to transform our present knowledge and data into insights that can be put to use to enhance treatment outcomes [62]. Advances in sequencing and imaging generate terabytes of data every year, and this data is now accessible to the biomedical community in ever-increasing quantities. When it comes to quantitative data, technological frontiers and digital image analysis – instead of direct, patient-related records – account for the vast bulk of data. The diagnosis of head and neck malignancies poses a distinct set of challenges because of their complex architecture and unpredictability. Nanomedicine may be able to solve these challenges [63]. Radiomics may be used to mine and extract medical imaging data with little danger and expenditure. These imaging qualities may be utilized as a surrogate for the overall tumor phenotypic traits in radiomics [64]. Stratification (or personalization) based on variations in (expected/predicted) mortality across patient groups and predictions of treatment outcomes is now possible because of radiation omics' prognostic and accurate machine learning algorithms. Some individuals may benefit from a reduction in systemic therapy and irradiation dosages [65].

2.3.6 FAIR data

As part of the FAIR (Findable; Accessible; Interoperable; Reusable) principles, data must be reusable in secondary research. Since their first publication in 2014, the FAIR data principles have been widely accepted [66]. According to the G20, the principles have been acknowledged and supported since 2016, as well as by G7 countries in 2017. The European Open Science Cloud, in the meanwhile, has placed FAIR data at the Center of its infrastructure (EOSC). To ensure the FAIRness of an information resource's metadata, it is crucial to have a permanent identity, well-defined access rules (including privacy requirements), and appropriate licensing. Data reusability (R) is also dependent on the correctness and fullness of the meta-data that accompany the data.

2.4 Archiving and Sharing of Biomolecular Patient Data in Repositories and Databases

Due to privacy and security concerns, many patients and hospitals are hesitant to share their personal information with researchers (such as those raised by the General Data Protection Regulation; GDPR). This is particularly true in the realm of biomedical research. In the case of genome sequences, for example, the data may be stored in a form that allows it to be utilized for future research while yet protecting the privacy of patients. One of the most pressing concerns is this: (www.phgfoundation.org). Big data in other fields, such as computer science, need more privacy protections than openness. To maintain the raw sequencing data, the European Genomics Archive was formed. Every study that makes use of EGA data relies on a data access committee (DAC), which is run by the research team that collects the data and has the authority to determine whether or not to provide access to the EGA data to other researchers [67]. Secondly, researchers have a dilemma if they want to study processed biomaterials data without even being able to track back particular markers. To overcome this problem, other repositories may utilize fine-grained access control to summarize the data without revealing specific markers [68]. At this point, it is just a matter of time until attempts are made to trace the exact computations performed on the data and integrate them in a manner that respects individual privacy.

Head and neck cancer and life sciences may benefit greatly from the use of large amounts of data. It might also have a significant influence on the way clinical and scientific data is communicated. With so much data and the near-real-time/streaming nature of data, individuals and organizations will be unable to continue sharing datasets the way they do today. Instead of relying on a single database, big data users may establish organic, decentralized virtual networks like those proposed by the Dutch Techcenter for Life Sciences (DTL) [69–70]. As nodes in these networks, databases are accessible to the public, provided they meet certain criteria. Complexity will necessitate new ways of interpreting data and translating it back to patients as the number of connections grows. Big data sets must be transformed into "little data" environments for each patient who relies on them to make this work. Even at the following step, which incorporates intuitive and emotional components, medical expertise is still needed. As far as bedside manners are concerned, machine learning and big data are not going to be of much use in the near future.

2.5 Data Sources of Big Data in Medicine

There are a plethora of various kinds of sources for big data. In oncology, patient-derived data is the most evident. Most of these are kept in computerized patient files for therapeutic reasons and comprise a variety of data points and subjects. In addition to a wide range of information on patients and tumors (such as demographics, comorbidities such as cancer-related symptoms, family history, and genetic predisposition), as well as imaging data (CT, MRI, PET, and ultrasound) and tissue-based analysis (such as immunohistochemistry, DNA/RNA sequencing experiments and whole-genome BAM files which contain oncogene sequences), these files also include radiological data [68]. However, *in vitro* research may also provide valuable information. Computational study of this data is a secondary source of big data. Radiomics and digital image analysis, genetic expression and mutation analysis, and other indirect and calculated data are included in the processed data. Machine learning is a growing source of this processed data, which often includes big computer data files including structured data. PREMs and PROMs are only two examples of patient-generated data that may be used to generate big data. Patients can use applications on computers and mobile devices to record various types of data, either supplied by their caregivers (eHealth, telehealth) or on their own initiative. Publications are the fourth and last source of data (IBM project). Biomedical articles, textbooks and other on-line resources are so vast that only few doctor of the globe can read it. However, the depth (quantity) of data per individual is the most essential part of the application of big data in medical research.

Oncologists frequently collect and store hundreds or even millions of observables per patient, despite the fact that patient cohorts tend to be tiny. When it comes to uncommon cancers like head and neck cancer, the disparity between the amount of data collected on individual patients and the cohort as a whole is even more pronounced.

Machine learning and neural networks, on the other hand, have made significant progress in recent years thanks to new methods. It takes tens of thousands to millions of photographs to train a computer to recognize objects in images. The more samples we have, the more accurate our results will be in terms of personalizing therapy. Effective data administration and standards are thus required in the field of head and neck cancer. To put it simply, data sharing, provenance, and protocols for exchanging data all play a role.

2.5.1 Integration of big data in head and neck cancer (HNC)

As a result, uniformity in data capture is necessary since there are so many sources and so many quantities. If this standard is implemented, datasets will be easy to connect to other data repositories. Data integration is crucial for data interpretation and the creation of value from the data, which requires standardization [70]. By analyzing the case mix of patients after surgery, for example, it is possible to assess the quality of care received by those who have had certain surgical procedures and those who have undergone other treatments (e.g., chemotherapy). In recent years, the Netherlands has created national archives for clinicopathologic data, genetic and genomic data, and PROM/PREMs data (Table 2.3). Several illness categories have been subclassified by the Dutch Institute for Clinical Auditing (DICA) since 2014, based on data from the Dutch Head and Neck Audit (DHNA) (cancer and non-cancer). Data (pathologic and genetic) has been gathered from 20 distinct kinds of tumor tissue and is now being synoptically presented

Table 2.3 Data sources for most common cancer types, including HNC.

Tumor type	Clinical	Pathological	Genetic/genomic	PROM/PREM
HNSCC	DHNA	PALGA	PALGA/HMF	NET-QUIBIC
Breast cancer	NBCA	PALGA	PALGA/HMF	
Lung cancer	DLCA/NVALT	PALGA	PALGA/HMF	
Prostate cancer		PALGA	PALGA/HMF	
CRC	DSCA	PALGA	PALGA/HMF	
Melanoma	DMTR	PALGA	PALGA/HMF	

throughout the country. Radiological data is still deficient in structure. NET-QUBIC has built a nationwide website for registering PREMs/PROMs and other forms of cancer as well. The great majority of all patient-derived data sources have now been included in HNC in Holland. The HNSCC collection and the TCIA are two examples of similar efforts taking place throughout the world (Cancer Imaging Archive). The most critical next step is to collect all of this data as rapidly as possible and to do it for each patient individually if possible. You should travel to one of eight Dutch institutes, each one of which has a separate team of partners suggested for head and neck cancer patients (NWHHT). We will all be better able to exchange information and improve data input consistency if all of these various databases are brought together. This organization is also in a unique position to connect these databases throughout the country and develop algorithms for integrated information analytics.

For the vast majority of tumor forms, data on clinical, pathological, genomic/genetic, and radiological results have been properly structured. DHNA: Dutch Head and Neck Audit; NBCA: National Breast Cancer Audit; DLCA: Dutch Lung Cancer Audit. DSCA: Dutch Surgical Colorectal Audit; DMTR: Dutch Melanoma Treatment Registry; PALGA: Pathologisch Anatomisch Landelijk Gegevens Archief; HMF: Hartwig Medical Foundation.

2.5.2 Challenges and future perspectives

However, despite the fact that big data is being used in medical care and has great potential, data producers in life science confront major obstacles. New technologies like next-generation genotyping (particularly whole exome/entire genome decoding) and radiology create an ever-increasing volume of data. These massive volumes of data may restrict the ability to expand the complexity of data analysis. When data volume and speed increase at the same time as data heterogeneity (variability) increases, it becomes more difficult to draw firm conclusions from the data. This is especially true when different study designs and methods, as well as different analytical methods as well as interpretation pipelines all contribute to data heterogeneity (variability). A second problem with data governance occurs because of the necessity to connect data from different sources. Data ownership and usage are questions that need to be answered. It is still up to the patient to make choices based on their information. It may also be a matter of researchers determining how it is done. If that is the case, please tell us who is in control and who is in charge of the data.

2.5.3 Archiving and sharing of biomolecular patient data in repositories and databases

As a result of privacy and security concerns, many patients and institutions are hesitant to reveal their personal information (such as those raised by the General Data Protection Regulation; GDPR). In biomedical research, this is particularly true. Keeping genetic sequences usable while still respecting the privacy of the persons from whose data they were obtained is difficult, to say the least (www.phgfoundation.org). Other professions such as computer engineering are bound by strict privacy regulations when it comes to large data volumes. EGA was created to hold raw sequencing data. A data access committee (DAC) may authorize or deny access to EGA-stored datasets based on the amount of work that was put into acquiring the data. Another issue arises when researchers want to evaluate bimolecular data that has been handled without being able to track back particular markers. For example, data in different repositories [68] may be summarized using fine-grained access control [69]. As a final point, a number of projects are aiming to integrate different data sources while keeping track of the specific computation that has been performed on the data [70, 71].

Biomedicine and hematology/oncology of the head and neck can benefit greatly from big data. The exchange of clinical and research data might also be transformed. Because of the vast amount of data and the (near) real-time/streaming nature of data, individuals and organizations will be unable to exchange datasets in person. As an alternative to consolidating all data into a single database, the Dutch Techcentre for Life Sciences (DTL) proposes that individuals using big data build their own organic, decentralized virtual networks. Users may get access to databases linked to these networks through nodes if certain requirements are satisfied. Complexity will need new techniques of data interpretation and translation back to the individual patient as connections and connectivity increase. Providers of healthcare will face significant difficulties in dealing with this situation. Big data sets must be transformed into "small data" environments specific to each patient who relies on them in order for this to be effective. In order to integrate the emotional and intuitive aspects of this last phase, we still require medical specialists. It is unlikely that big data or machine learning can ever have a handle on bedside etiquette in the near future.

2.6 Conclusion

People's everyday routines have been transformed by the use of big data and artificial intelligence (AI). Healthcare systems are particularly benefiting from this trend. Healthcare has been altered from a traditional system to a modern system for cardiovascular, oncology, ear, and asthmatic illnesses. Using virtual and real-time systems, big data has made it easier to identify an illness. This paper's contribution is a review of current research relevant to mHealth and eHealth, where different methodologies and models that employ big data for diagnostics and healthcare systems are reviewed. The present exciting uses of AI and big data in medical health and electronic health are discussed in this study and might possibly bring value to illness diagnosis and patient treatment. The planned study will aid in the development of innovative healthcare solutions.

Acknowledgment

The authors are highly thankful to the Department of Pharmacy of Galgotias University, Greater Noida, for providing all necessary support for the completion of the work.

Funding

There is no source of funding

Conflict of Interest

There is no conflict of interest in the publication of the contents of this chapter.

References

[1] De Mauro A, Greco M and Grimaldi M. A formal definition of big data based on its essential features. Lib Rev 2016; 65(3): 122–135.

[2] Vogel C, Zwolinsky S, Griffiths C, et al. A Delphi study to build consensus on the definition and use of big data in obesity research. Int J Obes 2019; 43(12): 2573–2586. Mallappallil et al. 9

[3] Hashem I, Yaqoob I, Anuar N, et al. The rise of "big data" on cloud computing: review and open research issues. Inform Syst 2015; 47: 98–115.

[4] Hilbert M and López P. The world's technological capacity to store, communicate, and compute information. Science 2011; 332: 60–65.

[5] https://blog.google/products/photos/google-photos-one-year200-million/(accessedApril26,2019).

[6] https://www.infoworld.com/article/2616022/facebookpushes-the-limits-of-hadoop.html(accessedJuly26,2019).

[7] Stephens Z, Lee S, Faghri F, et al. Big data: astronomical or genomical? PLoS Biol 2015; 13(7): e1002195.

[8] https://unstats.un.org/bigdata/(accessed4/5/2019).

[9] Chen Z, Chen J, Collins R, et al. China Kadoorie Biobank (CKB) collaborative group. China Kadoorie Biobank of 0.5 million people: Survey Methods, Baseline Characteristics and Long-term Follow-up. Int J Epidemiol 2011; 40(6): 1652–1666.

[10] Vaught J and Henderson MK. Biological sample collection, processing, storage, and information management. IARC Sci Publ 2011; 163:23–42.

[11] Hewitt R. Biobanking: the foundation of personalized medicine. Curr Opin Oncol 2011; 23(1): 112–119.

[12] De Souza Y and Greenspan J. Biobanking past, present and future: responsibilities and benefits. AIDS 2013; 27(3): 303–312.

[13] Peisert S, Dart E, Barnett W, et al. The medical science DMZ: a network design pattern for data-intensive medical science. J Am Med Inform Assoc 2018; 25(3): 267–274.

[14] Doyle C, David R, Li J, et al. Using the web for science in the classroom: online citizen science participation in teaching and learning. 2019, https://doi.org/10.1145/3292522.3326022(accessedJuly6,2019).

[15] Henderson J. The yellow brick road to penicillin: a story of serendipity. Mayo Clin Proc 1997; 72(7): 683–687.

[16] https://www.fda.gov/patients/hiv-timeline-and-historyapprovals/hivaids-historical-time-line-1981-1990(accessedAugust2,2019).

[17] https://www.fda.gov/patients/hiv-timeline-and-historyapprovals/hivaids-historical-time-line-2000-2010(accessedAugust2,2019).
[18] https://www.mckinsey.com/industries/pharmaceuticals-andmedical-products/our-insights/pursuing-breakthroughs-incancer-drug-development(accessedJanuary2019).
[19] Collins F, Green E, Guttmacher A, et al. A vision for the future of genomics research. Nature 2003; 422(6934): 835–847.
[20] MacArthur J, Bowler E, Cerezo M, et al. The new NHGRI-EBI catalog of published genome-wide association studies (GWAS Catalog). Nucleic Acids Res 2017; 45(D1): D896–D901.
[21] Wayne M and McIntyre L. Combining mapping and arraying: an approach to candidate gene identification. PNAS 2002; 99(23): 14903–14906.
[22] Stratton M, Campbell P and Futreal P. The cancer genome. Nature 2009; 458: 719–724
[23] Ogino S, Lochhead P, Giovannucci E, et al. Discovery of colorectal cancer PIK3CA mutation as potential predictive biomarker: power and promise of molecular pathological epidemiology. Oncogene 2014; 33(23): 2949–2955.
[24] https://clinicaltrials.gov/ct2/show/NCT03645993(accessedJune2,2019).
[25] Beck A, Sangoi A, Leung S, et al. West systematic analysis of breast cancer morphology uncovers stromal features associated with survival. Sci Transl Med 2011; 3(108): 108ra113.
[26] Song J, Song T, Seo D, et al. Data mining of web-based documents on social networking sites that included suicide-related words among Korean adolescents. J Adolesc Health 2016; 59(6): 668–673.
[27] Patorno E, Schneeweiss S, Gopalakrishnan C, et al. Using real-world data to predict findings of an ongoing phase IV cardiovascular outcome trial: cardiovascular safety of linagliptin versus glimepiride. Diabetes Care 2019.
[28] Ferrarotto R, Redman M, Gandara D, et al. Lung-MAP– framework, overview, and design principles. Chin Clin Oncol 2015; 4(3): 36.
[29] Lambin P, Zindler J, Vanneste B, et al. How rapid learning health care and cohort multiple randomised clinical trials complement traditional evidence-based medicine. Acta Oncologica 2018; 54(9): 1289–1300.

[30] Montani S and Striani M. Artificial intelligence in clinical decision support: a focused literature survey. Yearb Med Inform 2019; 28(1): 120–127.
[31] Magrabi F, Ammenwerth E, McNair JB, et al. Artificial intelligence in clinical decision support: challenges for evaluating AI and practical implications. Yearb Med Inform 2019; 28(1): 128–134.
[32] Perez-Gandia C, Garcia-Saez G, Subias D, et al. Decision support in diabetes care: the challenge of supporting patients in their daily living using a mobile glucose predictor. J Diabetes Sci Technol 2018; 12(2): 243–250
[33] Park JJH, Siden E, Zoratti MJ, et al. Systematic review of basket trials, umbrella trials, and platform trials: a landscape analysis of master protocols. Trials 2019; 20: 572.
[34] Boyd N, Dancey J, Gilks C, et al. Rare cancers: a sea of opportunity. Lancet Oncol 2016; 17(2): e52–e61.
[35] Drilon A, Laetsch T, Kummar S, et al. Efficacy of larotrectinib in TRK fusion-positive cancers in adults and children. N Engl J Med 2018; 378(8): 731–739.
[36] https://www.fda.gov/drugs/resources-information-approveddrugs/drug-information-soundcast-clinical-oncology-disco(accessed 2 August 2019)
[37] Qin B, Jiao X, Liu K, et al. Basket trials for intractable cancer. Front Oncol 2019; 9: 229.
[38] Berry DA. The brave new world of clinical cancer research: adaptive biomarker-driven trials integrating clinical practice with clinical research. Mol Oncol 2015; 9(4): 951–959.
[39] Tafti A, Badger J, LaRose E, et al. Adverse drug event discovery using biomedical literature: a big data neural network adventure. JMIR Med Inform 2017; 5(4): e51.
[40] Visvanathan K, Levit L, Raghavan D, et al. Untapped potential of observational research to inform clinical decision making: American Society of Clinical Oncology research statement. J Clin Oncol 2017; 35(16): 1845–1854.
[41] https://www.govinfo.gov/content/pkg/PLAW-114publ255/html/PLAW-114publ255.htm
[42] Köpcke F and Prokosch H. Employing computers for the recruitment into clinical trials: a comprehensive systematic review. J Med Intern Res 2014; 16(7): e161.

[43] Liu Y and Harbison S. A review of bioinformatic methods for forensic DNA analyses. Forensic Sci Int Genet 2018; 33: 117–128.
[44] Wang X, Williams C, Liu Z, et al. Big data management challenges in health research-a literature review. Brief Bioinformat 2019; 20(1): 156–167.
[45] Essa YM, Hemdan EE, El-Mahalawy A, et al. IFHDS: intelligent framework for securing healthcare bigdata. J Med Syst 2019; 43(5): 124.
[46] Cheng X, Chen F, Xie D, et al. Design of a secure medical data sharing scheme based on blockchain. J Med Syst 2020; 44(2): 52.
[47] Galetsi P, Katsaliaki K and Kumar S. Values, challenges and future directions of big data analytics in healthcare: a systematic review. Soc Sci Med 2019; 241: 112533.
[48] Shaikh AR, Butte AJ, Schully SD, et al. Collaborative biomedicine in the age of big data: the case of cancer. J Med Internet Res 2014;16(4):e101. https://doi.org/10.2196/jmir.2496.Apr7.
[49] . Roman-Belmonte JM, De la Corte-Rodriguez H, Rodriguez-Merchan EC. How blockchain technology can change medicine. Postgrad Med 2018;130(4):420–7.
[50] Bourne PE. What Big Data means to me. J Am Med Inform Assoc 2014;21(2):194. https://doi.org/10.1136/amiajnl-2014-002651.
[51] Zhang C, Bijlard J, Staiger C, Scollen S, et al. Systematically linking tranSMART, S.M. Willems, et al. Oral Oncology 98 (2019) 8–12 11 galaxy and EGA for reusing human translational research data. F1000Res. 2017;6. https://doi.org/10.12688/f1000research.12168.1 .Aug16ELIXIR-1488.
[52] Grossberg AJ, Mohamed ASR, El Halawani H, et al. Sci. Data 2018;5:180173. https://doi.org/10.1038/sdata.2018.173.Sep4.
[53] Prior F, Smith K, Sharma A, et al. The public cancer radiology imaging collections of The Cancer Imaging Archive. Sci Data. 2017;19(4):170124https://doi.org/10.1038/sdata.2017.124.Sep.
[54] Bousfield D, McEntyre J, Velankar S, et al. Patterns of database citation in articles and patents indicate long-term scientific and industry value of biological data resources. F1000Research 2016;5(160). https://doi.org/10.12688/f1000research.7911.1.
[55] Ooft ML, van Ipenburg J, Braunius WW, et al. A nation-wide epidemiological study on the risk of developing second malignancies in patients with different histological subtypes of nasopharyngeal carcinoma. Oral Oncol 2016;56:40–6.

[56] Datema FR, Ferrier MB, Vergouwe Y, et al. Update and external validation of a head and neck cancer prognostic model. Head Neck 2013;35(9):1232–7.

[57] Datema FR, Moya A, Krause P, et al. Novel head and neck cancer survival analysis approach: random survival forests versus Cox proportional hazards regression. Head Neck 2012;34(1):50–8.

[58] Barlesi F, Mazieres J, Merlio JP, et al. Routine molecular profiling of patients with advanced non-small-cell lung cancer: results of a 1-year nationwide programme of the French Cooperative Thoracic Intergroup (IFCT). Lancet 2016;387(10026):1415–26.

[59] Petersen JF, Timmermans AJ, van Dijk BAC. Trends in treatment, incidence and survival of hypopharynx cancer: a 20-year population-based study in the Netherlands. Eur Arch Otorhinolaryngol 2018;275(1):181–9.

[60] Timmermans AJ, van Dijk BA, Overbeek LI, et al. Trends in treatment and survival for advanced laryngeal cancer: A 20-year population-based study in The Netherlands. Head Neck 2016;38(Suppl 1):E1247–55.

[61] de Ridder M, Balm AJ, Smeele LE, et al. An epidemiological evaluation of salivary gland cancer in the Netherlands (1989–2010). Cancer Epidemiol 2015;39(1):14–20. Feb.

[62] Govers TM, Rovers MM, Brands MT, et al. Integrated prediction and decision models are valuable in informing personalized decision making. J Clin Epidemiol 2018. Aug 28 pii: S0895-4356(18)30447-5.

[63] Wong AJ, Kanwar A, Mohamed AS. Radiomics in head and neck cancer: from exploration to application. Transl Cancer Res 2016;5(4):371–82.

[64] Parmar C, Grossmann P, Rietveld D, et al. Radiomic machine-learning classifiers for prognostic biomarkers of head and neck cancer. Front Oncol 2015;3(5):272.

[65] Wilkinson MD, Dumontier M, Aalbersberg IJ, et al. The FAIR guiding principles for scientific data management and stewardship. Sci Data 2016;15(3):160018https://doi.org/10.1038/sdata.2016.18.

[66] Lappalainen I, Almeida-King J, Kumanduri V, et al. The European genome-phenome archive of human data consented for biomedical research. Nat Genet. 2015;47(7):692–5.

[67] Klonowska K, Czubak K, Wojciechowska M, et al. Oncogenomic portals for the visualization and analysis of genome-wide cancer data. Oncotarget 2016;7(1):176–92. Jan 5.

[68] Christoph J, Knell C, Bosserhoff A, et al. Usability and suitability of the omicsintegrating analysis platform tranSMART for translational research and education. Appl Clin Inform. 2017;8(4):1173–83.
[69] He S, Yong M, Matthews PM, et al. TranSMART-XNAT connector tranSMART-XNAT connector-image selection based on clinical phenotypes and genetic profiles. Bioinformatics 2017;33(5):787–8. Mar 1.
[70] Hoogstrate Y, Zhang C, Senf A, et al. Integration of EGA secure data access into galaxy. F1000Res 2016;5. https://doi.org/10.12688/f1000research.10221.1.Dec12pii:ELIXIR-2841.eCollection2016.
[71] Eijssen L, Evelo C, Kok R, et al. The Dutch tech centre for life sciences: enabling data intensive life science research in the Netherlands. F1000Research 2015. https://doi.org/10.12688/f1000research.6009.2.

3

Big Data and Precision Oncology in Healthcare

Arul Prakash Francis[1], Shah Alam Khan[2], Shivkanya Fuloria[3], Neeraj Kumar Fuloria[3], and Dhanalekshmi Unnikrishnan Meenakshi[2*]

[1]Centre for Molecular Medicine and Diagnostics (COMManD), Saveetha Dental College and Hospitals, Saveetha Institute of Medical and Technical Sciences, Saveetha University, India
[2]College of Pharmacy, National University of Science and Technology, Sultanate of Oman
[3]Faculty of Pharmacy, AIMST University, Malaysia
*Corresponding Author: College of Pharmacy, National University of Science and Technology, P.O. Box 620, P.C. 130 Muscat, Oman, E Mail: dhanalekshmi@nu.edu.om, Phone: [+968] 24235000, Fax: [+968] 24504820, ORCID ID: 0000-0002-2689-4079.

Abstract

Cancer is one of the most prevalent chronic diseases and is incredibly complex in nature. Big data (BD) is helpful in the precision and tailoring of treatment approaches for cancer patients that were never before possible, by linking together diverse datasets. Healthcare scientists are working toward the revolution of more comprehensive cancer treatment, intending to discover accurate, precise, and targeted cancer treatment using BD analytics tools to make a real impact in the fight against cancer. This chapter highlights the prominent role of BD analytics in clinical decisions based on patient health records for the management of cancer. It also spotlights the impact of BD in personalized interactions and its sustenance to diagnose cancer at an earlier phase. Furthermore, precision biomarkers and their function in forecasting the likelihood of a patient developing a malignancy, based on BD, are also addressed. The outcome of BD and predictive analytics are also briefed. The importance of BD in the context of EHR systems has also been emphasized. It also outlines the subset of ongoing challenges in computing large datasets by combining expertise from various fields. In addition to recognizing the potential advances in BD and precision oncology, related alarms about potential risks, complex interplay of techniques, and economic forces are briefly deliberated. Evidently, BD has tremendous potential in precision oncology. However, successful exploitation will be contingent on overcoming the hurdles associated with generating and integrating datasets and data analytics.

Keywords: Precision, Big Data, Oncology, Cancer, Datasets, Data analytics, Biomarkers, Healthcare, Patients.

3.1 Introduction

3.1.1 Precision medicine

For treating cancer patients, oncologists derive meaningful insights that should definitely suit the particular patient in front of them by considering all contextual aspects. Molecular profiling leads to more precise treatment, replacing the historical one-size-fits-all approaches. Enormous patient data are available for assisting oncologists in therapeutic decisions, but analysis of these data by individual oncologists is very difficult and this reveals the importance of big data analysis tools. It is essential to scrutinize all of the available cancer data to find unambiguous signatures for precise and successful cancer treatment. Precision Medicine (PM) in oncology relates to a specific treatment for one particular patient based on evidence of linking targeted intermediaries to cancerous molecular anomalies using biomarkers, molecular signatures, and phenotypes linked to patient diet and lifestyle [1, 2] In cancer, early diagnosis is crucial as it manifests to be heterotype. Cancer treatment prioritizes both prevention and therapy for the rehabilitation of the ailment [3]. PM aims to bring personalized solutions in terms of therapy. The main objective of PM is to improve overall patient care and determine when and how to personalize patients' treatments, that is, "the right patient, the right medicine, and the right time." Personalized healthcare can be possible only by measuring every single key factor that impacts a patient's health condition such as genetic, physical, dietary, environmental, and socioeconomic factors. On the other hand, PM helps physicians in streamlining diagnoses and risk evaluations for specific patients related to their health patterns and genomic origins.

3.1.2 Big data and its metaphors in healthcare

Human physiology is a complicated web of interconnections of interrelated configurations and data. Precision oncology and medicine also depend on managing and analyzing large sets of data, including genomics and imaging. Bio samples and data are progressively being shared in the management of cancer. To grasp the complexities of the cancer disease, it is essential to collect massive amounts of data that can be integrated quickly, using recently advanced information technologies. To enhance the performance of a potential medication to molecular target association, refined insights from individual patient data and the treatment response data from a large patient population should be analyzed and correlated. Machine learning (ML) tools are comparatively better than traditional models as these are more precise, accurate, and less costly risk assessments at the individual level. Precision oncology unwinds many strands of data using ML algorithms and helps to harness big data to turn knowledge into action. Advancements and approaches in cancer treatment and care are creating a new era of precision oncology. In the medical field, big data (BD) denotes medical records of a huge set of the patient population, considering the clinical information and medical history of the patient, genomic patterns, clinical trial records, and the billing information for economic analyses [4]. BD uses considerable amounts of medical data to hunt for patterns or relationships that are not visible in smaller datasets. BD has evolved to include massive data volumes and the growing ability to analyze and interpret data. A lot of data is not synonymous with BD, but BD's description of healthcare is related to the 5Vs and is described in Figure 3.1 [5]. BD and machine learning (ML) are used to construct algorithms as good as human physicians, which is a fascinating application of BD in precision oncology. It has been reported that BD and ML may appear incomprehensible at first, but they are closely related to classic statistical models that most doctors are familiar with [6]. As the amount of available data expands, hypotheticals like data analytics and data science must be materialized to address how to deal with it. Improved data collection, storage, cleansing, processing, and interpretation methods are still being developed in most fields. Starting from intelligent medication design and personalized therapy to population screening and electronic health record (EHR) mining, novel techniques are used to extract the clinical significance from enormous volumes of data to generate meaningful transformation in oncology medical practice [7].

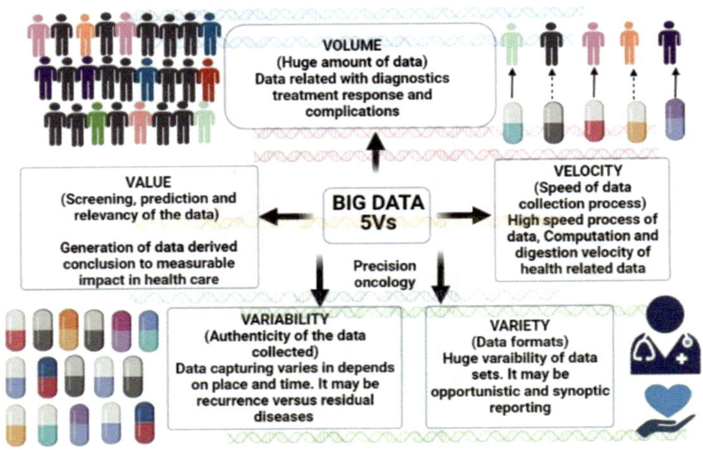

Figure 3.1 Description of innate characteristics and 5Vs of handoop BD.

EHRs, which contain sensitive personal data of the patients, are closely secured and not publicly accessible. These EHRs are typically reserved in the hospital charts or clinics, with no central sharing of data required to leverage BD technologies. The EHR belongs to the patient, and it is only with their approval that it can be accessed and used outside of the clinical realm. This strangles the rapid use of the vast amounts of data. Current medical practice will need to adapt data-driven strategies to embrace this situation. The Cancer Genome Atlas (TCGA) has produced integrated molecular data extracted from 33 distinct tumor types using more than 10,000 samples and has made it available to the public. This massive, joint endeavor not only incorporates epigenetic, transcriptomic, genomic, proteomic, and comprehensive clinical data, but it also explores a user-friendly data portal that makes access easier [8]. Molecular profiling allows for more precise treatment rather than the one-size-fits-all approach. Most crucially, there is compelling evidence that this strategy can succeed with cancer-related mutations, leading to target identification and the PM approach [9].

Clinical data is also more complex and not of much value compared to the data delivered to giant corporations, necessitating processing to convert it into a format that can be used. Even the technological context needed for clinical data modification, transportation, and management is insufficient. Furthermore, advanced algorithms enabled by high-performance computing enable these massive datasets to be transformed into knowledge. The increased use of sensing technology is another significant driver for BD. Due to advancements in technology, such as sensors with more effective wireless communications, better bandwidth, and improved microelectronics, personalized and stratified healthcare is being substituted by uninterrupted sensing and simultaneous advancements in integrated care instead of episodic monitoring [5]. In patients undergoing chemotherapy cycles, white blood cell and neutrophil counts can be kept track of continuously so that the early symptom of neutropenia might be combined with activation of granulocyte to prevent sepsis. Despite these advancements, several obstacles remain in the way of achieving PM and accurate public health interventions for individuals and populations [10].

Appropriate methodologies and analytical approaches are essential to turn BD into meaningful and valuable information for successful cancer treatment and patient care. Based on the above-ground reality, this chapter focuses on the use of BD in personalized interactions and its possible utility in preventing and detecting cancer at an earlier stage. Precision biomarkers and their relevance in using BD to forecast a patient's risk of developing cancer are also discussed in brief, with specific importance to EHR systems. By merging experience from many domains, it also illustrates a subset of ongoing issues in computing massive datasets. Aside from recognizing the potential advancements in BD and precision oncology, predictive analytics, relevant concerns regarding potential hazards, the complexity of technique interplay, and economic forces are briefly discussed. Clearly, BD has enormous potential in precision oncology. However, its full potential will be realized only after overcoming the obstacles of creating and integrating datasets and data analytics. Advancement in information and communications expertise facilitates new concepts in personalized healthcare. The Internet of Things (IoT) is an add-on to the current internet technology that enables universal connection across the physical and virtual worlds. Smart devices with developed technologies like near field communication (NFC) and radio frequency identification (RFID) have become a mainstream element of our lives [11].

3.2 Precision Medicine (PM), Biomarkers, and Big Data (BD)

PM is dependent on BD as it is essential for turning accessible data into actionable insights to progress treatment outcomes [12]. In terms of quantity, most data obtained have come through computerized data analytics such as digital image analysis and radiomics, rather than direct records related to the patients, available in everyday clinical practice. BD is likely to play a significant part in pharmacogenetics and personalized healthcare as it links to PM. Patients with the same malignancy subtype responded in differently to similar chemotherapy treatment. For instance, CYP2D6, a polymorphic gene linked to tamoxifen response, BRAF mutations associated with dasatinib response in non-small cell lung cancer (NSCLC), and several gene signatures have recently been linked to rectal cancer response to chemo-radiotherapy. The observed variability of medication response is thought to be due to genomic instability. Recent research has focused on the complicated relationship between genomes and chemotherapeutic sensitivity, resistance, and toxicity [13, 14]. Hence, the enormous data related to cancer patients worldwide could play a crucial role in precision oncology. Similarly, programs like the Cancer Cell Line Encyclopedia (CCLE) and the Genomics of Drug Sensitivity in Cancer (GDSC) create enormous genomic datasets to study the link between genetic biomarkers and drug sensitivity in different cancer cell lines. Data mentions that using computational algorithms, drug prediction for specific cell lines could be enhanced based on genomic profiles and drug-response data. They demonstrate the opportunity of harnessing massive pharmacogenetic datasets to envisage drug sensitivity. Such computational forecasts on clinical response and specific toxicity should be confirmed in cancer patients undergoing chemotherapy [14].

3.2.1 BD's influence and predictions in precision oncology

PM is designed to act on a precise target with the help of predictive biomarkers, revealing the maximum efficacy in response rate as well as endurance [15]. The clear understanding of inter-patient heterogeneity in various cancer subcategories has been simplified due to large-scale collaborative sequencing assignments [16]. A small-to-moderate proportion of patients reported druggable genomic aberrations highlighting the importance of multicenter association in primary drug development for effective clinical trial enrolment [17]. Clinical use of developed molecular targeted drugs benefitted only specific patient populations. The recent studies reported a few significant obstacles in PM, that is, intratumoral heterogeneity and clonal development following therapy. These hindrances, in turn, question the significance of a single-needle biopsy in capturing a patient's entire genetic background [18, 19]. Because BD refers to a large number of variables and a large sample size or fine-grained sampling, the focus of BD research is on population-wide genome sequencing, but other fields are currently being explored, such as merging traditional surveillance with geographic modeling [10]. It is predicted that integrating big genomic and environmental datasets will aid in predicting individuals or groups at the possibility of experiencing particular long-lasting diseases, especially cancer. It could drive targeted efforts to change environmental as well behavioral factors that contribute to health risks in specific groups. BD could assess current preventative efforts and uncover new insights that can help the physicians improve the therapeutic approach. BD can also be used to track the effects of specific/precise medicines to one particular patient, such as expensive and effective chemotherapeutics in connection to the patient's genetic features, in a therapeutic setting, with a focus on the economic status of the patient [5]. Novel integrative data models beyond the predictive, preventive, personalized, and participatory (P4) have emerged, representing the cornerstone in precision oncology [20, 21]. Multiple criteria must be considered when evaluating cancer therapy methods that include observing and estimating the immune reaction and quantifying response indications, taking account of environmental and behavioral variations. These applaud additional precise elements, describing multifariousness intrinsic to sub-clonality and relapse. Clones offer insights into tumorigenesis and treatment response by explaining the phenotypic heterogeneity [22]. However, most sub-clones from the primary tumor and metastasis lesion have characterized variability in genomic patterns due to the spectator mutations originally present [23].

Consequently, treatments directed toward high-variant genomic variations are predicted to deliver substantial tumor responses based on the available data. Since the development of drug resistance is known for the nullification of clinical responses to targeted agents, repetitive tumor biopsies of regular lesions and investigation of existing biomarkers are vital components of patient care. This, in turn, allows the identification of resistance mechanisms and potentially directs toward precise treatment. The complexity of data has increased as the volume of generated data increases exponentially. Moreover, it is not sufficient to sequence all types of the variant in a human genome, that is, protein levels, metabolites, transcript levels, and phenotypic traits. The remodeling of single-cell data compared with the massive investigation of diverse populations of cells provides considerably more vision into biological processes [24].

3.2.2 Impact of BDs in radiology and biomarker-related datasets

Radiology-related data in precision cancer care enables predictive and trustworthy ML approaches for personalization, such as categorizing disparities in sustenance among patients and predicting therapeutic output(s) to decide the right treatment option for patients with different cancers. It is possible for oncologists to (de)escalate systemically and apply radiation therapy in precise patient groups [5, 25]. Patients are categorized into cohorts using various criteria, including the ailment variations in susceptibility, prognosis, and response degrees. This information plays a crucial role in identifying patients who specifically require aggressive treatments to enhance clinical outcomes [26]. Genomic instability is a unique cancer feature categorized by genetic and epigenetic heterogeneity, and is unique in humans. As a result, people with almost a similar form of cancer may have different prognoses and require different treatments [27]. BD plays a crucial part in this arena of cancer diagnosis and treatment.

The primary goal of PM in oncology is the usage of biomarkers to regulate the treatment approach to each specific patient. Cancer biomarkers are diagnostic, prognostic, and predictive, and monitor therapeutic responses. Prognostic biomarkers deliver the overall cancer outcome of patients irrespective of therapy. Prognostic biomarkers can provide only the data information related to patients at high risk who require more aggressive treatments and not the information about a specific treatment. On the other hand, predictive biomarkers deliver the therapeutic benefit achieved by the patients from a particular treatment [28]. Radiomics, a developing field in PM, describe the extraction of quantitative data from medical images that stimulate the specific treatment. Radiomic data have been interrelated through breast cancer (BC) clinical data such as the disease's phase, lymph node contribution, and hormone receptor status [29]. Moreover, it can distinguish between malignant and benign lesions [30]. Microarray gene expression investigation is now clinically employed for BC molecular grouping that identifies several intrinsic BC molecular isoforms that differ in therapeutic outcomes and predict the patients' survival [31]. Unfortunately, microarray analysis or genome sequencing usage in routine clinical assessments is limited as it is too expensive, and every patient cannot afford the facilities. The clinically-available tests such as BC index [32], end predict [33], oncotype DX 21-gene recurrence score [34], breastOncPx 14-gene distant metastasis signature [35], and the MammaPrint 70-gene prognosis signature [36] are utilized to overcome the difficulty in expensive microarray analysis. Though these clinically validated genetic tests and data reports have significantly changed how patients are selected for chemotherapy, treatment plans still depend on standardized therapeutic approaches. The availability of authenticated radiosensitivity biomarkers for a typical application in the clinic is negligible. Biomarkers decide the radiation dose schedules for different patients depending on tumor biology or the risk of localized tissue toxicity. The development of data tools to predict radiotherapy response is slower due to the low financial capacity of radiation oncology suppliers to fund genome-oriented companies.

The Cancer Data Science Laboratory (CDSL) was established in 2018 to create computational algorithms for investigating and incorporating cancer omics from data of patients worldwide [4]. It is one of the large-scale publicly accessible and collaborative genomic datasets. It provides a report on fundamental research queries about genetic predisposition, vulnerability, and responsiveness to the management and treatment of cancer. Clinical Proteomic Tumor Analysis Consortium research projects and their repository data are available to international scientists. On other hand, a TARGET program (Therapeutically Applicable Research to Generate Effective Treatments) provides information about clinical and accrued tissue materials to create, analyze, and interpret genomics data and is considered the multi-omic approach. These data will help researchers in better understanding the genetic composition of different cancers (uniqueness) and hence could pave the way for identifying the effective therapeutic target (precise approach) [4, 37]. However, these approaches are in the developing stage, and an immense amount of data remains to be assembled and decoded. The algorithm to identify methods to make these data consortiums and the techniques to analyze them should be the focus of future research. As algorithms gain more authority, it is crucial to remember that these new algorithmic decision-making tools are not guaranteed to be fair, equitable, or even accurate. Even with the most advanced ML algorithms, the adage "garbage in, trash out" holds true. Regardless of where an algorithm falls on the ML scale, best analytic practices must be followed to ensure that the end output is reliable and accurate. This is particularly important in healthcare, where these algorithms can impact millions of patients' lives [6].

3.3 Electronic Health Records (EHR) and Precision Oncology

Evidence-based medicine, the conventional treatment strategy, depends on several randomized controlled trial results considered to address specific effects like therapeutic efficacy or toxicity. Moreover, while considering PM to treat patients, numerous clinical and biological parameters should be measured, making it almost awkward to strategize the dedicated trials [38]. Innovative technologies are required to describe different treatment strategies using the comprehensive phenotypic profiles of an enormous number of patients. The clinicians could analyze the unique set of defects responsible for the disease or associated with the therapeutic response from each patient and clinical outcome using the data generated to execute the personalized treatment. The idea of customized treatments relies on identifying the deviation of abnormalities for patients. The shortage of large phenotyped cohorts hinders discovering factors influencing the disease even though computing power has steadily increased [39, 40]. An adequate amount of phenotype data could be created by generalizing EHRs. The EHR systematically gathers and preserves the individual's health information, including demographic statistics, personal and family medical histories including vaccination records, laboratory tests, and imaging results [41]. Data science is crucial in predicting outcomes, and guides treatments through the models developed from large databases [42]. Technological innovation creates opportunities to speed up research and improve patient care. BD technologies link the data sources of the different patients to recognize patterns, define ideal therapy, and enhance PM outputs [43]. Computer models are needed to help doctors organize data, define designs, evaluate results, and set action criteria to provide a continuous learning infrastructure with real-time knowledge. There are already examples of BD analytics generating new information, improving clinical care, and streamlining public health surveillance [44]. EHR has been effectively mined for post-market pharmaceutical management and better pharmacovigilance [14].

EHR is an inherently significant resource that contains the information of each individual, including diagnoses, laboratory test results, and imaging. Besides, EHR has been used classically in all patient-care features, including effectiveness, patient-centeredness, communication, education, timeliness, safety, equity, billing, and auditing [45]. Lifestyle decisions such as food and exercise, family history, and linkages between individuals, race and ethnicity, adherence to prescribed drugs, allergies, and data from wearable technology would all improve EHR data. EHR, through virtual learning, provides opportunities to know more about the disease and risk factor-induced pleiotropic effects such as genetic variation, especially related to oncology. The data stored in the EHR is generally not complete as it was collected in a cohort-based study. EHRs solve the problems involving BD *viz.*

reliability and standardization of data and the accuracy of EHR phenotyping. EHR data is becoming more widely available for academic research, but it offers a number of computational issues that have yet to be resolved. On the other hand, scrutiny and quantification in the background are required as some of the algorithms could provide proof of identity of suspicious areas in complex images. Besides, quantitative data could be matched with large populations to deliver predictive data highlighting the drug response, progression, and outcome [7]. Missing information abounds in these records, relying on narrative text-mining rather than laboratory tests or genomic sequencing. Data standardization is also an issue, and it comprises both organized and unstructured data. Scalability is greatly enhanced by standardization across multiple nations and EHR software solutions [45].

Teamwork between industry and researchers will be required to warrant the relevancy of newly developed techniques and evaluation in real-life scenarios before being used in the clinical departments [7]. Physical data integration of EHRs takes a lot of time and work, but it is currently the most successful way to health data linkage because it is backed up by strict governance rules and a solid infrastructure. One noteworthy example is the national patient-centered clinical research network [10, 46]. Technological developments support physicians in interpreting the clinical images and choosing the precise treatment, but it is time-consuming. For example, an extended period was required to review and scrutinize the data, because in a year, 75 million mammograms are done worldwide [47]. Incorporating new BD techniques in the current clinical work environment is a significant barrier in its application and is likely to be disruptive. Incorporating new data-driven strategies requires a change in the current clinical practice. Testing new techniques, especially those that substitute human action, needs some time. Industry will be the major player that enables hardware and software to upkeep BD management. However, the data access and analysis require interruption of the standard clinical process as the uptake is slow or non-existent. Automation of programs provoked by image content and specific imaging examinations have resulted in tremendous success.

3.4 BD Predictive Analytics and Analytical Techniques

Predictive analytics has been mainly related to megabytes of data generated by EHRs. With its abundance of data, the field of oncology appears to be a good fit for predictive analytics. Regardless of the need for healthier life expectancy projections, acute care utilization, side effects, genetic and molecular risk, predictive analytics applications in cancer are limited [48]. If a patient is diagnosed with lung cancer, risk stratification can help determine if surgery and chemotherapy should be employed. Finally, reliable long-term consequence prediction is critical for communicating with family members and making medical decisions. As a result, clinical prediction literature has seen a tremendous rise in recent years [49].

A learning health system (LHS) combines clinically obtained data with ML to promote PM projects. The LHS is a complete tool that could be used for clinical decision-making, discovery, and hypothesis generation. These new applications may positively impact patients' long-term care and treatment [50]. A lack of relevant prognostic data hampers oncology risk stratification, with the requirement for laborious manual data entry, a deficiency of complete data, and, in some situations, an overreliance on clinical intuition. Even among patients with advanced solid tumors, prospective data suggest that doctors are bad at predicting outcomes. Failures in identifying individuals at high risk of mortality might result in unduly hostile end-of-life treatment in cancer patients [51]. Oncologists are increasingly asked to modify treatment plans based on a patient's risk of particular consequences. Data about the individual, episodes of treatment, and specific medical problems must be captured and interpreted. As part of the Oncology Care Model, the Centers for Medicare and Medicaid Services has compiled complete datasets of beneficiaries and worked with EHR suppliers to increase data collection [48, 52]. The future applications of BD in cancer will face problems such as obtaining enough datasets and targets, a paucity of future verification of predictive analysis, and the possibility of automated prejudice in descriptive datasets [48]. Developments in computational approaches for clinical risk assessment of cancer patients are expected to benefit oncology. Advanced algorithms that anticipate the risk of usage, costs, and clinical results will almost certainly play a more prominent role in how oncology patients are treated in the future. Oncology practitioners are increasingly using predictive analytics techniques to impact common elements of patient care. This cuts down on the time to respond to patients with unusual medical conditions [53]. Envisaging adverse events from chemotherapy, expected chemotherapeutic effects, risk of recurrence, and total life expectancy could be potential applications of analytics that help in decision-making. Numerous real-time EHR-based procedures were established as a proof of concept to evaluate cancer patients' risk of short-term death before starting chemotherapy. These algorithms are potentially applicable to any EHR and are based on organized and unstructured EHR data. Although the future applications of these algorithms are unknown, oncologists could find accurate mortality forecasts incredibly valuable at the point of care [54, 55].

Combining medical, hereditary, and molecular-based forecast concepts could introduce a new age of high-accuracy risk categorization in cancer, enabling a true analysis of BD relevant to precision oncology. There are practical techniques for leveraging BD to undertake predictive analytics. Using increasingly modern computational and information technology, a significant volume of patient- and disease-related data may be obtained with ease. As a result, there is a growing interest in using BD to advance healthcare. Precision oncology specifically relies on predictive analytics. Although there are numerous hurdles in using BD to advance healthcare, there is also considerable potential [56].

3.5 Sources of Data for BD

The volume of data required to predict personalized treatment is rapidly increasing, hence a large amount of data is to be collected and managed. Biomedical data becomes more complex in two ways: the number of samples, and heterogeneity. Short reads from next-generation sequencers must be combined into genomes. While most genomes are sequenced at 30X, a new study claims that high-quality genomes may need 126X (referred to as deep sequencing). Furthermore, an increasing number of samples are taken for the same individual; data might be collected across multiple organs, via single-cell genomics, or under different settings. Finally, the length of time available for sampling is growing. Gene expression, for example, can be tracked over time to gauge treatment effectiveness. Recent advancements in non-intrusive recording techniques will enable data collection over an individual's whole life cycle, paving the path for individualized care from womb to tomb. Some datasets are highly diverse; data of the same type can be gathered using many methods with varied coverage, bias, and noise robustness, and the same can be said for data of different sorts. Furthermore, due to the lack of a standard format in data repositories and the so-called data-extraction difficulty in BD, many data sources create data-gathering challenges [41]. The development in advanced techniques estimated that the preliminary data for a single patient would be around 7–10 GB, which includes the raw genomic data, clinical features, blood tests, administrative data, imaging data, and radiation oncology data.

BD sources are plentiful, and all are from different settings. Patient-derived data are the most evident in oncology and are frequently stored in computerized patient files for therapeutic purposes, including various data points/subjects. These data files enclose clinical history and information of patients, including tumor status, treatment, and outcomes, as well as demographic information like gender and age, family history, symptoms, co-morbidity, radiological data, and tissue-based analysis [5]. However, one crucial component in cancer is the depth of data received from a single patient. Many observables are created frequently and kept in oncology, even though average patient cohort sizes are tiny. The disparity between the depths of data per patient is more pronounced in infrequent malignancies like cancer of the head and neck. A variety of data sources of BD in oncology medicine are represented in Figure 3.2. Some private big data firms have emerged to collect and aggregate real-world data from clinical notes, medical records, and billing data to provide real-time feedback on treatments and outcomes to cancer care stakeholders. Flatiron Health, for example, has established Oncology Cloud specifically for data management. Flatiron's network consists of approximately 1.5 million active patients from 250 cancer clinics, and all are connected through a cloud-based EHR technology into a single data system. Their investigative tools can streamline a specific EMR system, examine individual patient care costs, compile quality indicators, and find candidates for clinical trials. Food and Drug Administration (FDA) utilizes Flatiron's technology to explore the role of real-world evidence along with clinical genomics data from Foundation Medicine. The Kaiser Permanente Clinical Research Networks and Market Scan by Truven Health are other private sector databanks [4].

3.6 BD Exchange in Radiation Oncology

Radiation oncology offers a variety of data from different sources to construct a base for BD research, and to improve cancer care. It provides patient demographics, and clinical baseline factors including family history and personal health status, constituting a vital source for BD analysis. The effectiveness of cancer interventions is valued using BD of clinical observations and treatment-related outcomes. Furthermore, radiological and diagnostic imaging modalities involving CT, PET, and MRI are the primary sources that generate an increased volume and variety of big data. Daily verification imaging with the patient under treatment rapidly increases the volume of image-based data. The picture archiving and communication system (PACS) is the most significant data repository in radiation oncology. Radiotherapy Treatment Planning System (TPS) is a computationally-intensive process that

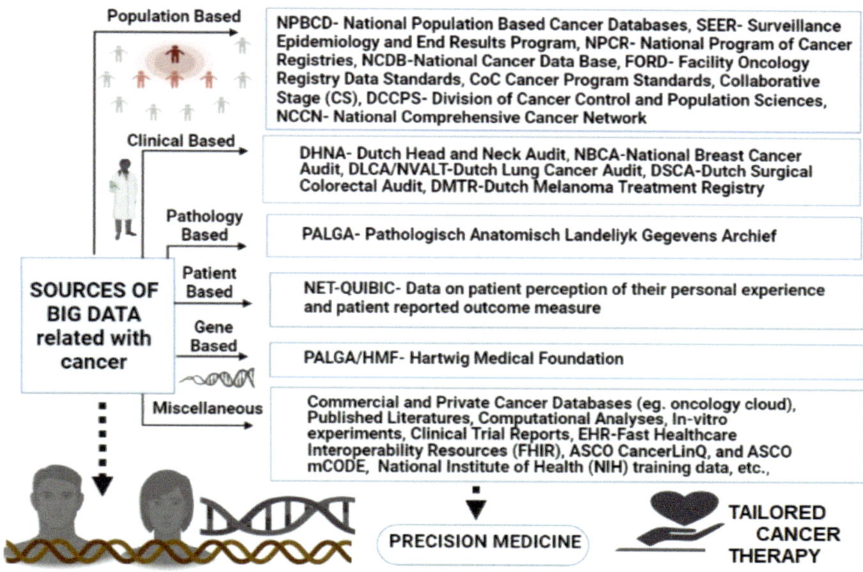

Figure 3.2 Description of variety of data sources of BD in Cancer for the benefit of treating patients.

produces big data regarding various parameters *viz* organ delineations, beam geometry, radiation energy, collimation settings, and spatial dose distribution. Adaptive radiotherapy responding to real-time imaging at the point of treatment delivery also increases the volume of data. Digital pathology and high-throughput specimen analysis from medical laboratories generate a rapidly developing data source, including genomics, proteomics, metabolomics, and histologic and hematologic data. Digital imaging and communications in medicine (DICOM) are widely used for medical imaging supported by all imaging systems [57]. DICOM consists of two components: a type of file format and a network communication protocol. Patient information (e.g., name, identifier, sex, date of birth) and images generated through medical image systems are stored as DICOM files. Followed by the storage, the DICOM protocol can exchange data (e.g., image or patient information) between various systems connected to the hospital network. HL7 is a widely accepted interoperability standard-setting organization providing standards to define the protocol, language, and data type used for information communication among different systems [58, 59]. HL7 FHIR, the current standard receiving positive attention from the community, has resulted in real-world implementations by medical vendors. To standardize the data exchange between medical information systems and widely accepted standard data communicating standards, necessary steps are being taken.

3.6.1 Data integration

Data integration in the radiation oncology field, that is, data processing, storage, management, and exchange within an institution and between multiple organizations, has been achieved through data pooling architecture that includes centralized, decentralized, and hybrid architecture. A centralized architecture physically controls the pooled data in a centralized repository and simplifies the data access and management. Though it is simple, several issues are raised in privacy and anonymization, duplication of data, mapping the local data to the central data model, and intellectual property (IP) rights. For example, data within one institution cannot be accessed by another institution due to the data privacy of patients or local laws. A decentralized architecture is a project-based architecture that directly exchanges the communication data to multiple institutions without intermediaries. But the infrastructure should comply with a standard exchange protocol that is mandatory at each site to permit data exchange. Shared data in decentralized architecture may be stored or not stored after the data transfer. On the other hand, hybrid architecture is focused to couple the strengths of both types of architecture, that is, transferring the data across multiple sites through direct communication [60].

3.6.2 Data interoperability

Data interoperability deals with the capacity of a system to read and understand data relocated from another system and exhibits a vital role in data exchange between multiple centers. To execute complex and comprehensive data analyses, it is essential to transform data sources into fully interoperable sources across multiple information technology (IT) systems. Data interoperability involves two key subprinciples (1) data representation should be identical between all institutions for writing and reading the information, and (2) syntactic and semantic interoperability should be suitably placed [61]. In the current scenario, the privacy of patient data, local policy, and even technical issues play a crucial role in achieving data interoperability. Data interoperability among different IT platforms and certain technologies (e.g., Semantic Web and ontology) have been applied to BD exchange in the field of radiation oncology.

3.7 Clinical Trial Updates on Precision Medicine

PM is defined as the customization of medical treatment to each patient's unique traits, as well as the classification of individuals into subpopulations that differ in their susceptibility to disease or response to treatment. PM offers remarkable opportunities to develop the future of healthcare, and currently, it is most advanced in oncology. Beyond oncology and late-stage diseases, PM also has extensive applications in genetic disorders. Though it promises to treat Covid-19, connecting PM into healthcare is complicated. Healthcare providers will be equipped with digital tools to recognize the complex data received from the precise technique [62]. Some pharmaceutical companies often use these data to treat broad populations undergoing clinical trials on some drugs and treatments. Subsequently, they work for some patients but not for others. Clinical trials are effective ways for evaluating diagnostic and therapeutic tactics, as well as in-depth data mining, which enable the use of BD for cancer detection and treatment. New evidence reveals that computational algorithms can be used to identify features in computed tomography pictures, allowing for more accurate result assessments. By considering each patient as an individual, PM can overcome the restrictions of traditional medicine by shifting the importance from the reaction towards prevention. Pharmacogenomics was possibly the most primitive application of PM. Trials of genotyping *VKORC1* and *CYP2C9* to augment warfarin dose brought about success such as methods to programmed dose assessment [63]. FDA included the likelihood of testing with black box warnings. Indeed, it is believed that ongoing sequencing efforts would identify more protective variants that drugs can target. These medicines, which include targeted therapy drugs, would benefit only a restricted group of patients. Clinical trial schemes for various diagnostics and therapeutics are shown in Table 3.1.

All investors (business, regulatory, government, patient representatives, and academics) must be collaborated to build frameworks to analyze assay performance before regulatory approval and to harmonize the certification process [66]. Improved discussion among physicians and patients is essential in delivering a precise forecast of possible benefits and side effects of the PM approach, while not demoralizing clinical trial participation, thereby avoiding the current therapeutic chaos in which patients and clinicians abandon standard therapy in favor of unconfirmed therapeutic targets. This necessitates the development of standard teaching frameworks for doctors, laboratory technicians, and patients alike. The lack of training and knowledge between primary care physicians is another challenge in PM. Primary care physicians have

Table 3.1 Precision medicine trials in various fields.

Trail designs	Applications	Examples
Adaptive studies: Adaptive trials act as a compact trial design and cumulative evidence to alter features of the trial without jeopardizing its feasibility and credibility [64] .	Uncertainty arises regarding the best trial design. Data could be collected and evaluated in order to ascertain particular aspects of the trial.	Adaptive designs require bio-markers. These designs may be helpful in type I diabetic patients to lower the dose of IL-2. [65].
Umbrella protocols: In an umbrella trial design, several personalized therapies are tested in parallel.	In case of multiple treatment options, a sufficient number of patients were directed toward various treatments in an umbrella trial.	Plasma MATCH is a multi-center study that looks into five distinct therapy options for BC. Therapies were divided into five therapy groups based on molecular markers. Patients with estrogen receptor gene 1 mutation (group A), HER2 mutation (group B), AKT (serine/threonine-specific protein kinase B) mutation (group C), AKT activation (group D), or triple-negative status (group E) were divided into five categories [66]. Five different targeted therapeutic options for advanced breast cancer were examined in terms of clinical usefulness.
Basket trials: Basket trials concentrate mostly on the primary target rather than the condition or clinical syndrome [64].	Basket trials are used in individuals with a variety of medical disorders or syndromes who have a well-defined therapy goal.	This trial is based on EEG biomarkers to simplify the process of identifying medications for cross-diagnostically related ailment categories such as cognitive impairment. For example, patients with bipolar illness could be excluded from a unique pro-cognitive trial testing a new Drug X if they have MMN malleability [66]. Furthermore, a basket trial could evaluate the efficacy of a therapeutic strategy without being constrained by established criteria that are applicable to a variety of conditions.

to create the clinical context progressively around the patient's test results as a curiosity in genetic testing rises among consumers; still, most practitioners have not had in-depth training in genomics or genetics. Healthcare educators should teach PM techniques by including genomics and genetics in continuing professional development training courses. Human genetics was propelled forward by the identification of genetic factors, healthcare, and health and illness. Clinical trial approaches such as adaptive designs, umbrella designs, and basket designs cross-examine the efficacy of drugs and will be crucial to progressing PM. Overall, PM has massive and valuable potential to increase patient outcomes and challenge the future of healthcare. But, to see the actual value and capabilities of precision medicine techniques, the industry must overcome the obstacles around infrastructure, inequalities, and knowledge gaps.

3.8 Challenges and Future Perspectives

Even with advanced technological development, the increases in data volume surpass hospitals' capability to handle the ultimatum for the storage of data [67]. Establishing external storage for the oldest and largest data files is the solution to overcome difficulty in data storage. Digital data can be conserved and easily accessed by storing the voluminous data separately from the query platform. Furthermore, healthcare data that is only known by physicians and patients should be transferred to a platform that is under the jurisdiction of separate data governance organizations and is accessible for future study and analysis. Hadoop is a software that provides a framework for circulated storage and treating BD, using the MapReduce programming model. However, the existing systems are inefficient in importing data in global exchange standards [68]. In a (bio) medical research situation, the goal is to gather vast datasets from several hospitals and clinics as much as feasible; however, privacy concerns and security and protection measures (GDPR) frequently prevent this. One of the most difficult challenges is storing identifiable patient data, such as genome categorizations, that allows the data to be reused for future research although protecting the privacy from where the data was composed. While free access to large datasets may be preferred in other areas, such as computer science, privacy apprehensions must obviously be more important than a desire for complete openness in health sciences [69]. When researchers want to browse processed biomolecular data without tracking back specific markers, they face an additional challenge [5]. Because of the increased interconnection and (hence) complexity, new techniques to data analysis and data and interpretation transfer back to the individual patient will be required. This involves the translation of BD-derived information into the context of the individual care-dependent patient's small data. Medical professionals are still required for this final step, which includes merging intuitive and emotional factors. BD and ML technologies will not help with bedside manners for the foreseeable future. In addition, the investigation podiums should be linked within the institutions for effective data collection and integration. Custom-made pipelines (extract, transform, and load (ETL)) with different architecture designed by IT teams are not used again in other institutions. On the other hand, FAIR data principles increase data accessibility for research purposes [70]. BD's critical challenges for any organization are health data security and accessibility. Data should be accessible easily from any place without compromising safety. An architecture that considers high-security constraints, including strong user authentication and methods that guarantee the traceability of all data processing steps, is essential for the remote accessibility of the data. Medical record linkage and data anonymization are frequently needed steps in data sharing for research. A responsible third party is a prerequisite to taking care of these procedures. Using remote solutions with potential advantages, that is, cost-effectiveness, computing power, and flexibility, is often problematic as the final user does not control the server – creating issues with data privacy regulations. Producers of BD confront problems in making them ideally usable in precision oncology, despite their current usage in clinical practice. The volume of data continues to grow dramatically as new technologies emerge, such as NGS and radionics. These BD add a layer of complexity to the data, making it difficult to analyze. This is especially true when the increase in data (volume and velocity) is accompanied by an increase in data heterogeneity (variability), which includes treatments, outcomes, differences in study design, analytical methods, and data interpretation pipelines, all of which make it difficult to draw firm conclusions from the data [5]. In many ways, PM- and BD- related enhancements are required to validate its routine and effective use in healthcare. Interpretation of the relationship among the environment, microbiome, and genetics; treatment response prediction; improved drug design and delivery; and a more user-friendly method for accessing and using electronic health records and databases must be validated [63].

3.9 Conclusion

BD and ML have an impact on practically every aspect of modern life in some way. PM, as addressed in this chapter, necessitates multidisciplinary competence that extends beyond omics, particularly in oncology care, innovative research, and development paradigms. The role of BD in precision oncology to enhance health status is briefly summarized in this chapter. Oncology stands to improve based on the developments in computational tools for clinical risk assessment. However, a number of fundamental roadblocks in BD and PM remain; predictive analysis is not yet accurately reported; modeling interoperability; lack of contextualization and integration with ethics; trust in the new paradigm shift; data management, data integration, data sharing, and data sources are all issues that must be addressed. With the rapid growth of technology, certainly, data for each and every treatment step, genomes, resistance, recurrence, financial, and other parameters from every cancer patient will be available. Because the ultimate goal of BD in precision oncology is to combine available individuals' databases and compile wide-ranging data reports on all cancer patients for further in-depth analysis, the BD technique will undoubtedly aid in bridging the gap between clinical trials and real-world therapeutic interventions. BD analysis holds a lot of promise for improving healthcare and transforming population health. However, realizing this promise will necessitate addressing issues such as data privacy, security, ownership, and governance. With meticulous data gathering, reporting, quality assurance, and analysis, BD will aid in the delivery of precision oncology treatment and patient care.

Acknowledgment

The authors are thankful to their respective Universities for supporting the successful completion of this chapter.

Conflicts of Interest

The authors declare no conflict of interest.

Funding Information

Nothing to disclose.

References

[1] Hoelder S, Clarke P A and Workman P 2012 Discovery of small molecule cancer drugs: successes, challenges and opportunities *Mol Oncol* **6** 155–76

[2] Ghasemi M, Nabipour I, Omrani A, Alipour Z and Assadi M 2016 Precision medicine and molecular imaging: new targeted approaches toward cancer therapeutic and diagnosis *Am J Nucl Med Mol Imaging* **6** 310–27

[3] Teare H J A, Hogg J, Kaye J, Luqmani R, Rush E, Turner A, Watts L, Williams M and Javaid M K 2017 The RUDY study: using digital technologies to enable a research partnership *Eur J Hum Genet* **25** 816–22

[4] Tsai C J, Riaz N and Gomez S L 2019 Big Data in Cancer Research: Real-World Resources for Precision Oncology to Improve Cancer Care Delivery *Semin Radiat Oncol* **29** 306–10

[5] Willems S M, Abeln S, Feenstra K A, de Bree R, van der Poel E F, Baatenburg de Jong R J, Heringa J and van den Brekel M W M 2019 The potential use of big data in oncology *Oral Oncol* **98** 8–12

[6] Beam A L and Kohane I S 2018 Big Data and Machine Learning in Health Care *JAMA* **319** 1317–8

[7] Hulsen T, Jamuar S, Moody A R, Karnes J H, Varga O, Hedensted S, Spreafico R, Hafler D A and McKinney E F 2019 From big data to precision medicine *Front Med* **6** 34

[8] Hoadley K A, Yau C, Hinoue T, Wolf D M, Lazar A J, Drill E, Shen R, Taylor A M, Cherniack A D, Thorsson V, Akbani R, Bowlby R, Wong C K, Wiznerowicz M, Sanchez-Vega F, Robertson A G, Schneider B G, Lawrence M S, Noushmehr H, Malta T M, Caesar-Johnson S J, Demchok J A, Felau I, Kasapi M, Ferguson M L, Hutter C M, Sofia H J, Tarnuzzer R, Wang Z, Yang L, Zenklusen J C, Zhang J (Julia), Chudamani S, Liu J, Lolla L, Naresh R, Pihl T, Sun Q, Wan Y, Wu Y, Cho J, DeFreitas T, Frazer S, Gehlenborg N, Getz G, Heiman D I, Kim J, Lawrence M S, Lin P, Meier S, Noble M S, Saksena G, Voet D, Zhang H, Bernard B, Chambwe N, Dhankani V, Knijnenburg T, Kramer R, Leinonen K, Liu Y, Miller M, Reynolds S, Shmulevich I, Thorsson V, Zhang W, Akbani R, Broom B M, Hegde A M, Ju Z, Kanchi R S, Korkut A, Li J, Liang H, Ling S, Liu W, Lu Y, Mills G B, Ng K S, Rao A, Ryan M, Wang J, Weinstein J N, Zhang J, Abeshouse A, Armenia J, Chakravarty D, Chatila W K, de Bruijn I, Gao J, Gross B E, Heins Z J, Kundra R, La K, Ladanyi M, Luna A, Nissan M G, Ochoa A, et al 2018 Cell-of-Origin Patterns Dominate the Molecular Classification of 10,000 Tumors from 33 Types of Cancer *Cell* **173** 291-304.e6

[9] Parikh A R and Corcoran R B 2017 Fast-TRKing Drug Development for Rare Molecular Targets *Cancer Discov* **7** 934–6

[10] Prosperi M, Min J S, Bian J and Modave F 2018 Big data hurdles in precision medicine and precision public health *BMC Med Inform Decis Mak* **18**

[11] Schreier G 2014 The Internet of Things for Personalized Health *Stud Health Technol Inform* **200** 22–31

[12] Govers T M, Rovers M, Brands M T, Dronkers E A C, Baatenburg de Jong R J, Merkx M A W, Takes R P and Grutters J P C 2018 Integrated prediction and decision models are valuable in informing personalized decision making *J Clin Epidemiol* **104** 73–83

[13] Garnett M J, Edelman E J, Heidorn S J, Greenman C D, Dastur A, Lau K W, Greninger P, Thompson I R, Luo X, Soares J, Liu Q, Iorio F, Surdez D, Chen L, Milano R J, Bignell G R, Tam A T, Davies H, Stevenson J A, Barthorpe S, Lutz S R, Kogera F, Lawrence K, McLaren-Douglas A, Mitropoulos X, Mironenko T, Thi H, Richardson L, Zhou W, Jewitt F, Zhang T, O'Brien P, Boisvert J L, Price S, Hur W, Yang W, Deng X, Butler A, Choi H G, Chang J W, Baselga J, Stamenkovic I, Engelman J A, Sharma S v., Delattre O, Saez-Rodriguez J, Gray N S, Settleman J, Futreal P A, Haber D A, Stratton M R, Ramaswamy S, McDermott U and Benes C H 2012 Systematic identification of genomic markers of drug sensitivity in cancer cells *Nature* **483** 570–5

[14] Leff D R and Yang G Z 2015 Big Data for Precision Medicine *Engineering* **1** 277–9

[15] Ocana A, Amir E, Vera-Badillo F, Seruga B and Tannock I F 2013 Phase III Trials of Targeted Anticancer Therapies: Redesigning the Concept *Clin Cancer Res* **19** 4931–40

[16] Dienstmann R, Rodon J, Barretina J and Tabernero J 2013 Genomic medicine frontier in human solid tumors: prospects and challenges *J Clin Oncol* **31** 1874–84

[17] Dienstmann R, Rodon J and Tabernero J 2015 Optimal design of trials to demonstrate the utility of genomically-guided therapy: Putting Precision Cancer Medicine to the test *Mol Oncol* **9** 940–50

[18] Hunter D J 2016 Uncertainty in the Era of Precision Medicine *N Engl J Med* **375** 711–3

[19] Bedard P L, Hansen A R, Ratain M J and Siu L 2013 Tumour heterogeneity in the clinic *Nature* **501** 355–64

[20] Hood L and Flores M 2012 A personal view on systems medicine and the emergence of proactive P4 medicine: predictive, preventive, personalized and participatory *N Biotechnol* **29** 613–24

[21] Schellekens H, Aldosari M, Talsma H and Mastrobattista E 2017 Making individualized drugs a reality *Nat Biotechnol* **35** 507–13

[22] Tabassum D P and Polyak K 2015 Tumorigenesis: it takes a village *Nat Rev Cancer* **15** 473–83

[23] Yap T A, Gerlinger M, Futreal P A, Pusztai L and Swanton C 2012 Intratumor heterogeneity: seeing the wood for the trees *Sci Transl Med* **4** 127ps10
[24] Ritchie M D, Holzinger E R, Li R, Pendergrass S A and Kim D 2015 Methods of integrating data to uncover genotype-phenotype interactions *Nat Rev Genet* **16** 85–97
[25] Forghani R, Savadjiev P, Chatterjee A, Muthukrishnan N, Reinhold C and Forghani B 2019 Radiomics and Artificial Intelligence for Biomarker and Prediction Model Development in Oncology *Comput Struct Biotechnol J* **17** 995–1008
[26] Penet M F, Krishnamachary B, Chen Z, Jin J and Bhujwalla Z M 2014 Molecular Imaging of the Tumor Microenvironment for Precision Medicine and Theranostics *Adv Cancer Res* **124** 235
[27] Burrell R A, McGranahan N, Bartek J and Swanton C 2013 The causes and consequences of genetic heterogeneity in cancer evolution *Nature* **501** 338–45
[28] Polley M Y C, Freidlin B, Korn E L, Conley B A, Abrams J S and McShane L M 2013 Statistical and practical considerations for clinical evaluation of predictive biomarkers *J Natl Cancer Inst* **105** 1677–83
[29] Guo W, Li H, Zhu Y, Lan L, Yang S, Drukker K, Morris E, Burnside E, Whitman G, Giger M L, Ji Y and Group T B P R 2015 Prediction of clinical phenotypes in invasive breast carcinomas from the integration of radiomics and genomics data *J Med Imaging* **2** 041007
[30] Bickelhaupt S, Paech D, Kickingereder P, Steudle F, Lederer W, Daniel H, Götz M, Gählert N, Tichy D, Wiesenfarth M, Laun F B, Maier-Hein K H, Schlemmer H P and Bonekamp D 2017 Prediction of malignancy by a radiomic signature from contrast agent-free diffusion MRI in suspicious breast lesions found on screening mammography *J Magn Reson Imaging* **46** 604–16
[31] Sotiriou C, Neo S Y, McShane L M, Korn E L, Long P M, Jazaeri A, Martiat P, Fox S B, Harris A L and Liu E T 2003 Breast cancer classification and prognosis based on gene expression profiles from a population-based study *Proc Natl Acad Sci U S A* **100** 10393–8
[32] Sgroi D C, Sestak I, Cuzick J, Zhang Y, Schnabel C A, Schroeder B, Erlander M G, Dunbier A, Sidhu K, Lopez-Knowles E, Goss P E and Dowsett M 2013 Prediction of late distant recurrence in patients with oestrogen-receptor-positive breast cancer: a prospective comparison of the breast-cancer index (BCI) assay, 21-gene recurrence score, and IHC4 in the TransATAC study population *Lancet Oncol* **14** 1067–76

[33] Dubsky P, Brase J C, Jakesz R, Rudas M, Singer C F, Greil R, Dietze O, Luisser I, Klug E, Sedivy R, Bachner M, Mayr D, Schmidt M, Gehrmann M C, Petry C, Weber K E, Fisch K, Kronenwett R, Gnant M and Filipits M 2013 The EndoPredict score provides prognostic information on late distant metastases in ER+/HER2- breast cancer patients *Br J Cancer* **109** 2959–64

[34] Paik S, Shak S, Tang G, Kim C, Baker J, Cronin M, Baehner F L, Walker M G, Watson D, Park T, Hiller W, Fisher E R, Wickerham D L, Bryant J and Wolmark N 2004 A multigene assay to predict recurrence of tamoxifen-treated, node-negative breast cancer *N Engl J Med* **351** 2817–26

[35] Tutt A, Wang A, Rowland C, Gillett C, Lau K, Chew K, Dai H, Kwok S, Ryder K, Shu H, Springall R, Cane P, McCallie B, Kam-Morgan L, Anderson S, Buerger H, Gray J, Bennington J, Esserman L, Hastie T, Broder S, Sninsky J, Brandt B and Waldman F 2008 Risk estimation of distant metastasis in node-negative, estrogen receptor-positive breast cancer patients using an RT-PCR based prognostic expression signature *BMC Cancer* **8** 339

[36] van de Vijver M J, He Y D, van 't Veer L J, Dai H, Hart A M, Voskuil D W, Schreiber G J, Peterse J L, Roberts C, Marton M J, Parrish M, Atsma D, Witteveen A, Glas A, Delahaye L, van der Velde T, Bartelink H, Rodenhuis S, Rutgers E T, Friend S H and Bernards R 2002 A gene-expression signature as a predictor of survival in breast cancer *N Engl J Med* **347** 1999–2009

[37] Singal G, Miller P G, Agarwala V, Li G, Kaushik G, Backenroth D, Gossai A, Frampton G M, Torres A Z, Lehnert E M, Bourque D, O'Connell C, Bowser B, Caron T, Baydur E, Seidl-Rathkopf K, Ivanov I, Alpha-Cobb G, Guria A, He J, Frank S, Nunnally A C, Bailey M, Jaskiw A, Feuchtbaum D, Nussbaum N, Abernethy A P and Miller V A 2019 Association of Patient Characteristics and Tumor Genomics With Clinical Outcomes Among Patients With Non-Small Cell Lung Cancer Using a Clinicogenomic Database *JAMA* **321** 1391–9

[38] Chen C, He M, Zhu Y, Shi L and Wang X 2015 Five critical elements to ensure the precision medicine *Cancer Metastasis Rev* **34** 313–8

[39] Mardis E R 2011 A decade's perspective on DNA sequencing technology *Nature* **470** 198–203

[40] Metzker M L 2009 Sequencing technologies — the next generation *Nat Rev Genet* **11** 31–46

[41] Gligorijević V, Malod-Dognin N and Pržulj N 2016 Integrative methods for analyzing big data in precision medicine *Proteomics* **16** 741–58
[42] Bibault J-E and Xing L 2020 The Role of Big Data in Personalized Medicine *Precis Med Oncol* ed B Aydogan and J A Radosevich (John Wiley & Sons, Ltd) pp 229–47
[43] Yarchoan M, Hopkins A and Jaffee E M 2017 Tumor Mutational Burden and Response Rate to PD-1 Inhibition *N Engl J Med* **377** 2500–1
[44] Cirillo D and Valencia A 2019 Big data analytics for personalized medicine *Curr Opin Biotechnol* **58** 161–7
[45] Hemingway H, Asselbergs F W, Danesh J, Dobson R, Maniadakis N, Maggioni A, van Thiel G J M, Cronin M, Brobert G, Vardas P, Anker S D, Grobbee Di E and Denaxas S 2018 Big data from electronic health records for early and late translational cardiovascular research: challenges and potential *Eur Heart J* **39** 1481–95
[46] Evans R S 2016 Electronic Health Records: Then, Now, and in the Future *Yearb med inform* S48–61
[47] Jourquin J, Reffey S B, Jernigan C, Levy M, Zinser G, Sabelko K, Pietenpol J and Sledge G 2019 Susan G. Komen Big Data for Breast Cancer Initiative: How Patient Advocacy Organizations Can Facilitate Using Big Data to Improve Patient Outcomes. *JCO Precis Oncol* **3** 1–9
[48] Parikh R B, Gdowski A, Patt D A, Hertler A, Mermel C and Bekelman J E 2019 Using Big Data and Predictive Analytics to Determine Patient Risk in Oncology *Am Soc Clin Oncol Educ Book* **39** e53–8
[49] Zhang Z 2020 Predictive analytics in the era of big data: opportunities and challenges *Ann Transl Med* **8** 68–68
[50] McNutt T R, Benedict S H, Low D A, Moore K, Shpitser I, Jiang W, Lakshminarayanan P, Cheng Z, Han P, Hui X, Nakatsugawa M, Lee J, Moore J A, Robertson S P, Shah V, Taylor R, Quon H, Wong J and DeWeese T 2018 Using Big Data Analytics to Advance Precision Radiation Oncology *Int J Radiat Oncol Biol Phys* **101** 285–91
[51] Sborov K, Giaretta S, Koong A, Aggarwal S, Aslakson R, Gensheimer M F, Chang D T and Pollom E L 2019 Impact of Accuracy of Survival Predictions on Quality of End-of-Life Care Among Patients With Metastatic Cancer Who Receive Radiation Therapy *J Oncol Pract* **15** E262–70
[52] Kline R, Adelson K, Kirshner J J, Strawbridge L M, Devita M, Sinanis N, Conway P H and Basch E 2017 The Oncology Care Model: Perspectives From the Centers for Medicare & Medicaid Services and

Participating Oncology Practices in Academia and the Community *Am Soc Clin Oncol Educ Book* **37** 460–6

[53] Rosen M A, DiazGranados D, Dietz A S, Benishek L E, Thompson D, Pronovost P J and Weaver S J 2018 Teamwork in Healthcare: Key Discoveries Enabling Safer, High-Quality Care *Am Psychol* **73** 433–50

[54] Elfiky A A, Pany M J, Parikh R B and Obermeyer Z 2018 Development and Application of a Machine Learning Approach to Assess Short-term Mortality Risk Among Patients With Cancer Starting Chemotherapy *JAMA Netw Open* **1** e180926

[55] Bertsimas D, Dunn J, Pawlowski C, Silberholz J, Weinstein A, Zhuo Y D, Chen E and Elfiky A A 2018 Applied Informatics Decision Support Tool for Mortality Predictions in Patients With Cancer *JCO Clin Cancer Inform* **2** 1–11

[56] Gill J and Prasad V 2018 Improving observational studies in the era of big data *The Lancet* **392** 716–7

[57] Mildenberger P, Eichelberg M and Martin E 2002 Introduction to the DICOM standard *Eur Radiol* **12** 920–7

[58] Dolin R H, Alschuler L, Boyer S, Beebe C, Behlen F M, Biron P v. and Shabo A 2006 HL7 Clinical Document Architecture, Release 2 *J Am Med Inform Assoc* **13** 30–9

[59] Dolin R H, Alschuler L, Beebe C, Biron P v., Boyer S L, Essin D, Kimber E, Lincoln T and Mattison J E 2001 The HL7 Clinical Document Architecture *J Am Med Inform Assoc* **8** 552–69

[60] Skripcak T, Belka C, Bosch W, Brink C, Brunner T, Budach V, Büttner D, Debus J, Dekker A, Grau C, Gulliford S, Hurkmans C, Just U, Krause M, Lambin P, Langendijk J A, Lewensohn R, Lühr A, Maingon P, Masucci M, Niyazi M, Poortmans P, Simon M, Schmidberger H, Spezi E, Stuschke M, Valentini V, Verheij M, Whitfield G, Zackrisson B, Zips D and Baumann M 2014 Creating a data exchange strategy for radiotherapy research: towards federated databases and anonymised public datasets *Radiother Oncol* **113** 303–9

[61] Valentini V, Schmoll H J and van de Velde C J H 2011 *Multidisciplinary management of rectal cancer: Questions and answers* (Springer-Verlag Berlin Heidelberg)

[62] Dash S, Shakyawar S K, Sharma M and Kaushik S 2019 Big data in healthcare: management, analysis and future prospects *J Big Data* **6** 54

[63] Hasanzad M, Larijani B, Reza H, Meybodi A, Hasanzad M and Larijani B 2017 Path to Personalized Medicine for Type 2 Diabetes Mellitus: Reality and Hope *Acta Med Iran* **55** 166–74

[64] Barker A D, Sigman C C, Kelloff G J, Hylton N M, Berry D A and Esserman L J 2009 I-SPY 2: an adaptive breast cancer trial design in the setting of neoadjuvant chemotherapy *Clin Pharmacol Ther* **86** 97–100

[65] Zardavas D and Piccart-Gebhart M 2016 New generation of breast cancer clinical trials implementing molecular profiling *Cancer Biol Med* **13** 226–35

[66] Salgado R, Moore H, Martens J W M, Lively T, Malik S, McDermott U, Michiels S, Moscow J A, Tejpar S, McKee T, Lacombe D, Becker R, Beer P, Bergh J, Bogaerts J, Dovedi S, Fojo A T, Gerstung M, Golfinopoulos V, Hewitt S, Hochhauser D, Juhl H, Kinders R, Lillie T, Herbert K L, Maheswaran S, Mesri M, Nagai S, Norstedt I, O'Connor D, Oliver K, Oyen W J G, Pignatti F, Polley E, Rosenfeld N, Schellens J, Schilsky R, Schneider E, Senderowicz A, Tenhunen O, van Dongen A, Vietz C and Wilking N 2017 Societal challenges of precision medicine: Bringing order to chaos *Eur J Cancer* **84** 325–34

[67] Huser V and Cimino J 2016 Impending Challenges for the Use of Big Data *Int J Radiat Oncol Biol Phys* **95** 890–4

[68] Dolin R H, Rogers B and Jaffe C 2015 Health level seven interoperability strategy: big data, incrementally structured *Methods Inf Med* **54** 75–82

[69] Lappalainen I, Almeida-King J, Kumanduri V, Senf A, Spalding J D, Ur-Rehman S, Saunders G, Kandasamy J, Caccamo M, Leinonen R, Vaughan B, Laurent T, Rowland F, Marin-Garcia P, Barker J, Jokinen P, Torres A C, de Argila J R, Llobet O M, Medina I, Puy M S, Alberich M, de La Torre S, Navarro A, Paschall J and Flicek P 2015 The European Genome-phenome Archive of human data consented for biomedical research *Nat Genet* **47** 692–5

[70] Wilkinson M D, Dumontier M, Aalbersberg Ij J, Appleton G, Axton M, Baak A, Blomberg N, Boiten J W, da Silva Santos L B, Bourne P E, Bouwman J, Brookes A J, Clark T, Crosas M, Dillo I, Dumon O, Edmunds S, Evelo C T, Finkers R, Gonzalez-Beltran A, Gray A J G, Groth P, Goble C, Grethe J S, Heringa J, t Hoen P A C, Hooft R, Kuhn T, Kok R, Kok J, Lusher S J, Martone M E, Mons A, Packer A L, Persson B, Rocca-Serra P, Roos M, van Schaik R, Sansone S A, Schultes E, Sengstag T, Slater T, Strawn G, Swertz M A, Thompson M, van der Lei J, van Mulligen E, Velterop J, Waagmeester A, Wittenburg P, Wolstencroft K, Zhao J and Mons B 2016 The FAIR Guiding Principles for scientific data management and stewardship *Sci Data* **3** 160018

AUTHOR BIOGRAPHY

Dhanalekshmi Unnikrishnan Meenakshi holds a doctorate in Pharmacology (specialization in Nanomedicine) from the Council of Scientific and Industrial Research (CSIR)-CLRI, India. She worked as a Scientist in the Council of Scientific and Industrial Research - NEIST, India, and was involved in the government-funded projects for North East Exploration of Pharmaceuticals from natural sources. She has over 10 years of research and teaching experience with leading National and International Organizations. She has been working on an array of projects relating to cancer and gene therapy, nanotechnology, and pharmacology. Her research vicinity focuses on preclinical and clinical trials, mechanism of intoxication, etc. She uses state-of-the-art technology, for systematic evaluation of the efficiency of novel polymeric nanoparticles encapsulated with biologically active agents. Currently, she is a faculty member in the College of Pharmacy, National University of Science and Technology, Muscat, Sultanate of Oman. She has published extensively in nanomedicine, drug delivery, and formulation technology in peer-reviewed reputed journals and books. She bagged various awards and fellowships for her excellent contributions in her respective fields.

4

Big Data in Oncology: Extracting Knowledge from Machine Learning

Arun Kumar Singh[1], Rishabha Malviya[1], Sonali Sundram[1*], and Vetriselvan Subramaniyan[2]

[1]Department of Pharmacy, School of Medical & Allied Science, Galgotias University, Greater Noida, India
[2]Faculty of Medicine, Bioscience and Nursing, MAHSA University, Malaysia
*Corresponding Author: Department of Pharmacy, School of Medical and Allied Sciences, Galgotias University, Greater Noida, India, EMail: sonaliaim13@gmail.com

Abstract

Large amounts of data, rising healthcare expenses, and a focus on individualized treatment have all conspired to push the use of big data in healthcare to new heights in the last few years. Data that is too large or complicated for traditional data processing technologies to decipher is referred to as "big data" in the healthcare industry. Electronic medical records and electronic health records (EMR/EHR) contain data on patient history, diagnosis and treatment, medications, treatment plans, allergies, research center and test results, genomic sequencing, medical imaging, health insurers, and other clinical information. The Internet of Things (IoT) and EMR/EHRs are two examples of big data sources for healthcare. A variety of machine learning algorithms were tested on a variety of healthcare datasets and the results are presented in this study. Big data processing, management, and application issues are also discussed. For this article, machine learning techniques and the necessity to deal with and use large data from a new viewpoint will be discussed.

Keywords: Big Data, Machine Learning, IoT, Electronic Health Record.

4.1 Introduction

Information such as a patient's demographic trends, diagnosis, surgical treatments, prescription drugs, vital signs and vaccinations, laboratory results, and medical tests can be gleaned from large datasets, which depict the information on the quantity, velocity, variety, and time of data production that was originally recorded by the healthcare providers. Sensors, streaming machines, and high-throughput equipment are among the electronic health sources that are amassing more data as medical data collection advances. There are a variety of uses for this healthcare big data, including diagnostics, drug development, precision medicine, illness prediction, and more [1]. There is a wide range of applications for big data, from healthcare to scientific research to business to using the internet for business purposes, and governmental administration [1]. The 5Vs of big data may be summarized as follows:

- **Volume**: Big data is unquestionably represented by the large volume. As the volume of data grows, so does the need for conventional database platforms and procedures to be bolstered to handle it. Biometric sensor readings and genomes, among many other things, are being added to healthcare databases together with personal information and x-ray images. The database grows in size and complexity as a result of all this information.
- **Velocity**: The volume of data that is being created in a short amount of time is referred to as "big data." As a result of social media's data explosion, there is now a greater diversity of data. The amount of health information stored in paper documents, x-ray films, and texts continues to expand at an exponential rate.
- **Variety**: Big data is evident in the wide range of information that may be gleaned from it. For example, a simple text file can hold a database, an Excel spreadsheet, or a CSV file. Structured, unstructured, and semi-structured variations of health data are already present. There are two types of structured information: data that is organized, such as clinical data, and data that is unstructured or semi-structured.
- **Veracity**: This data's accuracy is certain; it is straight out of a bug report. In other words, it is an indicator of data accessibility rather than quality. This feature provides a certification of the accurate diagnosis, treatment, prescription, procedure, and outcome in healthcare data.
- **Value**: Big data reflects the value of data. Considering the costs and advantages of big data, analytics is more important than other concerns.

Healthcare providers and other stakeholders should be compensated based on the value they provide to patients. The primary purpose of healthcare delivery should be to provide great value to patients.

4.2 Application of Healthcare Big Data

Big data applications open up new avenues for the discovery of previously undiscovered information and the development of previously untried approaches to improving healthcare. Patients' profiles, evidence-based care, remote monitoring, and public health are just a few of the many useful applications. The following diagrams depict the big data frameworks and storage systems used in healthcare.

4.2.1 Internet of Things (IoT)

With the help of IoT, healthcare monitoring systems are now a reality, unleashing the potential for people to live healthy lives and enabling doctors to provide superior care. In addition, it has improved patient participation and satisfaction because of the simplicity and speed of communication with doctors. Reducing hospital stays and the number of re-admissions is also a benefit of remote monitoring of patients' health. Healthcare expenditures can be greatly reduced and treatment outcomes improved thanks, to the Internet of Things (IoT).

4.2.2 Digital epidemiology

has been characterized as using digital approaches, from data collecting to evaluate data in the field of digital epidemiology. An array of investigations ranging from environmental studies to cross-sectional surveys may benefit from the usage of this technology. Non-health-related and health-related data sources are also incorporated into the system.

FRED (a framework for reconstructing epidemiological dynamics) is an approachable framework for epidemic modeling in contrast to particular disease models. FRED uses geographic regions to represent each person as an agent. For each person, there is a collection of sociodemographic data such as sex, employment position, and location of the home, as well as a set of the social active network in which they are involved. To represent the illness epidemic in FRED, a synthetic population dataset is used.

A distributed NoSQL database for storing massive amounts of data. The relational schema is not used in a NoSQL database. Significant stores, unit groups, graphs, and document repositories are all types of NoSQL databases. With the key-value system, data is encrypted and saved in digital certificates. The wide-column database is also capable of storing huge amounts of data in a variety of columns. This document is used for semi-structured data, and it stores a lot of information about the document type. Edges between nodes with trace and query links are also stored in the graph database.

Healthcare, video surveillance, social media, computer data, and sensor readings are just a few of the many potential sources of big data. Structured and unstructured data are both acceptable forms of information Large datasets need the use of a big data platform. If you have less than 100 GB of data, you do not need a big data platform, but if you have more than this, you do. Big data Projects and other big data sources demand a big data architecture. Traditional database systems are not capable of handling data that is too big or complicated to be stored and analyzed in a big data architecture [2]. Big data architecture is designed to manage the various types of activities:

- Big data processing in batches
- Massive amounts of data may now be processed in real-time.
- Machine learning and predictive analytics

The data source layer, substrate layer, manage layer, and big data platforms have two core components: an analytic and visualization layer. Semantic-based architecture for mixed multimodal retrieval, extensive security monitoring, and modular software are a few examples of contemporary designs. There are pros and cons to every design.

4.3 Big Data Analytics in Healthcare

From terabytes to trillions of terabytes, sophisticated analytic methods are utilized on structured, semi-structured, and unstructured data in big data analytics. Big data analytics in healthcare can combine and evaluate a great quantity of complex and heterogeneous data, including genomics, epigenetics, transcriptomic, proteomic, metabolic, specialist, and personalized medicine and biological data (Electronic Health Record). At the top of the list is electronic health records (EHRs), which have been implemented in many countries. The basic purpose of EHR is to extract usable information from the health process by analyzing enormous volumes of big data.

An EHR is a computer-based database that stores a patient's medical records in electronic form. An electronic health record (EHR) stores all of a patient's medical information, including their diagnosis, medicines, tests, allergies, and vaccinations. A patient's electronic health record (EHR) is accessible to all healthcare clinicians involved in their care, allowing them to make informed decisions about the patient's treatment. EMR is another name for EHR (Electronic Medical Record). More and more data is being created every second, from a variety of sources, including internet browsing, social media, mobile transactions, and online shopping. The big data paradigm has indeed expanded, and the amount of this type of organized and unstructured data has allowed for fresh viewpoints to be offered. It is now possible to gain a deeper insight into a person's motivations and behavior by using new data sources such as social media and location-based services. Understanding and extracting hidden information that may be used to enhance the user's experience from a large and diverse set of data is made possible by gaining useful insights from this data [3]. An organization's ability to innovate has a great deal to do with the type of EHR system it chooses. Researchers are being compelled by the advent of big data to broaden their perspectives and make strategic decisions about the future. To assist healthcare professionals in making decisions and providing better care for their patients, clinical decision support systems (CDSS) use data analysis. To derive clinical advice from various patient-related data, knowledge management is employed by the CDSS. Healthcare systems provide integrated workflows, help at the site of therapy, and offer care-plan recommendations. To diagnose and enhance patient care, doctors use CDSSs to reduce the number of non-essential tests that are performed, increase patient safety, and prevent potentially harmful and expensive complications. Cost savings, elimination of risk factors, disease prediction, and improved preventative care are only a few of the benefits of using big data in healthcare. What are some of the healthcare industry's most difficult challenges? How do you choose the best treatment for a specific disease? (ii) What are the effects of certain policies on spending and behavior? (iii) In what ways are future healthcare costs projected to rise. (iv) How is it possible to tell if a claim is false? In terms of healthcare costs, is there a regional variation? [4].

Big data analysis tools and methodologies can help answer these questions. Quality healthcare has four major foundations. As an example, there is real-time monitoring of patients as well as patient-centered care, treatment technique improvement, and disease prediction analytics. With descriptive, predictive, and prescriptive big data tools, these four pillars of great healthcare can be effectively handled.

4.3.1 Machine learning for healthcare big data

The phrase "machine learning," which is a subfield of AI, refers to the ability of computer systems to recognize patterns in databases and use that information to solve problems on their own. Based on pre-existing algorithms and datasets, machine learning helps IT systems discover patterns and create appropriate solution concepts. This is why artificial knowledge is developed in machine learning based on experience. In machine learning, datasets are utilized to learn by applying statistical and mathematical techniques to them. There are two primary systems: symbolic and subsymbolic. In contrast to symbolic systems, sub-symbolic systems provide knowledge content in the form of artificial neural networks rather than explicit rules and instances.

These are based on the human brain's principle of implicit representation of knowledge content. Large-scale data, the variety of the kinds of data, high-speed streaming data, and unclear or partial data are all major challenges in machine learning for big data [5]. Supervised, unsupervised, and reinforcement learning are the three main types of machine learning. Table 4.1 shows the various types of machine learning, with categories and examples, as well as a brief description of each.

First, supervised learning is an issue that necessitates a model that can be used to create a mapping between the implementation of an existing system and the goal output, therefore it needs training using labeled data to accomplish this task. As a second benefit, unsupervised learning does not necessitate the use of labeled training data and instead relies on the model to explain or extract associations in data. Reward learning is a type of problem where an agent acts in a context and learns through feedback. The following are some more lessons that must be learned to solve the big data problems:

- learning through the use of representations,
- learning by doing is a popular method of instruction,
- extensive training,
- transfer of knowledge,
- learning in a distributed and concurrent manner, and
- learning based on kernels.

Deep learning plays a crucial part in healthcare, among the many other fields. Deep learning is a sort of machine learning, and addresses the issues

Table 4.1 Different machine learning techniques

S. no	Learning types	Categories	Examples
1	Supervised learning	Learning problems	Support vector machine (SVM), naïve Bayes (NB), neural networks (NN)
2	Unsupervised learning	Learning problems	K-means, Gaussian mixture model, Dirichlet process mixture model
3	Reinforcement learning	Learning problems	Q-learning, R-learning TD learning
4	Semi-supervised learning	Hybrid learning problems	Speech analysis, internet content classification, protein sequence classification
5	Self-supervised learning	Hybrid learning problems	Generative adversarial networks (GANs), autoencoders
6	Multi-instance learning	Hybrid learning problems	Categorization of medical images, action of molecules
7	Inductive learning	Statistical inference	Weather prediction
8	Transductive learning	Statistical inference	K-nearest neighbors (KNN)
9	Online learning	Learning techniques	A decrease in gradient descent
10	Ensemble learning	Learning techniques	Stacking, bagging
11	Deep learning	Learning technique	Voice recognition software, analysis of medical images
12	Transfer learning	Learning techniques	Image classification

that were previously unsolvable by machine learning. Deep learning makes use of Neural Networks to speed up computations while maintaining high accuracy. To better serve the healthcare business, deep learning is helping medical practitioners and academics uncover previously untapped potential in data.

4.3.2 Deep learning for healthcare big data

For everything from defense and monitoring to human–computer interface (HCI) and question–answer systems, DNNs (deep neural networks) are at the cutting edge of machine learning and data-learning data analytics.

DNN architectures fall into three categories: FNN (feed-forward neural network), CNN (convolution neural network), and RNN (recurrent neural network) [6].

Deep learning in healthcare enables clinicians to do more accurate analyses of diseases and to make better treatment choices as a consequence. Hospital management information systems can benefit from the implementation of deep learning technologies to reduce costs, shorten hospital stays and lengthen their duration, control insurance fraud, detect patterns of disease change, provide high-quality healthcare and improve the efficiency with which medical resources are allocated. There are many other sorts of biomedical data that may be utilized, described in the following paragraphs, such as images from medical research, time signals from medical research, and data from wearable devices.

4.3.3 Drug discovery

It is possible to find and create new drugs with the aid of deep learning in the healthcare industry. Patients' medical histories are taken into account while developing a treatment plan using this cutting-edge technology. In addition, this technology accumulates knowledge from the symptoms and tests of patients.

4.3.4 Medical imaging

Heart disease, cancer, and brain tumors may all be detected with the use of medical imaging procedures like MRIs, CT scans, and ECGs. Because of this, physicians can better understand the condition and give the best possible therapy to their patients.

4.3.5 Alzheimer's disease

Alzheimer's disease presents a major challenge to the medical community. Deep learning can be used to diagnose Alzheimer's disease in its early stages.

4.3.6 Genome

Patients may learn more about the diseases that may impact them if a deep learning approach is applied to decipher a genome. The genetics and insurance industries both stand to benefit greatly from deep learning. Parents may monitor their children's health through a digital device in real-time using

Cells Cope's deep learning capabilities, which reduces trips to the doctor. Doctors and patients will be able to benefit greatly from the use of deep learning in the healthcare industry.

4.4 Tools and Tasks

Machine learning (ML) includes a wide variety of activities and approaches [7]. The result of supervised learning tasks may be predicted in advance, such as the existence of a tumor, the duration of survival, or the response to therapy. When there is no obvious result to forecast, unsupervised learning may identify patterns and subgroups in the data. It is often used for more exploratory analysis. As a third kind of machine learning technique, reinforcement-based learning is used to develop strategies for sequential decision-making, such as selecting the most effective treatment plans for cancer patients [8, 9]. The emphasis of this study is on learning environments that are both supervised and unsupervised.

4.4.1 Supervised learning

In this part, we will go over a few of the most prevalent supervised learning strategies used in oncology. It is possible to use these algorithms to predict a continuous (regression) or a discrete (exponential) result based on a collection of input characteristics (classification). Various approaches for supervised learning are tabulated (Table 4.2).

4.4.2 Linear models

Using a linear equation, a linear model connects the different factors to the final result of interest. The coefficients for each of the characteristics are found using a linear regression model. An observation's prediction is then given by a weighted combination of these features (i.e., $\beta_0 + \beta_1 x_1 + \ldots + \beta_p x_p$, where a patient's features are given by variables $x_1 \ldots x_p$).

According to linear regression, the result is directly proportional to feature values, and this connection is both linear and additive. Logistic and Cox regression models, for example, both assume an additive connection between features, but the linear function is transformed dependent on the prediction objective. This is the case for both models.

Because of the ease with which they may be understood and used, linear techniques have long been favored for use in mathematical modeling [10].

Table 4.2 Approaches for supervised learning.

Model type	**Overview**	**Benefits**	**Drawbacks**	**Algorithms examples**
Linear models	Additive models use a linear function of patient characteristics to compute risk.	Intuitive: Coefficients reveal the link between attributes and outcomes in an explicit manner.	In the absence of naturally occurring interactions between variables, something is said to be additive.	Linear regression, logistic regression [9]
Decision trees	Subpopulations of the feature space with similar result predictions can be partitioned using a single decision tree.	Variable interactions can be represented in a nonlinear way. Interpretable: Decision routes identify high- and low-risk feature combinations.	Does not reflect the continuous links between factors and results naturally.	CART [10],optimal classification trees [11, 12]
Ensemble methods	Methods that use a large number of decision trees to create predictions. The final forecast is based on the outcomes of all the trees that were built on a portion of the data and features.	Complex interactions can be captured more effectively when multiple models are combined into a nonlinear ensemble.	There is not a single final model to relate traits to outcomes, which makes it more difficult to interpret. Model interpretations must be generated through post-processing.	Random forests [13], XGBoost [14]
Neural networks	Models with a large number of weighted modifications of the input features that produce highly nonlinear predictions	Complex interactions are captured in a non-linear fashion. High-dimensional, unstructured data can be processed (e.g., images)	The method in a "black box," meaning it is tough to decipher. The complexity of training: a large number of parameters must be tweaked to build models	Convolutional and recurrent neural networks [15]

There are a large number of current risk ratings and prediction models built on top of these types of models. However, the properties of the results are generally nonlinear. There may be a difference in the impact of tumor size on tumor recurrence risk according to age. Such interconnections between variables are not easily captured by a linear model. As an example of how to model a combined impact between age and tumor size, one may design an extracted feature that combines the two into an interaction variable. This is performed on an ad hoc basis since it is difficult to analyze all potential changes of pairs or larger groups of variables. In the following sections, we will describe nonlinear techniques that take into consideration variable interactions from the beginning.

4.4.3 Decision tree models

Leo Breiman first introduced classification and regression trees (CARTs) as an alternative to linear models [10]. Data are divided into smaller groups by feature splits in a decision tree, and the leaves of the tree hold the final subgroups of observations. As a result of these feature splits, the final tree can divide the population into discrete leaves [11]. Each leaf generates a single forecast. When dealing with classification, the prediction takes the form of a probability, which is often computed using the leaf's periodicity of the most frequent result, and a numerical value which is typically based on the leaf's overall average of the outcomes. A schematic representation of binary classification is shown as Figure 4.1.

The model's tree-based structure enables it to capture correlations between features that are not linearly related. For example, it is capable of distinguishing between patients of varying ages to their risk levels. Some comorbidities may only be relevant to male patients, for example, because of interdependencies between variables.

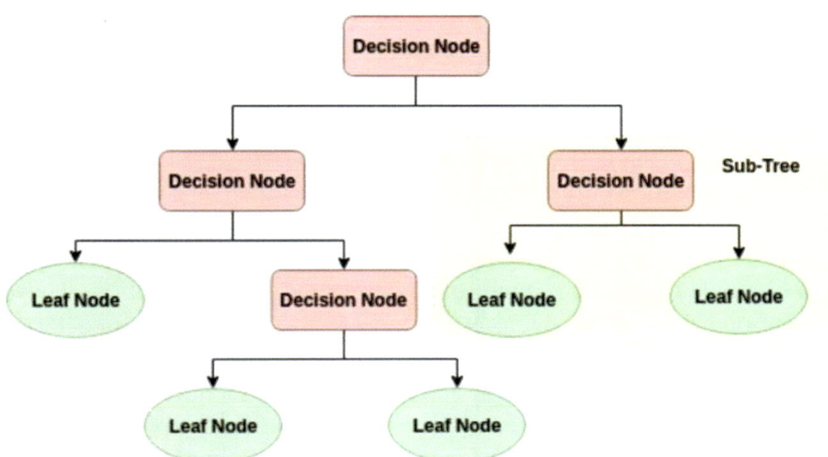

Figure 4.1 A decision tree for binary classification.

Loss functions, which assess the prediction error, are used to identify characteristic breaks in decision trees that minimize it. Observations that were wrongly categorized in a classification task would have a misclassification rate, but observations that were successfully classified in a regression task would have an absolute mean difference. A decision tree is constructed using CART by performing greedy recursive splits. Using the split that minimizes error, it first divides up the data into two subsets, and then splits those subsets again, going on to the next level without altering the previous ones. Allowing just those splits that satisfy an error improvement criterion regulates the splitting. Last but not least, the clustering algorithm is edited to remove splits that do not provide a significant improvement in model error. OCT, a newer decision tree algorithm, has been proposed as an alternative in the last several years. Among the OCTs, an optimization framework takes into account the tree's structure while considering possible splits [12]. Data partitions that cannot be found using a greedy technique may be recovered using a local search process. A complexity parameter is also used to limit the depth of the tree. In general, OCTs outperform CARTs while retaining CART's high level of interpretability.

4.4.4 Ensemble models

Ensemble techniques, such as random woods and gradient-boosted machines, go beyond the structure of a decision tree. These methods build a huge decision tree and then utilize the models to forecast [13–17]. When a random forest is used to build models, each tree is trained on a random set of data and attributes. To conclude, all tree projections are integrated. Each consecutive tree is generated to give greater weight to observations that had a high degree of error in the previous three. To improve, this is done in stages. In terms of speed, this error-correction technique surpasses random forests.

Since ensemble algorithms aggregate multiple unique trees, no one model can unambiguously relate the input attributes to the final prediction. They are harder to understand than linear programming and decision trees because they lack coefficients or feature divisions. There is a lack of openness in the application area, which is critical for adoption. Others, such as Shapley, are also available. It has been suggested that the use of additional explanation and feature-importance metrics calculated from the models might provide insights [18, 19].

4.4.5 Neural networks

Neural networks use a multilayer network of mathematical changes to relate information to projected outcomes. With just one layer, the feedforward neural network shown in Figure 4.2 is easy to understand. The model uses linear functions to transform input characteristics into vertices in a hidden unit. Non-linear activation functions are employed to map these nodes into a result. These network dynamics enable neural network models to capture intricate connections between the parameters and the result.

Some of the more recent developments in neural networks include recurrent neural networks, convolutional neural networks, and generative adversarial networks. Everything has been shown in Schmidhuber's comprehensive report. Deep learning, a kind of machine learning based on neural networks, is built on techniques like these [20].

The capacity of neural networks to synthesize unprocessed photos and unstructured text has made them particularly appealing. Whenever the number of data characteristics is higher than the number of occurrences in a high-dimensional situation, they may scale to unstructured data types. Interpretability is sacrificed for model power and complexity. For this reason, neural networks have come to be referred to as "black box" approaches. The lack of interpretability restricts their usefulness in certain therapeutic situations, as it does with ensemble approaches.

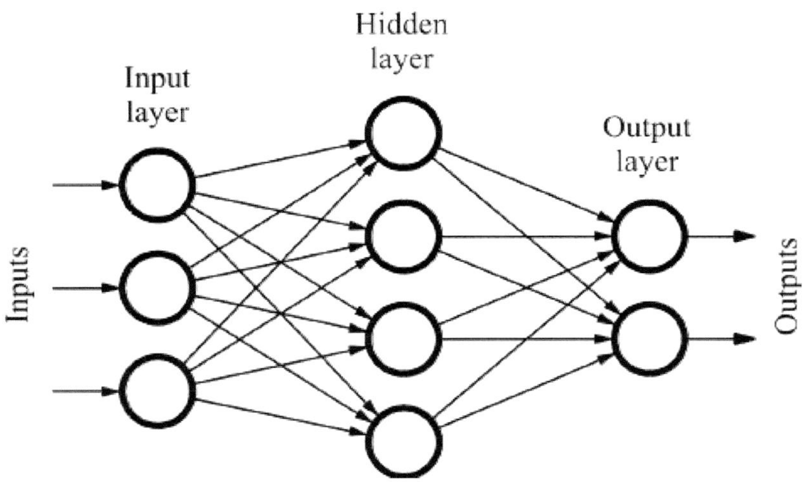

Figure 4.2 An example of a feedforward neural network.

4.4.6 Unsupervised learning

Unsupervised learning, on the other hand, is less focused on predicting a certain result; rather, it aims to uncover the underlying patterns in data. Non-task-specific (i.e., based on an anticipated result such as survival) outputs from these methodologies give broad knowledge. Table 4.3 lists the many unsupervised learning versions.

This evaluation focuses on clustering since it has the most obvious meaning in a healthcare context. Patient EMR data clusters for a certain illness type, for example, can include information on several patient profiles within that condition. With just two variables, such as age and BMI, as shown in Figure 4.3, we may see an example of basic clustering (BMI). If you want to maximize similarity and separation across groups, you divide your data into K distinct clusters. There should be clusters that are homogeneous and different from each other in a proper cluster assignment. Patients are more comparable if they are of similar age and BMI.

Most people use K-means and hierarchical clustering for clustering data. To identify the optimal K, K-means clustering employs a heuristic. Hierarchical clustering starts with a single cluster for each observation and grows slowly as the distance between the clusters increases. For any K clusters, a tree-like structure is built, making it easy to discover the correct assignment. Because exploratory data analysis relies so heavily on grouping, unsupervised learning offers a big obstacle in comprehending clusters. It is common for users to want to know the specifics of the clusters they have picked. As an example, a cluster may include people with a large number of comorbidities, whereas an older patient may make up another. The mean and variance of

Table 4.3 Overview of unsupervised learning tasks.

Tasks	Objective	Algorithm examples
Clustering	Organize a collection of data into groups based on shared characteristics.	K-means [20], hierarchical clustering [21], DBScan [22]
Matrix factorization	Reduce the dimensionality of highly linked data by identifying the underlying feature structure.	Principal component analysis [23], singular value decomposition [24], A priori algorithm [25]
Association analysis	Automatically extract "A implies B"-style rules and dependencies between features.	

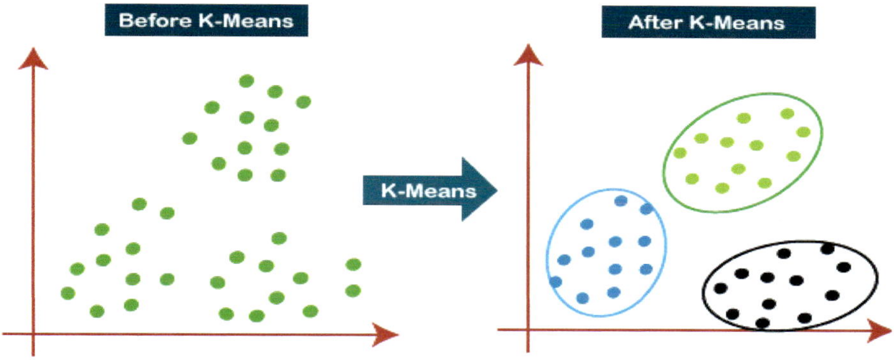

Figure 4.3 A two-dimensional example of a cluster with a K of **3.**

each attribute for all clusters may be examined to see which features differ the most across groups. However, with such large feature spaces [21], this may be challenging. As an alternative to using a clustering approach, a multiclass classification model such as CART or OCT may be employed [21]. Using the output tree, we can predict which nodes belong to which cluster to learn more about the distinctive features of each group [22–28]. Clustering algorithms do not take into consideration the interpretation of the data they create, even if this is a post-processing step. As additional clustering algorithms have been developed, they have produced clusters that can be directly analyzed [29–31].

4.4.7 Medical data resources

Scalable and objective data analysis is made possible by ML models. In terms of structure and content, healthcare data sources are vastly different. Data encoding strategies, increased computer power, and algorithmic advancements have made it possible for ML to make use of a wealth of new data sources. While clinical data is our primary emphasis, other sources such as financial claims records or data from national registries may also give useful information.

4.4.8 EMR

Recorded patient experiences with a medical system are the basis for electronic medical records (EMR). These records include the patient's basic demographics, health and psychosocial history, and a history of their treatment, including prescriptions, diagnoses, operations, vitals, laboratory findings, and their outcomes of tests. For large-scale analysis, this is a priceless resource since it offers structured data that is very uniform among patients in a medical system and generally consistent between systems that employ the same EMR provider. Additional benefits include providing a full and long-term perspective of an individual patient's health history in the EMR. The EMR data format, which naturally lends itself to ML algorithms, allows for the direct use of many clinical variables as model inputs. Unstructured elements like free-text notes and reports are critical to the success of electronic medical records (EMRs), even if their structure makes them appealing. A radiological report may reveal information about a tumor not found in the patient's electronic medical record [32]. Structured data limits the accessibility of information. This difficulty was addressed by the development of techniques in the field of natural language processing (NLP) for transforming raw text into characteristics that can be used as inputs to ML algorithms These approaches range from simple word frequency counting to more complicated systems like GloVe33, which encodes words as vectors based on their contextual meaning [33].

4.4.9 Data curation challenges

Quality data collection and translation into clinical features are critical for ML. Particular difficulties arise in the extraction and purification of EMR data. It is difficult to merge datasets from various organizations because of the great variety of data-gathering techniques employed by EMR systems. Even within a single EMR, it is challenging to merge patient information into a single data collection. The three most challenging aspects of the process are data collection and transmission, data imputation, and medical validation.

4.4.10 Data extraction and transfer

The first step in data curation is to get the original electronic medical records (EMRs). Cancer-specific data, such as stage and therapy, is also needed for oncology studies. Cancer-specific software and registries are typically used to store this data. Hospital electronic medical records are also used. Furthermore, the methods used by various hospital departments while collecting data sometimes diverge. As an example, a researcher may be interested in obtaining the total neutrophil count (ANC). The desired trait may not be captured by a single standardized field. Different regions of the hospital, separate equipment, or even different labs may all be used to collect ANC samples through time. To achieve the most correct answer, one must combine all of the different meanings into a single field. Data uniformity against physician freedom is a fundamental trade-off. There are numerous factors to consider when combining (roughly) similar areas of study from a research standpoint, and doctors generally prefer to design their workflows. The physician's capacity to communicate information is not affected, but data aggregation is more challenging. To meet this problem, clinical ontologies must relate data pieces to clinically significant domains, as shown by this example. In this way, it would be easier for researchers or doctors to organize clinical characteristics. Codes like Logical Names and Codes or Anatomic Therapeutic Chemical Classification Scheme exist, however, they are typically inadequate and not uniformly implemented across EMRs [34, 35].

4.4.11 Data imputation

Even after successful extraction, certain clinical data will be absent. Even if all of a patient's lab data are documented, their medical history may be incomplete. There are a variety of ways to deal with data that is missing. The term "full case analysis" refers to the exclusion of observations with any missing data. However, this might be deceptive. There may be a systematic lack of data in certain sectors. Removing patients with incomplete data may bias the sample toward nonsmokers, for example, since smokers may be hesitant to reveal their smoking status. Missing data may be replaced by the mean, median, or mode for continuous variables, or the mode for categorical variables. Although this may lead to prejudice, it is also possible. Large-scale data analysis has led to the introduction of a variety of increasingly complex imputation algorithms. Multiple imputations by linked equation teach from the entire feature space rather than looking at each variable individually [36]. For optimal imputation, the global structure of the data is used to optimize imputation [37]. This approach has been used in the medical field to account for datasets with several instances of the same patient with time (MedImpute) [38].

4.4.12 Clinical validation

Data must be assessed for clinical validity after extraction and organization into an initial feature space. Big datasets need repeatable and scalable validation tests, which can be performed on small cohorts, but machine learning thrives on large ones. Medical applications offer extra complications

Table 4.4 Key elements for clinical data validation.

Challenges	Approach
A patient's clinically reasonable (even though rare) numbers may indicate an error with data entry or unit selection.	Make sure that the variable boundaries are clinically validated: There should be suitable minimum and maximum values. Impute all out-of-bounds entries with NAs using this approach.
Internal consistency refers to the uniformity of observations made within a single study. Is it possible, for example, to have both a height and weight column?	BMI, critical laboratory ratios, and other derived fields must be calculated and their limits checked. Consider the relationship between variables as well, such as that a patient can only be diagnosed with stage IV cancer if they have metastases. Approach: the patient may be omitted if the data in these fields cannot be resolved by a chart review.
Is there any consistency in the entries made over time? As an example, does the variation in blood pressure over time appear to be logical, or does it indicate data entry errors?	Values with a large relative rise or drop between chart review visits should be flagged. For imputation, NA values should be used to replace any errant spikes. Since relative change % will be more affected by noise, it is preferable to discretize the term "change in X."

to data cleansing and verification, which are two of the key challenges to ML [39, 40]. Distinguishing between incorrect values and genuine rare situations is also a difficulty. However, despite the need to clean data to reduce value mistakes or inconsistencies caused by incorrect data input, severe situations should not be inadvertently adjusted. Conditional functional dependencies, statistical estimate methods, and crowdsourcing have been used to clean data, among other things [41–43]. Table 4.4 outlines three major challenges in data validation: value feasibility, internal consistency, and temporal consistency [44–46].

4.5 Applications

4.5.1 Diagnosis and early detection

To make an accurate cancer diagnosis, a combination of clinical data, such as gene expression levels, x-ray imaging, and histology, is required. Gene expression profiles have been utilized since the early 2000s to find cancer biomarkers [47–50]. Computer vision advancements have switched the emphasis from raw pictures to diagnosis. In light of the importance of mammography in detecting cancer, breast cancer has long been a pioneer in this area. Since 1995, significant progress has been made in the diagnosis of breast cancer via mammography [51–53]. CT scans, which are often used to detect lung cancer, have used similar methods for making the diagnosis [54]. A comprehensive evaluation of imaging-based diagnostic applications is provided by Hu et al. Image-based diagnostics has also been used for histology In addition to breast cancer lymph node metastases, convolutional neural networks have effectively detected prostate cancer and bladder cancer [55–59].

Scaled trend analysis over a large number of cancer patients may also be used to identify malignancies before they have spread to other organs. Detection of cancer at an early stage is commonly accepted, but it is difficult because the features that are indicative of cancer appearance are typically subtle and vary from patient to patient [60]. Mammograms and CT scans have been used to forecast the likelihood of breast cancer or lung cancer in the future [61, 62]. Researchers have used EMR data in addition to gene expression data to predict the probability of pancreatic cancer in a high-risk group of patients [63, 34]. Policies and procedures for cancer screening could benefit from these early warning systems. They have the greatest promise for early intervention and better patient outcomes.

4.5.2 Cancer classification and staging

Several criteria, including the kind of cancer and its stage, dictate the classification of many forms of cancer. Clinical trial eligibility and prognosis prediction are commonly based on this classification, therefore its utilization might have a substantial influence on treatment recommendations and patient care. Staging criteria developed by the American Joint Committee on Cancer (AJCC) have been regarded as the gold standard in oncology since they were first published in 1977 [65, 66]. Only the initial tumor size (T), the number of affected lymph nodes (N), and the presence of metastasis (M) may be used to determine a patient's disease stage (M). Classification techniques for cancer benefit from their simplicity since they only need a minimal quantity of data to be collected. These approaches, on the other hand, rely on cutoff values derived from clinical data and ignore potentially important clinical aspects. The shortcomings of the existing system have prompted an investigation into possible alternatives. To put it another way, cancer staging is an attempt to categorize patients into distinct risk categories. ML might enhance prognostic differentiation across phases by allowing the definition of staging criteria directly from data. It is possible to classify patients according to a model that predicts disease-free survival. In a sense, this is a cancer classification system.

Pancreatic cancer and intrahepatic cholangiocarcinoma are two examples of cancers that researchers have used this strategy to treat [67, 68]. Although the methodologies differ, they all use huge datasets to uncover new variables and provide superior patient stratification to the AJCC scheme. When the AJCC was founded, it was impossible to concurrently analyze thousands of patients in a manner that can now be done with these studies.

A supervised learning framework is often used in cancer staging to provide survival predictions, and these predictions are subsequently examined to develop staging criteria. Unsupervised learning, on the other hand, has been successful in detecting unique cancer cohorts. For lung and breast cancer, clustering has been used to identify discrete subgroups with better prognoses, despite the method not explicitly taking survival into account [69–71]. Due to the noise and difficulties in assessing survival, it may be desirable to generate subgroups without a stated result. Patients may now be divided into more distinct clinical categories thanks to a fresh viewpoint provided by clustering in cancer classification. Also, cancer-related gene signatures have been discovered by using unsupervised learning, as well. Cancer researchers have used this method to identify unique subtypes of the disease. Better disease knowledge and treatment options may be made with the identification of these cohorts [72].

4.5.3 Evaluation and prediction of treatment reactions

ML may also provide prescriptive insights. Personalized predictions of treatment response and anticipated adverse effects of alternative medications may help doctors and patients make treatment decisions and monitor patients more effectively. The increased availability of cell line data has made large-scale drug sensitivity estimates based on genetic profiles conceivable [73–76]. Clinical response to leucovorin, fluorouracil, and oxaliplatin in colorectal cancer patients has been predicted using genetic information in pan-cancer analysis as well as more specialized interactions [77–79]. Breast cancer neoadjuvant chemotherapy patients who have non–small-cell lung cancer (NSCLC) radiomics and the combination of clinical and imaging features have been utilized to predict treatment response. [80]. At the medicine or patient level, further research has been done to discover potential side effects of therapies. To evaluate tumor response, two-dimensional tumor measures using RECIST may be replaced with ML [81, 82]. To employ characteristics that could be measured by radiologists, it was necessary to rely on two-dimensional measurements [83, 84]. There are certain issues with this technique, as studies have discovered that RECIST may not correctly correspond with improvement in outcomes [85]. NSCLC patients have had RECIST criteria automatically detected using ML, much as in diagnostic imaging [86]. For NSCLC and brain cancers, several studies have used CT scan sequences and volumetric data from magnetic resonance imaging to evaluate response [87, 88].

4.6 Conclusion

Patients' health profiles and prediction models are improved through the use of big data in healthcare, allowing us to better detect and treat illness. Today's healthcare and the pharmaceutical business are hampered by a lack of knowledge about disease biology. Agamaegating more and more information on numerous scales for what makes an illness, from DNA, proteins, and metabolites, to cells and tissues, is where big data comes into play. Big data integration can help us better understand the scales of biology that we need to be modeling. A variety of machine learning approaches were used in this work to analyze and handle large volumes of data. Machine learning models are evaluated using metrics derived from large datasets. In addition, machine learning aids decision-making by using a variety of strategies to anticipate illness and make accurate diagnoses on time, both of which have the potential to improve the health of patients. In the future, illnesses may be avoided at an early stage by predicting the information. Models may be built using a variety of methods to aid and forecast variables, and the accuracy is always flawless thanks to machine learning. It is been a boon to human health research because of the quick implementation of EHR, which has resulted in an abundance of fresh data on patients. In the future, the healthcare sector and healthcare organizations will rapidly and widely incorporate and exploit big data and machine learning. Big data analytics raises challenges about protecting privacy, setting standards, enforcing governance, and continually developing tools and technology. Amorphous as they may be now, healthcare big data analytics and applications may be accelerated by quick advancements in platforms and technologies.

Acknowledgment

The authors are highly thankful to the Department of Pharmacy of Galgotias University, Greater Noida, for providing all necessary support for the completion of the work.

Funding

There is no source of funding.

Conflict of Interest

There is no conflict of interest in the publication of this chapter's content.

References

[1] Manogaran G, Lopez D, Thota C, et al. (2017) Big data analytics in healthcare internet of things, In: Innovative Healthcare Systems for the 21st Century, Springer, Cham, 263-284.
[2] Lopez D and Manogaran G, (2017) A survey of big data architectures and machine learning algorithms in healthcare. Int J Biomed Eng Technol 25: 182.
[3] Khamlichi KY, Chaoui NEH and Khennou F, (2018) Improving the use of big data analytics within electronic health records: A case study based open HER. Procedia Compute Sci 127: 60-68.
[4] Sharma M, Kaur P and Mittal M, (2018) Big data and machine learning-based secure healthcare framework. Procedia Compute Sci 132: 1049-1059.
[5] Xu Y, Qiu J, Wu Q, et al. (2016) A survey of machine learning for big data processing. EURASIP J Adv Signal Proc 2016: 67.
[6] Gerrero-Curieses A, Munoz-Romero S, Bote-Curiel L, et al. (2019) Deep learning and big data in healthcare: A double review for critical beginners. Appl Sci 9: 2331.
[7] Hastie T, Tibshirani R, Friedman J: The Elements of Statistical Learning: Data Mining, Inference, and Prediction. New York, NY, Springer, 2013.
[8] Padmanabhan R, Meskin N, Haddad WM: Reinforcement learning-based control of drug dosing for cancer chemotherapy treatment. Math Biosci 293:11-20, 2017.

[9] Pedregosa F, Varoquaux G, Gramfort A, et al: Scikit-learn: Machine learning in Python. http://scikit-learn.sourceforge.net
[10] Breiman L, Friedman JH, Olshen RA, et al: Classification and Regression Trees. New York, NY, Routledge, 1984
[11] Interpretable AI: Interpretable AI documentation. https://docs.interpretable.ai/stable/(accessedon15September,2021)
[12] Bertsimas D, Dunn J: Optimal classification trees. Mach Learn 106:1039-1082, 2017
[13] Breiman L: Random forests. Mach Learn 45:5-32, 2001
[14] Chen T, He T, Benesty M, et al: xgboost: Extreme gradient boosting. https://cran.r-project.org/package=xgboost
[15] Schmidhuber J: Deep learning in neural networks: An overview. Neural Netw 61:85-117, 2015
[16] Friedman JH: Greedy function approximation: A gradient boosting machine. https://projecteuclid.org/download/pdf_1/euclid.aos/1013203451(accessedon25decemeber2021)
[17] Chen T, Guestrin C: XGBoost: A scalable tree boosting system. https://arxiv.org/pdf/1603.02754.pdf(accessedon15January,2022)
[18] Lundberg SM, Lee SI: A unified approach to interpreting model predictions. https://github.com/slundberg/shap(accessed on 15 January, 2022)
[19] Lundberg SM, Erion G, Chen H, et al: From local explanations to global understanding with explainable AI for trees. Nat Mach Intell 2:56-67, 2020
[20] MacQueen J: Some methods for classification and analysis of multivariate observations. https://www.cs.cmu.edu/~bhiksha/courses/mlsp.fall2010/class14/macqueen.pdf
[21] Sneath PHA, Sokal RR: Numerical Taxonomy: The Principles and Practice of Numerical Classification. San Francisco, CA, W. H. Freeman and Company, 1973
[22] Ester M, Kriegel H-P, Sander J, et al: A density-based algorithm for discovering clusters in large spatial databases with noise. https://www.aaai.org/Papers/KDD/1996/KDD96-037.pdf
[23] Rao CR: The use and interpretation of principal component analysis in applied research. Sankhya Indian J Stat Ser A 26:329-358, 1964
[24] Golub GH, Reinsch C: Handbook series linear algebra: Singular value decomposition and least squares solutions. http://people.duke.edu/~hpgavin/SystemID/References/Golub+Reinsch-NM-1970.pdf

[25] Agrawal R, Srikant R: Fast algorithms for mining association rules in large databases, in: Proceedings of the 20th International Conference on Very Large Data Bases. San Francisco, CA, Morgan Kaufmann Publishers, 1994, pp 487-499
[26] Radev DR, Jing H, Stys M, et al: Centroid-based summarization of multiple documents. Inf Process Manage 40:919-938, 2004
[27] Jain AK, Murty MN, Flynn PJ: Data clustering: A review. ACM Comput Surv 31:264-323, 1999
[28] Hancock TP, Coomans DH, Everingham YL: Supervised hierarchical clustering using CART. https://www.mssanz.org.au/MODSIM03/Volume_04/C07/06_Hancock.pdf(lastaccessedon25December2021)
[29] Bertsimas D, Orfanoudaki A, Wiberg H: Interpretable clustering: An optimization approach. Mach Learn 1-50, 2020
[30] Fraiman R, Ghattas B, Svarc M: Interpretable clustering using unsupervised binary trees. Adv Data Anal Classif 7:125-145, 2013
[31] Blockeel H, De Raedt L, Ramon J: Top-down induction of clustering trees. https://arxiv.org/pdf/cs/0011032.pdf(Lastaccessedon27december 2021)
[32] Manning C, Schutze H: Foundations of Statistical Natural Language Processing. Cambridge, MA, MIT Press, 1999
[33] Pennington J, Socher R, Manning CD: GloVe: Global vectors for word representation. https://nlp.stanford.edu/pubs/glove.pdf
[34] Stephens ZD, Lee SY, Faghri F, et al: Big data: Astronomical or genomical? PLoS Biol 13: e1002195, 2015
[35] Aerts HJWL, Velazquez ER, Leijenaar RTH, et al: Decoding tumour phenotype by noninvasive imaging using a quantitative radiomics approach. Nat Commun 5: 4006, 2014
[36] Miotto R, Wang F, Wang S, et al: Deep learning for healthcare: Review, opportunities and challenges. Brief Bioinform 19:1236-1246, 2018
[37] McDonald CJ, Huff SM, Suico JG, et al: LOINC, a universal standard for identifying laboratory observations: A 5-year update. Clin Chem 49:624-633, 2003
[38] WHO: The Anatomical Therapeutic Chemical Classification System with Defined Daily Doses (ATC/DDD). https://www.who.int/classifications/atcddd/en/
[39] van Buuren S, Groothuis-Oudshoorn K: mice: Multivariate imputation by chained equations in R. J Stat Softw 45:2-20, 2011

[40] Bertsimas D, Pawlowski C, Zhuo YD: From predictive methods to missing data imputation: An optimization approach. http://jmlr.org/papers/v18/17-073.html
[41] Bertsimas D, Orfanoudaki A, Pawlowski C: Imputation of clinical covariates in time series. http://arxiv.org/abs/1812.00418
[42] Rahm E, Do H: Data cleaning: Problems and current approaches. IEEE Data Eng Bull 23:3-13, 2000
[43] Chu X, Ilyas IF, Krishnan S, et al: Data cleaning: Overview and emerging challenges. Proceedings of the ACM SIGMOD International Conference on Management of Data, San Francisco, CA, June 26-July 1, 2016, pp 2201-2206
[44] Bohannon P, Fan W, Geerts F, et al: Conditional functional dependencies for data cleaning. https://ieeexplore.ieee.org/document/4221723/
[45] Krishnan S, Wang J, Franklin MJ, et al: SampleClean:Fast and reliable analytics on dirty data.http://sites.computer.org/debull/A15sept/p59.pdf
[46] Chu X, Morcos J, Ilyas IF, et al: Katara: A data cleaning system powered by knowledge bases and crowdsourcing. Proceedings of the ACM SIGMOD International Conference on Management of Data, Melbourne, Victoria, Australia, May 31-June 4, 2015, pp 1247-1261
[47] Tan AC, Gilbert D: Ensemble machine learning on gene expression data for cancer classification. Appl Bioinformatics 2:S75-S83, 2003 (suppl 3) http://europepmc.org/abstract/MED/15130820
[48] Hwang K-B, Cho D-Y, Park S-W, et al: Applying machine learning techniques to analysis of gene expression data: Cancer diagnosis, in Lin SM, Johnson KF (eds): Methods of Microarray Data Analysis: Papers from CAMDA '00. Boston, MA, Springer, 2002, pp 167-182
[49] Danaee P, Ghaeini R, Hendrix DA: A deep learning approach for cancer detection and relevant gene identification. Pac Symp Biocomput 22:219-229, 2017
[50] Ye QH, Qin LX, Forgues M, et al: Predicting hepatitis B virus-positive metastatic hepatocellular carcinomas using gene expression profiling and supervised machine learning. Nat Med 9:416-423, 2003
[51] Wolberg WH, Street WN, Mangasarian OL: Image analysis and machine learning applied to breast cancer diagnosis and prognosis. Anal Quant Cytol Histol 17:77-87, 1995
[52] Shen L, Margolies LR, Rothstein JH, et al: Deep learning to improve breast cancer detection on screening mammography. Sci Rep 9:12495, 2019

[53] Ramos-Pollan R, Guevara-L t' opez MA, Su t' arez-Ortega C, et al: Discovering mammography-based machine learning classi t' fiers for breast cancer diagnosis. J Med Syst 36:2259-2269, 2012
[54] Sun W, Zheng B, Qian W: Automatic feature learning using multichannel ROI based on deep structured algorithms for computerized lung cancer diagnosis. Comput Biol Med 89:530-539, 2017
[55] Hu Z, Tang J, Wang Z, et al: Deep learning for image-based cancer detection and diagnosis: A survey. Pattern Recognit 83:134-149, 2018
[56] Madabhushi A: Digital pathology image analysis: Opportunities and challenges. Imaging Med 1:7-10, 2009
[57] Litjens G, Sanchez CI, Timofeeva N, et al: Deep learning as a tool for increased accuracy and ef t' ficiency of histopathological diagnosis. Sci Rep 6:26286, 2016
[58] Zhang Z, Chen P, McGough M, et al: Pathologist-level interpretable whole-slide cancer diagnosis with deep learning. Nat Mach Intell 1:236-245, 2019
[59] Liu Y, Gadepalli K, Norouzi M, et al: Detecting cancer metastases on gigapixel pathology images. http://arxiv.org/abs/1703.02442
[60] Etzioni R, Urban N, Ramsey S, et al: The case for early detection. Nat Rev Cancer 3:243-252, 2003
[61] Yala A, Lehman C, Schuster T, et al: A deep learning mammography-based model for improved breast cancer risk prediction. Radiology 292:60-66, 2019
[62] Huang P, Lin CT, Li Y, et al: Prediction of lung cancer risk at follow-up screening with low-dose CT: A training and validation study of a deep learning method. Lancet Digit Health 1:e353-e362, 2019
[63] Kim BJ, Kim SH: Prediction of inherited genomic susceptibility to 20 common cancer types by a supervised machine-learning method. Proc Natl Acad Sci USA 115:1322-1327, 2018
[64] Boursi B, Finkelman B, Giantonio BJ, et al: A clinical prediction model to assess risk for pancreatic cancer among patients with new-onset diabetes. Gastroenterology 152:840-850.e3, 2017
[65] American Joint Committee on Cancer: Cancer Staging Manual. Chicago, IL, American Joint Committee on Cancer, 1977
[66] Amin MB, Greene FL, Edge SB, et al: The Eighth Edition AJCC Cancer Staging Manual: Continuing to build a bridge from a population-based to a more "personalized" approach to cancer staging. CA Cancer J Clin 67:93-99, 2017

[67] Das A, Ngamruengphong S: Machine learning based predictive models are more accurate than TNM staging in predicting survival in patients with pancreatic cancer. Am J Gastroenterol 114:S48, 2019

[68] Tsilimigras DI, Mehta R, Moris D, et al: A machine-based approach to preoperatively identify patients with the most and least benefit associated with resection for intrahepatic cholangiocarcinoma: An international multi-institutional analysis of 1146 patients. Ann Surg Oncol 27:1110-1119, 2020

[69] Chen D, Xing K, Henson D, et al: Developing prognostic systems of cancer patients by ensemble clustering. J Biomed Biotechnol 2009:632786, 2009

[70] Lynch CM, Van Berkel VH, Frieboes HB: Application of unsupervised analysis techniques to lung cancer patient data. PLoS One 12:e0184370, 2017

[71] Aure MR, Vitelli V, Jernstrom S, et al: Integrative clustering reveals a novel split in the luminal A subtype of breast cancer with impact on outcome. Breast Cancer Í Res 19:44, 2017

[72] Kakushadze Z, Yu W: *K-means and cluster models for cancer signatures. Biomol Detect Quantif 13:7-31, 2017

[73] Menden MP, Iorio F, Garnett M, et al: Machine learning prediction of cancer cell sensitivity to drugs based on genomic and chemical properties. PLoS One 8: 61318, 2013

[74] Garnett MJ, Edelman EJ, Heidorn SJ, et al: Systematic identification of genomic markers of drug sensitivity in cancer cells. Nature 483:570-575, 2012

[75] Huang C, Mezencev R, McDonald JF, et al: Open source machine-learning algorithms for the prediction of optimal cancer drug therapies. PLoS One 12: e0186906, 2017

[76] Ding MQ, Chen L, Cooper GF, et al: Precision oncology beyond targeted therapy: Combining omics data with machine learning matches the majority of cancer cells to effective therapeutics. Mol Cancer Res 16:269-278, 2018

[77] Huang C, Clayton EA, Matyunina LV, et al: Machine learning predicts individual cancer patient responses to therapeutic drugs with high accuracy. Sci Rep 8: 16444, 2018

[78] Lu W, Fu D, Kong X, et al: FOLFOX treatment response prediction in metastatic or recurrent colorectal cancer patients via machine learning algorithms. Cancer Med 9:1419-1429, 2020

[79] Coroller TP, Agrawal V, Narayan V, et al: Radiomic phenotype features predict pathological response in non-small cell lung cancer. Radiother Oncol 119:480-486, 2016

[80] Mani S, Chen Y, Arlinghaus LR, et al: Early prediction of the response of breast tumors to neoadjuvant chemotherapy using quantitative MRI and machine learning. AMIA Annu Symp Proc 2011:868-877, 2011

[81] Bloomingdale P, Mager DE: Machine learning models for the prediction of chemotherapy-induced peripheral neuropathy. Pharm Res 36:35, 2019

[82] Abajian A, Murali N, Savic LJ, et al: Predicting treatment response to intra-arterial therapies for hepatocellular carcinoma with the use of supervised machine learning: An artificial intelligence concept. J Vasc Interv Radiol 29:850-857.e1, 2018

[83] Therasse P, Arbuck SG, Eisenhauer EA, et al: New guidelines to evaluate the response to treatment in solid tumors. J Natl Cancer Inst 92:205-216, 2000

[84] Eisenhauer EA, Therasse P, Bogaerts J, et al: New response evaluation criteria in solid tumours: revised RECIST guideline (version 1.1). Eur J Cancer 45:228-247, 2009

[85] Villaruz LC, Socinski MA: The clinical viewpoint: definitions, limitations of RECIST, practical considerations of measurement. Clin Cancer Res 19:2629-2636, 2013

[86] Arbour KC, Anh Tuan L, Rizvi H, et al: ml-RECIST: Machine learning to estimate RECIST in patients with NSCLC treated with PD-(L)1 blockade. J Clin Oncol 37:9052-9052, 2019 (suppl; abst 9052)

[87] Xu Y, Hosny A, Zeleznik R, et al: Deep learning predicts lung cancer treatment response from serial medical imaging. Clin Cancer Res 25:3266-3275, 2019

[88] Kickingereder P, Isensee F, Tursunova I, et al: Automated quantitative tumour response assessment of MRI in neuro-oncology with artificial neural networks: A multicentre, retrospective study. Lancet Oncol 20:728-740, 2019

5

Impact of Big Data on Cancer Care and Research

Nitu Singh[1*], Urvashi Sharma[2], Deepika Bairagee[1], Neelam Jain[3], Surendra Jain[4], and Neelam Khan[1]

[1]RGS College of Pharmacy, Lucknow, India
[2]Medi-Caps University, India
[3]Pacific college of Pharmacy, Pacific University, India
[4]Department of Biomolecular Sciences, School of Pharmacy, University of Mississippi, USA
*Corresponding Author.

Abstract

Despite the fact that normal patient cohort sizes are small, oncology regularly creates and maintains a large number of observables (thousands to millions). The difference between the depth of data/patient and the mass of the cohort is much greater. Hospitals can better monitor intensive care patients using big data. Pandemic outbreaks can be predicted and the efficacy of (expensive) medicine may be examined. There is a huge and rising demand for evidence in the development of cancer therapies today. Clinical trials are critical in the discovery of new drugs, but they have limitations. They are both expensive and time-consuming, for example. Only a small percentage of patients take part, and those who do are not usually representative of the wider public. Molecular features are currently being utilized to divide tumors into sub-groups that are too small to be studied in a randomized trial. The past method of evidence production, which relied only on prospective clinical trials, will not be able to close the large evidence gap. Big data analysis can provide evidence to answer many of the problems that are currently being raised in cancer care and medicinal research. This article will provide an overview of how big data approaches can be utilized in cancer research, as well as how they may be utilized to translate statistics into innovative ways to make better cancer care and delivery decisions. CGHub, COSMIC, cBioPortal, CPRG, GDAC, canEvolve, UCSC cancer, CGWB, MethyCancer, SomamiR, NONCODE, GDSC, and CanSAR are only a few of the web-based resources included in this study. This document also includes a number of additional relevant sites for cancer research.

Keywords: Big Data, Cancer Therapy, Clinical Trials, Oncology, Web-based Resources.

5.1 Introduction

Cancer is a comprehensive term that encompasses a wide range of ailments that can affect any organ of the living body. Malignant tumor and neoplasm are the other synonyms used. Cancer is accompanied by the fast growth of aberrant cells that nurture outside their normal borders, allowing them to infect neighboring sections of the living organism and migrate to the other body parts; this is known as metastasis. Cancer metastasis is the primary reason for disease spread. It is the top cause of death due to the disease, with an expected ten million losses estimated by 2020 [1]. The below have been the most common types of fresh reports of malignancy in 2020:

- Mammary gland (2.26 million cases);
- Bronchi (2.21 million cases);
- Colon and rectum (1.93 million cases);
- Prostate (1.41 million cases);
- Skin (non-melanoma) (1.20 million cases); and
- Stomach (1.09 million cases).

The communal bases of malignancy demise in 2020 were:

- Lung (1.80 million deaths);
- Colon and rectum (935 000 deaths);
- hepatic (830 000 deaths);
- Stomach (769 000 deaths); and
- Breast (685 000 deaths) [1].

Malignancy kills over 70% of people in low- and middle-income nations. Tobacco use, a high BMI, alcohol usage, low consumption of fruits and herbs, and lack of exercise account for almost one-third of the fatalities of cancer. Carcinogenic illnesses such as inflammation of the liver and the papillomavirus (HPV) contribute to around thirty percent of malignancy incidences in poor- and relatively low-earnings countries [2]. In low- and medium-income nations, late-stage manifestations and an absence of availability of assessment and treatment are widespread. More than 90 percent of high-income regions have comprehensive therapy, but just 15% of low-income countries have it, according to studies [3]. Cancer has a major and growing economic impact. Cancer's total yearly economic cost was projected to be US$ 1.16 trillion in 2010 [4].

Oncologists face various obstacles today, including uniformity of treatment, staying updated in the interdisciplinary treatment of all cancers, gathering data on a variety of scales to determine what defines a disease, and more. According to the Times of India, cancer appears to be tightening its grip, with a million new cases diagnosed each year. According to some analysts, the prevalence of the deadly disease would increase five-fold by 2025. According to statistics, cancer cases have risen from 700 to almost 1000 per million people. Cancer is one of the most difficult and fast-developing illnesses. It is the globe's second major cause of death. Each year, the World Health Organization estimates that 14 million new instances of cancer are reported. Over the next 20 years, that figure is predicted to climb by almost 70%. Although the battle against cancer has made significant progress in the last 30 years, and the survival rate has doubled, a universal treatment remains elusive [5].

Oncology risk stratification is hampered by a lack of relevant prognostic data, the requirement for lengthy physical data entry, a deficiency of complete statistics, and, in certain situations, an over-dependence on clinical perception. Clinical prognosis estimation is poor, especially among patients with advanced solid tumors, according to prospective studies [6]. Failures in recognizing patients with a low survival risk, resulting in unnecessary end-of-life care is an extreme or wasteful use of critical care in cancer patients [7]. Although prognostic aids for tumors exist in oncology, they are seldom utilized since they do not apply to most malignancies [8, 9], do not incorporate genetic information, do not predict who might die in under a year [10], and require time-consuming manual data entry [11]. Besides, interclinician heterogeneity and bias exist in the assessment of prognostic factors such as performance status [12]. Even fewer available facts exist for estimating the risk of other key results in cancer patients, including hospitalization or side effects. The desire for more patient-centered care is a motivating force behind improved risk stratification models in oncology. Furthermore, shifting reimbursement mechanisms, such as alternative payment models, will encourage the exact care for the correct patient rather than solely motivating the size of facilities provided [13]. Oncologists are being requested more and more to change treatment strategies depending on the formal risk of a range of complications for a client. Data on demographics, treatment episodes, and specific clinical issues must be collected and analyzed. The Centers for Medicare and Medicaid Services have accumulated substantial datasets on Medicare beneficiaries as the portion of the Oncology Attention Exemplary and are already collaborating with EHR providers to facilitate information collection and data requirements 14–16]. Despite the increasing availability of rich data combining clinical and use characteristics, powerful prediction methods are required to predict future risks of acute usage or other harmful outcomes. In domains like re-entry risk mitigation in the general inpatient context, and decision aids, it has been discovered that anticipatory analytics can help with value-based healthcare decision-making [17, 18].

Now the question is, how can oncologists address these issues? This is where big data comes into the scene. The big data market is predicted to touch $6.6 billion by 2021, according to studies. As technology develops, it will be able to supplement traditional healthcare by recognizing trends and improving disease detection. In fact, in some parts of the world, it is already being used for more reliable and early detection of diseases such as cancer. Big data is endowed with powerful capabilities and is best suited to deal with vast volumes of facts. The purpose of big data in medicine is to create better health profiles and predictive models for individuals in order to detect and treat disease more effectively. IBM Watson Health, for example, has used big data to solve some of the most critical issues. It combines human professionals with augmented intelligence to aid healthcare and examiners all across the domain, to decode statistics and information to perceptions so they can deliver better results. Big data is among the ten leading revolutions in the future decade, along with the digital technology to analyze it [19]. Its influence is expected to be similar to that of the internet, the cloud, and more recently, the bitcoin [20]. The big data phenomenon is affecting almost every industry. Information power firms (IBM, Google, Facebook, and Amazon) were the first to make extensive use of them. These massive IT firms have developed algorithms that use computational models and machine-learning procedures to forecast individual choices and use that data for personalized marketing. Big data breakthroughs have also piqued the interest of health insurance firms and governments, and big data has made its way into the life sciences [21].

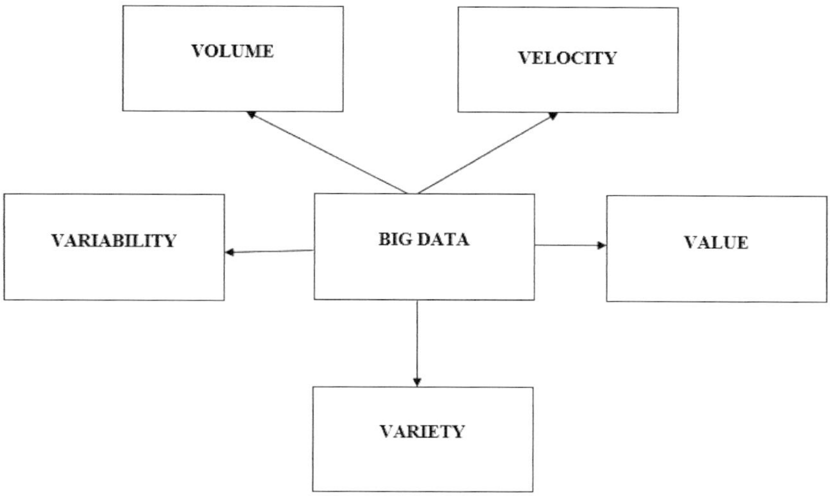

Figure 5.1 Five Vs of big data [19, 20].

5.2 What Is Big Data?

Despite the statistic that the word "big data" is used by many individuals and corporations, it does not necessarily imply the same thing or have the same meaning. Big data is more than simply "a lot of data," as the majority of us have a foggy sense of what it is ("everything which would not appear on an Excel sheet"). The five Vs is a phrase used to refer to large data in health care [22] Figure 5.1.

Big data, according to this definition, have the following characteristics:

- Volume: Big data are large in magnitude, having a large number of data points/records from various subjects. Diagnostic data (clinical, radiological, and pathological), treatment records, response data, and consequences are some of these.
- Velocity: It has two types: (i) big data is generated at a constantly increasing rate, and (ii) it is computed reasonably quickly.

Cancer is becoming more common over the world. With the advancement of technology and monitoring equipment, a large amount of data will need to be handled at a similar period.

- Variety: The term "big data" encompasses a broad range of data types. This diversity presents both opportunities (numerous distinct data formats enhance the quality and utility of the data) and challenges (the heterogeneity of the statistics necessitates standardization).
- Variability: It is critical to distinguish that data collection happens at different places and at different times. Getting the majority of (synoptic) reported data requires capturing a (predefined) obligatory minimal dataset.
- Value: Creating data-driven assumptions or quantities based on accurate records that may aim to demonstrate changes or outcomes in healthcare is only advantageous if it allows for the production of data-driven assumptions or quantities. While the sheer volume of data collected is frequently a source of concern, a more major concern is that data are usually dispersed around the earth and maintained that make data integration tough. Because the facts must be sent over the internet, this may cause scalability issues, but it also necessitates harmonization and standardization efforts [19, 20].

Hospitals can better monitor intensive care patients using big data. The efficacy of expensive medication can be assessed, and pandemic breakouts can be predicted in advance. Big data could be useful in designing and reforming ailment anticipation methods, especially in the medical field. Combining big genomic and environmental datasets will aid in predicting whether a single person or large groups are in danger for particular (enduring) ailments, including tumors. This could lead to targeted measures aimed at influencing environmental and behavioral factors that result in health hazards in Hospitals. Big data could also be used to evaluate existing anticipatory initiatives and uncover new insights that can be used to improve them. Big data can also be used in a therapeutic setting to track the impact of specialized medicines, such as pricey oncolytics, particularly in connection to patient and tumor (genetic) features. This will aid in the advancement of precision medicine and provide critical information for calculating the cost effectiveness of specific treatment regimens. Big data in medicine data sources are abundant and come in a variety of shapes and sizes. Patient-derived data are the most evident in oncology. These are frequently stored in computerized patient files for therapeutic purposes and include a variety of data points/subjects. These data include sex, age, present symptoms, family history, comorbidities, radiographic data, solid and liquid tissue-based analyses, and whole-genome BAM files, averaging 100 GB in size. Experiments conducted in vitro, on the other hand, are useful and can be a beneficial source. The computational analysis of these data is the next foundation of big data. Radiomics and internet picture processing, as well as gene regulation and transformation studies, are examples for indirect and computed data are among the processed records. Machine learning is a growing source of processed data, which typically consists of big computer data files including structured data. Patients, including patient-related outcome measurements (PROMs), and patient-related experience measurements (PREMs), are the third font of big data, through software running on PCs and other devices to document all kinds of measured data, either as a result of their attendants' efforts (eHealth, telemedicine) or with their own. The published literature is a fourth pillar (IBM project). No clinician on the planet can read even a percent of the information gained each year, because over 1 million biomedical publications are published every year. Nonetheless, in the case of cancer, one essential component stands out: the amount of data per patient. Countless observations (thousands to millions) are frequently generated and maintained in oncology, despite the fact that average patient cohort sizes are tiny. The mismatch between the volume of data per individual and the size of the group is extra obvious for exceptional diseases like head and neck cancer. But, if there are examples to learn over, current methodological developments in instrument knowledge and neural networks are very potent. Object recognition in photos, for example, works well, but optimizing such systems requires thousands to millions of samples. As a result, if we want to use this to improve tailored therapies, we will need more data depth in terms of sample size [23].This can be achieved by big data, and cancer researchers now have influential novel tools for extracting value from a range of data sources. For any patient, these sources include a large quantity of data. Carcinoma is a biochemically complicated illness with high intra- and inter-tumoral variability among cancer types and even people; oncologists may use omics data to develop customized treatment methods by providing each patient with a distinct genetic profile [24]. The strategy of combining various data sources is employed in Comprehensive Cancer Centers (CCCs) [25]. The MASTER (Molecularly Aided Stratification for Tumor Eradication Research) project is housed in the National Center of Tumor Diseases, one of Germany's 13 CCCs. In a current review study [26], the following experiment was highlighted as an example of a very effective approach for addressing molecular profiles in cancer patients. The MASTER project collects, analyzes, and discusses information related to the identification of lower-age patients with locally innovative malignancy illnesses using whole exome or whole genetic

Table 5.1 In chronological order, EU-funded initiatives in Europe that integrate the usage of big data in oncology [28].

Project acronym	Full title	Coordinator country	Start date	End date
ClouDx-i	A cloud-based software system for next-generation infectious disease diagnosis	Ireland	07/01/2013	06/01/2017
iManageCancer	Increasing patient empowerment and self-management in cancer patients	Germany	01/02/2015	31/07/2018
MedBioinformatics	Creating medically-driven integrative bioinformatics applications focused on oncology, CNS disorders, and their comorbidities	Spain	01/05/2015	30/04/2018
MOCHA	Models of child health appraised	United Kingdom	01/06/2015	30/11/2018
ENLIGHT-TEN	European network linking informatics and genomics of helper T cells	Germany	01/10/2015	30/09/2019
PERMIDES	Personalized medicine innovation through digital enterprise solutions	Germany	01/09/2016	31/08/2018
IASIS	Integration and analysis of heterogeneous big data for precision medicine and suggested treatments for different types of patients	Greece	01/04/2017	31/05/2020

analysis, as well as RNA sequencing. The INFORM (Individualized Therapy for Relapsed Malignancies in Children) registry (mostly for sectors 1, 2, and 3), intends to relieve high-risk tumor recurrence in children, and is another success story mentioned in the study. To find patient-specific therapeutic targets, whole-exome fingerprinting, minimal complete genetic sequencing, RNA-seq, and microarray-grounded DNA methylation monitoring were all used by the researchers. The INFORM database started as a German project and has now grown to include eight European countries and Australia. Various other initiatives are concentrating on the effectiveness of big data in the study of cancer, in addition to the programs mentioned; Europe itself funds over 90 projects in this field (projects with funding greater than Ă499.999 are included in Table 5.1). Cancer Core Europe might act as a hub for bringing national efforts like the ones listed above to the European level [27].

The amount of data obtained for a single patient is huge, while conversely, the sum of patients that a solitary hospital may see is inadequate to yield statistically meaningful effects. Therefore, collaborations are especially important in the case of pediatric or other uncommon cancers. Gaining access to the statistics, as well as the capacity to quickly investigate the vast volumes of facts, is one of the key challenges of these collaborations. Physicians, academics, and informatics specialists may be able to take advantage of the acquired statistics and skilled understanding only if they have rapid and easy access toward their own or partners' data. Tools are being developed at the German Cancer Research Center, for example, to allow users to have right of entry and examine their personal or a partner's data.

Both The Cancer Genome Atlas (TCGA) and the International Cancer Genome Consortium (ICGC), which give examiners access to a number of sequenced individuals with diverse carcinomas, are international examples for transmitting intelligence with partners or the wider public. Due to the availability of these data, enormous meta-analyses using appliance knowledge algorithms are often used to combine different cancers, resulting in the discovery of new malignancy genetic variants that correspond to precise ways and could be employed as promising treatments. A variety of database searches, such as COSMIC (Catalogue of Somatic Mutations in Cancer), also allow access to cancer-related mutation catalogs. All of these data sources, in combination with the establishment and funding of CCCs around Europe, have the opportunity for an increase in the number of patients who can benefit through genetic profiling and customized medication, as an outcome of big data analysis.

5.3 The Outcome of Big Data on the Cancer Care/Research

There are various reported outcomes of Big Data in cancer care and research. Few examples of currently available applications are given in Figure 5.2.

5.3.1 Daily diagnostics

Big data has already been shown to be useful in therapeutic treatment. One instance is Dutch pathologists' location-specific nationwide access to each patient's histological follow-up. Since 1971, the PALGA organization has managed all electronic histological recordings in the Netherlands. PALGA is one of the world's leading biological databanks. It holds the country's 55 forensic labs. When a Dutch physician signs a histological outcome, one duplicate is stored in the native healthcare domain, while another is stored in the PALGA database. As a result, this catalog encompasses compulsory follow-up statistics for every patient in real-time, which any PALGA member may access (pathologist or molecular biologist). This has a lot of potential in terms of recognizing relevant patients' (oncological) history, such as excluding pathological data on past significant pathological findings (such as amputation margins and favorable lymph system) in situations where pathology had been done in another lab, or a recent malignancy in instances of a suspect malignancy of unknown origin. This database may also be used to investigate the simultaneous diseases or unexplained connections in low-prevalence disorders that appear unrelated at first glance [29]. Electronic patient files provide a massive quantity of medical data that may

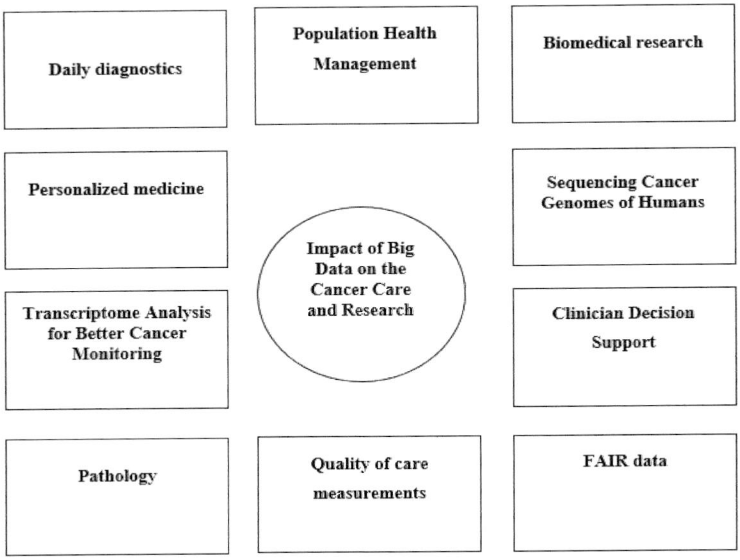

Figure 5.2 The outcome of big data on the cancer care/research.

be utilized to estimate prognosis. www.oncologiq.nl [30] has one of the first predictive models for head and neck cancer patients getting care in medical centers in industrialized nations. When statistical prognostication procedures are automated, models may be updated automatically as fresh data is collected [31]. These data might potentially be designed to make clinical ruling tools for better patient advice and non-binary treatment outcome evaluations.

5.3.2 Population health management

In order to reduce unwanted side effects, community health management includes targeting medications to high-risk individuals. Predictive algorithms can be used to recognize chemotherapy patients who are in danger of dying or need emergency care. This prediction might be utilized to affect clinicians. Following chemotherapy [32], carcinoma surgery [33, 34], or discharge preparation [35], are all examples of behavior throughout the disease spectrum. Prevalence with these elevated patients could help to cut down on resource consumption. Prognostic methods are used by Penn Medicine and New Century Health, for instance, to detect cancer patients who have a higher chance of being rushed to the hospital or visit the emergency care, and then use proactive phone calls or visits to target care management solutions [36, 37]. Even though electronic health record (EHR) data is extremely problematic to use, employees from Google [38] claim to have used the Fast Healthcare Interoperability Resources format to streamline the time-consuming practice of acquiring facts from EHRs. Deep knowledge systems were developed utilizing over 46 billion data points in the Fast Healthcare Interoperability Resources format to accurately predict a variety of medical outcomes, such as in-hospital mortality, readmissions, number of days stayed, and diagnoses of discharge. Big data analytics may be used to enhance patient-centered services, such as detecting illness outbreaks sooner, generating new understandings into disease causes, monitoring the class of medical and healthcare facilities, and developing better treatment approaches 39–42]. Furthermore, combining and analyzing data of various types, such as communal and technical data, results in innovative information and intellect, as well as the exploration of new hypotheses and the discovery of hidden patterns [43]. Smartphones are now a fantastic platform for sending personalized messages to patients in order to encourage them to make behavioral changes that will enhance their overall health and well-being. The delivery of medical and motivational counsel to patients can be replaced by text messaging [43].

5.3.3 Biomedical research

Big data will most likely provide the most value in the realm of exploration. The period of (GWAS) "genome-wide association studies" is giving way to a period of "data-wide association studies" (DWAS), and big data is taking center stage. Growing data, both as a result of increased usage of imaging and molecular studies and as a result of data combinations with other data, provide an unrivalled opportunity for every data researcher and bioinformatician. Big data reports a necessity in biomedical research. One of current medication's major faults is the lack of knowledge of disease biology. All crucial multisensor elements, such as DNA, RNA, protein, and metabolomics data, can only be gathered and included into far more reliable estimates to foresee how cancers will behave and which patients will get relief the most from various therapies, if massive amounts of big data are pooled.

5.3.4 Personalized medicine

Big data is very necessary for translating present knowledge and accessible data into functioning visions that may be utilized to enhance treatment outcomes in personalized medicine [44]. The volume of data available to the biomedical community is growing at an exponential rate, especially when new technologies, including sequencing and imaging, generate terabytes of data. In terms of quantity, the majority of data comes from calculated programmed information investigation such as radionics and alphanumeric image examination, rather than the straight, patient accounts available in everyday medical practice. Head and neck cancers provide a distinctive combination of analytical and treatment trials due to their complex architecture and unpredictability. Radiomics is capable of overcoming these challenges. Radiomics is a low-cost, non-invasive method of collecting and analyzing a wide variety of clinical imaging characteristics [45]. Radiomics is based on the idea of using imaging data to quantify the phenotypic properties of a full tumor. Radiomics in cancer detection and cancer care enables forecasting and dependable appliance knowledge techniques for stratified (or customization), i.e. recognizing distinctions in (initially supposed) subsistence among (groups of) patient populations, and treatment outcome(s) supposition to assist individuals with head and neck cancer in selecting the best possible management [46]. As a result of this, medical and radiation specialists could be allowed to (de-)escalate systemic medication and radiation dosages in specific patient groups.

5.3.5 Cancer genome sequencing

Every body cell includes the same set of chromosomes and roughly the same proportion of DNA. On the other hand, cancer cells contain unique chromosomal content and growth abnormalities that may be used in silico to extract information down to the DNA level. These discoveries might aid cell biologists, bioinformaticians, molecular biologists, and nano biotechnologists in developing more effective techniques for rectifying chromosomal defects, perhaps leading to new treatment options. The genome of the single-celled bacterium Escherichia coli was the first to be sequenced. Plants such as Arabidopsis thaliana and non-vertebrates such as worms, reptiles, and rodents had their genomes mapped. As these organisms got more intricate, genome sequencing became increasingly significant in terms of determining a single cell's potential to manage, sensitize, or fend off carcinomatous cells. There were also some working concepts on cancer triggers. To complement or enhance cancer detection techniques, researchers are actively analyzing actual information using mouse/chicken/hamster ovarian cancer cell lines, as well as increasing the accuracy of existing screening tests [47].

Strong algorithms that can estimate risk based on various genes found are necessary as the use of germline screening and next-generation carcinoma sequenced among cancer patients grows. Because technology in the next sequencing is too expensive to deploy as a randomly distributed screening technique for a whole population, prediction algorithms based on the patient clinical history appearances are able to be utilized to focus inherited evaluation on particular people. Machine learning algorithms have been shown to successfully stratify real variations from artifacts when used on targeted next-generation sequencing panels; this could be a useful prediction device as alternates of uncertain importance may result in a lot of misunderstanding among doctors and patients [48, 49]. Genetic risk stratification can also be used to determine if a person will benefit from a breast cancer screening. When compared to the existing age-based breast screening paradigm, proposing breast malignancy mammography to females who are at a higher hereditary threat of breast tumor reduced analysis and greater fee, according to research from the United Kingdom [50], when compared to the current age-based breast screening paradigm, suggesting breast cancer mammography to females who are at a higher genetic threat of breast tumour results in less analysis and a low cost.

5.3.6 Transcriptome analysis monitoring cancer better

Huge data of screening and experiment have been just established in the current period as a result of the cancer-related investigation. It has become more vital to uphold indicator genes, which are currently the principal instruments for oncogene surveillance, drug development, and biocompatibility research. Furthermore, several businesses extrapolate cancer genomic data to study transcriptomes and protein production. This is important for monitoring conservative and non-conservative mutations by detecting misplaced gene segments and their products [47].

5.3.7 Clinician decision support

Oncology practitioners will increasingly employ predictive analytics technology to impact routine sections of patient treatment after they achieve a performance level. In prospective trials, predictive algorithms have been demonstrated to reduce the time it takes to react to sepsis patients and to enable stroke patients to be treated more quickly [50, 51]. Among other things, analytics might be used to anticipate chemotherapy side effects, the projected length of response, reappearance hazard, and total life anticipation at the time of treatment [52]. Many real-time EHR-based algorithms have been created as a proof of concept to estimate cancer patients' danger of instant death preceding the start treatment [53]. These algorithms may be used with any EHR and are grounded on both structured and unstructured EHR data. While the future applications of these algorithms are uncertain, oncologists may find exact death estimates to be quite useful at the site of therapy.

5.3.8 Incorporating machine learning algorithms for diagnostic modeling

Massive amounts of data are stored in healthcare systems, which may now be accessed more simply owing to modern technologies. Machine learning techniques that can read data, communicate with data, and provide the highest level of accuracy inside the integration of large datasets are being used by biotech/interdisciplinary researchers to undertake complete assessments of such databases. To get a better understanding of tumors, researchers are merging cancer-related data from many sources using machine learning algorithms and high-tech data modeling tools. New ways to analyze cancer cell development and the death of healthy cells are being enabled by genetic data visualization tools, which are creating waves in cancer diagnosis. One of the most successful open-source research data management solutions for visualizing high throughput sequencing and screening data is the Genetic Modification and Clinical Research Information System, which has been put to excellent use [47].

5.3.9 Presenting greater clarity on disease prognosis

CancerLinQ, for example, is a healthcare data visualization software program that allows the medical practitioner to see superior patient therapeutic data. This is significant because it aids researchers in gaining a better understanding of previous cancer cases, disease development, and treatment regimens. Clinicians are utilizing screening tools to refer to patients' safe medical data and use it to promote clinical trials, give tailored treatment regimens, and better define the scope of cancer care. Tumor ID cards, which allow data to be centralized for clinical evaluation, have been implemented in hospitals with high cancer admission rates [47].

5.3.10 Feasible responses for cancer relapses through clinical data

To learn why some people have recurring malignancies and some do not, more doctors are turning to data analytics. Doctors are reviewing a significant number of case reports to analyze patient health issues from a much broader perspective than previously. Even though therapeutic event news has been around for long time, its use and accessibility have only recently increased. This means that rather than being subjected to standard identification approaches, laboratory data is reviewed after being compared to other internationally reported occurrences. As a result, data is the most critical requirement for personalized treatment. Our medicines are evolving at a quicker rate than the disease itself. As a result, if you wish to defeat, control, or avoid it, you must concentrate your efforts on target identification. Big data is an important tool for cancer researchers to employ to expand the excellence of their research and achieve results quicker [47].

5.3.11 Pathology

Pathology, which is critical in cancer treatment, is another area that will gain from predictive analytics. In areas such as determining the Gleason Score from prostate biopsies and diagnosing non–small cell tumors from bronchoscopic samples, pathologists have a lot of disagreement [54, 55]. Inaccurate biopsy results may result in treatment recommendations that are either unnecessary or inappropriate. Machine learning algorithms have the same sort of discrimination as pathologists in detecting metastatic breast cancer in sentinel lymph node biopsies [56]. Pathologists' capacity to scan large tissue sections for cancer cells is increased by these models, which may allow them to spend more time on other tasks and enhance their workflow.

5.3.12 Quality care measurements

Associating patient outcome records with data on patient characteristics and therapy has the potential to provide unparalleled feedback on care quality and efficacy. France's molecular testing landscape for helpful approaches in non-small cell lung cancer (NSCLC), also the treatment regimens based on it, was presented in a recent French study [57]. This provides immediate feedback on the best test–treatment connections. Expressively, it could provide a solid inducement for failing facilities to adapt their procedure in order to enhance overall care quality. The diversity in clinical care, which can be increased by merging data out from the national cancer registry (which includes clinical phase, therapy, and latest research findings) with the aforementioned PALGA database in the Netherlands, has been highlighted [58, 59]. Though transparency in such data is the best way to enhance treatment quality, it should be noted that providing commentary on such data, particularly consequence data and when compared, should be handled with great caution, since institutions and hospitals may be concerned regarding adverse publicity or naming-and-shaming [60]. In reality, most hospitals are willing to participate in such transparent feedback, as long as it is offered to the public confidentially and only shared with one person at a time. As a result, algorithms have been created in the Netherlands, such as those developed by the Dutch Institute for Clinical Auditing, which provide frequent automated feedback on pathology and treatment-related factors (www.dica.nl). If transparent data shows greater recurrence rates than comparable hospitals, there may be an enticement to identifying the underlying chain to fix any flaws.

5.3.13 FAIR data

The FAIR (Findable, Accessible, Interoperable, and Reusable) principles must be followed to guarantee that data may be utilized in secondary research. The concepts of FAIR data were initially stated in 2014 [61]. FAIR data is at the soul of Europe's European Open Science Cloud (EOSC). Findability (F) necessitates a permanent identifier; accessibility (A) necessitates explicitly specified access restrictions (data privacy requirements are incorporated in the specification); and interoperability (I) to describe the data, it is necessary to use a community-accepted ontology. Finally, for data reuse, the origin of the information, as well as the completeness and accuracy of the meta-data, is critical (R) [61].

5.4 Database for Cancer Research

The introduction of high-throughput sequencing technologies in the current period has resulted in a massive records outburst as well as a thorough examination of the tumor genome. It was a first where entire genome sequences, containing point mutations or structural changes, were made available to the public for a wide range of cancer types, allowing for unprecedented global cancer subtype differentiation.

5.4.1 Cancer Genomics Hub

The Cancer Genomics Hub (CGHub) is a single compendium for genetic data collected in the United States by three National Cancer Institute (NCI) projects: The Cancer Genome Atlas (TCGA), the Cancer Cell Line Encyclopedia (CCLE), and the Therapeutically Applicable Research to Generate Effective Treatments (TARGET). CGHub is presented by the University of California, Santa Cruz (UCSC), and data right of entry is restricted in order to safeguard patient privacy. CGHub has 1.9 PB of data from 42 distinct cancer types and healthy controls [62].

5.4.2 Catalog of Somatic Mutations in Cancer

COSMIC is the leading global dataset of human cancer genetic mutations and their implications. The database is hand-curated from available writings and contains exceptionally detailed disease categorization and patient data. With 2,710,499 code mutations, 10,567 gene hybrids, 61,232 genome rearrangements, 702,652 copy number variations (CNVs), and 118,886,698 aberrant expression Variants, the database contains 15,047 entire cancer genomes from 1,058,292 samples. Key phrases can be used to query the data, and registered users can download it. For reference, COSMIC has a database given in the following screens and whole-genome shotguns sequencing studies. The COSMIC database is a crucial resource for cancer research because of its large, manually vetted, and constantly updated dataset [63].

5.4.3 Cancer Program Resource Gateway

The Broad Institute is a well-known cancer research institute. Its Cancer Program aims to gain a better understanding of cancer's fundamental mechanisms and undertake research that leads to practical applications. The Broad Cancer Program Resource Gateway (CPRG) gives researchers access to a variety of datasets and tools. Broad's Genome Data Analysis Center (GDAC) will be given as an example of one of these resources.

5.4.4 Broad's GDAC

Analyzing terabytes of sequence data is vital, but it is time-consuming, for most of the laboratories. Broad's GDAC's Firehose system, on the other hand, is upending the existing status quo. The GDAC examines data from the TCGA pilot in a methodical manner, as well as data from other disorders. Firehose processes over 6000 pipelines each month while assembling 40 terabytes of TCGA data. Every two weeks, GDAC acquires and analyzes TCGA data before making them public [64]. Firehose provides a set of standardized genetic screening pipelines, and computer technology, that is free for the community. Using Broad's powerful processing environment, Firehose provides constantly updated data and findings on many layers, including edition-regulated datasets, statistical results, and scientist reports.

5.4.5 SNP500Cancer

The SNP500Cancer database is a resource for sequenced and genetic validation of Single nucleotide polymorphisms (SNPs) in malignancy and further complicated disorders and is associated with the Cancer Genome Anatomy Project (CGAP) (http://cgap.nci.nih.gov/Tools). The SNP500Cancer project's main goal is to rearrange standard examples in order to uncover existing or novel SNPs for cancer molecular epidemiology research. To get sequencing information for anonymized control DNA samples, the database can be searched by gene, gene ontology route, chromosome, SNP Database (dbSNP) ID, or SNP500Cancer SNP ID. Due to its high confidence, SNP500Cancer is a useful tool for researchers to choose SNPs for further examination, but its data volume is limited [65].

5.4.6 canEvolve

Genome-wide tumor profiling has dramatically increased in size and availability due to the fast expansion of biological tools. canEvolve satisfies the demand for data integration and analysis. canEvolve is a database that combines data from 90 trials involving over 10,000 participants. All of the information can be found in a graphical format. canEvolve is a practical genomics raised area that is available to the community that can be used for a variety of purposes [66].

5.4.7 MethyCancer

Oncogene initiation, chromosomal uncertainty, and tumor suppressor gene are all connected to DNA methylation and are all critical in the genesis of cancer. MethyCancer is a software that explains how DNA methylation, gene expression, and malignance are linked. MethyCancer is a library that comprises data on (1)DNA Methylation, (2) CNV and cancer, (3) CpG Island replicas, and (4)the interconnections among the data. One may easily search the database using the user-friendly MethyView interface [67].

5.4.8 SomamiR

Variations in miRNA sequence have been associated with a diversity of malignancies. SomamiR is a resource that was produced to explore the link across somatic and germline alterations, as well as miRNA function in cancer. GEO and the Pediatric Cancer Genome Project (PCGP) provided the mutation data, while miRNA data were obtained from miRBase version 18. SomamiR can be used to detect somatic mutations that alter miRNA target regions, as well as cancer GWAS, candidate gene connections in carcinoma, KEGG trails, and mutation facts in a genome website. Many human germline mutations in SomamiR have been linked to cancer and have a consequence on miRNA function. Investigators will be able to anticipate whether mutations will influence miRNA binding and, as a consequence, functional regulation because mutations data is contained in SomamiR [68].

5.4.9 cBioPortal

Researchers may use the cBioPortal for Cancer Genomics to explore, display, and analyze multimodal cancer genomics data. Data of numerous published malignancy research, together with CCLE and TCGA, are available on the site. It is worth noting that cBioPortal breaks down unique molecule profile facts of carcinoma-affected tissue and cell lines into smaller pieces. Researchers can utilize the cBioPortal web query comparison tool gene variation frequencies throughout studies, or summarize all significant genomic variants in a single tumor specimen interactively. It also offers capabilities for biological pathway study, model validation, and data download [69]. Consumers can inquire about sustaining a healthy ratio based on DNA mutation by looking at Kaplan–Meier diagrams [70], which show the findings of patient OS and DFS.

5.4.10 GEPIA Database

The Gene Expression Profiling Interactive Analysis (GEPIA) database (http://gepia.cancer-pku.cn/), a completely redesigned internet-grounded tool, offers vital collaborative and personalized features such as tumor vs. standard differential expression analysis, profiling plotting conferring to carcinomas of diverse medical levels, co-relation, treatment success analysis, similar DNA detection, and dimension decrease investigation, all based on the Cancer Genome Atlas [71]. GEPIA is an internet interactive program for analyzing gene expression-based survival data. Users can select whether they want to look at overall survival (OS) or disease-free survival (DFS). GEPIA permits two unique genomes to be input simultaneously for survival analysis key features of gene normalization. GEPIA also includes a list of the top genes linked to cancer patient survival. This feature is quite beneficial to users. Other uses for GEPIA include differential expression analysis based on distinct malignancy, numerous gene comparisons, and associated gene identification, in contrast to a study of survival rates [72].

5.4.11 Genomics of Drug Sensitivity in Cancer

The Genomics of Drug Sensitivity in Cancer (GDSC) databank is the world's biggest exposed knowledge base on cancer drug sensitivity and response. Experiments recording the activity of over 200 anticancer medicines across over 1000 tumor cell lines are part of the GDSC project. It integrates cell line treatment response data with massive genetic databases to uncover molecular indications of cancer therapy response. On the GDSC website, you can look up drug sensitivity by searching for a tumor gene or a tumor cell line and chemical. The data is offered in a number of graphical formats that may be seen and downloaded. The enormous assortment of cell lines, drug comparison, and genomic information has helped us better comprehend cancer cell genetic heterogeneity and find new patient-specific treatments [73].

5.4.12 canSAR

canSAR is a cross-discipline knowledge base focused on cancer that intends to make translational cancer research easier. Material from a range of disciplines can be found in the knowledge base. Researchers might gather data on current knowledge of a protein or treatment, such as the protein's conformation or alteration in cancer, cellular susceptibility patterns, and specific drug binding proteins. canSAR now has access to a massive number of human proteome data. Carcinoma statistics and descriptions, nontransformed cell lines, and protein 3D structural summaries are all available in this database [74].

5.4.13 NONCODE

Noncoding RNAs (ncRNAs) have a role in an extensive variety of tumors. NONCODE is a knowledge base for noncoding RNAs (excluding tRNAs and rRNAs) from a variety of species, including humans and mice [75]. According to *Science*, NONCODE was first introduced in 2005. NONCODE version

4.0 has 595,854 ncRNA records, with 210,831 (nearby 35%) being long ncRNAs (lncRNAs). NONCODE has over eighty percent experimentally derived ncRNA data, making it a very trustworthy resource for customers. NONCODE offers the iLncRNA online toolchain, which allows users to analyze customized RNA-seq data and obtain lncRNA expression profiles among tissues [76].

5.5 Bioinformatics Tools for Evaluating Cancer Prognosis

5.5.1 UCSC Cancer Genomics Browser

The UCSC Cancer Genomes Portal is a digital data storage, visualization, and analysis tool for cancer genomics and biomedical research data. Through the web, users can obtain data from a specific study as well as diagnostic features for a large number of samples. Moreover, two or more pieces of information may be displayed at the same time to allow for comparisons of gene expression and CNV across different data and cancer types. Data is downloaded and a clinical heat plot can be viewed. The browser now includes gene research on 71,870 specimens, the majority of which originate from across-the-board global tumor initiatives such as the TCGA, CCLE, and the Standing for the carcinomas project [77].

5.5.2 Cancer Genome Work Bench

The Cancer Genome Work Bench (CGWB) is a robust implementation that examines gene expression, mutations, copy number variation, and methylation in cancer patients. Using an automated analytic approach, users may evaluate and analyze genetic facts and inheritable factor expression dissimilarities in every single trial. The Heatmap viewer and the Genome browser are the two main viewers in CGWB. Users may move between the two to see information like the expression of genes, somatic mutations, and the process background whereby the mutations may be implicated. Medical studies and individual genotypes have been kept confidential by the CGWB, and access to this information needs permission [78].

5.5.3 GENT2

Differentially expressed analysis and prognosis analysis based on tumor subtypes are provided by GENT2. Consumers can compare genetic factor expression levels between tissue subtypes and search gene expression profiles of various tissues. This program includes a Kaplan–Meier plot with a log-rank test for existence examination, as well as a Cox proportional risk model for meta-analysis. It presently includes data on 27 malignancies and 46 variants of 19 forms of cancer [79].

5.5.4 PROGgeneV2

It is a digital application for researching genetic factor prediction in various malignancies. The database presently has 193 data points for 27 malignance categories. Operators may do single-gene, multi-gene, and two-gene expression ratio survival analyses, as well as use the shifting covariate survival models feature. Users can submit their genetic factor datasets for existence analysis and associate the findings to previous studies [80, 81].

5.5.5 SurvExpress

SurvExpress is a tool that allows risk scrutiny and survival analysis research, having around 29,000 illustrations from 26 different malignancy types, as well as medical data from 144 different databases. SurvExpress creates Kaplan–Meier graphs for each risk group, a gene expression heat map, and a visual representation of the data [82].

5.5.6 PRECOG

PRECOG is a strategy for combining genetic data and scientific data from 165 carcinoma gene databases, totaling around 19,000 specimens and survival rates data. It brings together around 19,000 illustrations with data on continued existence rates from 165 malignancy gene databases and spans 39 cancer types. It agrees with scientists to assess if patient survival is connected to gene expression. Thirty-nine distinct histologic kinds of tumors were split into 18 groups for easy viewing. Using univariate Cox regression, the link across an expression of genes and overall survival was explored. PRECOG also delivers prognostic analysis for all cancer genes. New users, on the other hand, must register and login [83].

5.5.7 Oncomine

Oncomine is a potential malignancy chip collection and facts gathering board dedicated to the study of cancer genes. Oncomine has a greater variety of cancer mutations, genomic data, and clinical information, many of which can help researchers find new biomarkers or therapeutic targets. However, the Kaplan–Meier plot is not displayed right away. Users can get the findings of changed expression, co-expression analysis, the molecular concept with multiple interface network, and gene function and survival status correlation through Oncomine. Meta-analysis could also be used to contrast research in order to reach more consistent and accurate results [84, 85].

5.5.8 PrognoScan

Researchers integrated a huge number of freely searchable cancer microarray data with clinical evidence to develop Prognoscan, a tool that predicts the link between both gene expression as well as patient survival. It covers 14 cancer kinds with a range of survival terms. One of its benefits is that it does survival analysis using the minimal P-value approach and provides an optimum cut-off [86].

5.5.9 GSCALite

GSCALite is a dynamic and graphical implement of assessing appearance/dissimilarity/medical association of genetic factor sets in malignancies. It offers three survival analysis modules grounded on TCGA cancer multi-omics data for a gene set. (1) Differences in mRNA genetic material set among malignancy and equivalent normal tissues, as well as gene function among malignancy subtypes, influence overall survival. (2) The effect of single nucleotide variants (SNVs) and the variety of inheritable factors set to change on complete continued existence in a certain malignancy category. (3) The effect of methylation expression discrepancies amongst tumor and healthy trials on life expectancies. At the transcriptome, epigenetic modification, and DNA mutation levels, users can hunt for prognostic indications. Researchers can check for cancer trail action interrelated with gene appearance, and also the relationship amongst genomes and drug compassion, to quickly investigate treatment resistance in tumors [87].

5.5.10 UALCAN

UALCAN is a web-based application that assesses the connotation between gene expression and patient existence by means of TCGA, RNA-seq, and clinical data. Users can do differential expression and fatality analyses on specific genes, as well as obtain expression and survival data for a single gene across 31 cancer types using pan-cancer analysis. Protein differentially expressed analysis, for breast cancer, colon cancer, and other cancer types, is currently available through UALCAN, however protein-based survival analysis is not. UALCAN also includes links to Pubmed, TargetScan, and DRUGBANK, which provide added statistics about the particular genes or aims, as well as other resources, allowing researchers to access even more critical data and information [88].

5.5.11 CAS-viewer

CAS-viewer is an internet program that integrates inter-data from several cancer types, including mRNA, miRNA, methylation, SNP, and clinical data to perform multiple level comprehensive analyses. It connects the differential transcriptional expression rate of 33 cancer types to methylation, miRNA, and splicing regulatory regions. Users may uncover possible transcripts connected to varied survival outcomes of each cancer type using the "Clinical correlation" module, which displays a Kaplan–Meier plot demonstrating the association between survival rate and percent spliced in (PSI) value [89].

5.5.12 MEXPRESS

MEXPRESS is an easy-to-use internet implement for analyzing gene expression, DNA methylation, and their relationships through medical data like patient survival. It has a one-of-a-kind visual interface that allows users to link certain genetic attributes (such as DNA methylation) to gene appearance and medical data. The MEXPRESS platform may be used by researchers to investigate the association with DNA methylation, gene expression, and variability of experimental factors [90].

5.5.13 CaPSSA

CaPSSA is a tool that helps users determine the patient subgroups with prognostic value involved in gene expressions, mutations, or genomic anomalies in search genes. It also allows for individualized histopathological data analysis in addition to clinical statistics, which is critical. For analyzing the prognostic significance of individual candidate genetic variants that allow participatory patient sections on gene expression profiles and chromosomal abnormalities, the consequences of the log-rank testing and Kaplan–Meier plots will also be supplied [91].

5.5.14 TCPAv3.0

TCPAv3.0 is an enhanced form of TCPA that uses TCGA RPPA data to explore and analyze protein expression. It combines protein information with additional TCGA information (somatic mutations, DNA methylation, mRNA and miRNA expression, and easy-going scientific data) to produce complete protein-centric analysis. Protein markers that are highly regarded with patient survival can be found using the Cox proportional analytics platform and the log-rank test. Pan-cancer analysis allows users to see which proteins are linked to the prognosis of various tumors and subtypes. In a number of malignancies, researchers can use the pan-cancer proper platform in connection with multi-omic TCGA data to assess different protein-driven multi-omic hypotheses [92, 93].

5.5.15 TRGAted

TRGAted is a user-friendly implement for examining the relationship between $\geq$ 200 proteins and cancer survival of 31 different cancer types. TRGAted contains RPPA data from the TCPA Portal. Gender, age, tumor phase, histological type, and reaction to therapy are all included in the cancer clinical information presented. The Cox proportional hazard model is being used to predict altogether proteins in a certain malignancy category or a specific protein across all malignancy categories. In respect of survival markers, TRGAted beats TCPAv3.0, and its capacity to see all molecules in a type of cancer can help researchers uncover life protein in a specific disease [94].

5.5.16 MethSurv

MethSurv is an internet survival analysis tool that uses DNA methylation statistics from the TCGA, which includes 7,358 samples from 25 cancer types. MethSurv offers a variety of long-term survival studies, covering single CpG, region-based analysis, all tumors, top indicators, and gene visibility, which are all accessible from the home page. Users may get CpG survival analysis results for specific sections of a chromosome, as well as exploration for a gene of curiosity to see all CpG survival data. Users may look at the top biomarkers listed by p-value for all CpG cancer types in the whole genome. Finally, MethSurv is an appropriate platform for assessing methylation cancer potential biomarkers in a preliminary study [95].

5.5.17 TransPRECISE and PRECISE

TransPRECISE is a program that uses data from 7,714 clinical specimens collected by means of the Cancer Proteome Atlas from 31 different cancer types. The data is combined with drug susceptibility statistics from the Genomic of Drug Sensitivity in Cancer model system and 640 tumor cell lines from the MD Anderson Cell Lines Project, which represents 481 medications. The new tool is based on the team's previous PRECISE model (personalized cancer-specific integrated network estimation model). The PRECISE model was created to investigate changes in the molecule arrangement of specific individual malignancies. Statistics from cell lines, along with drug sensitivity, are incorporated into TransPRECISE, making it easier for researchers to translate cancer cell biology into therapeutic development [96].

5.6 Conclusion

Cancer is a broad word that refers to a variety of illnesses that can occur in any part of the body. Cancer is one of the most difficult and fast-developing illnesses. It is the second biggest cause of death globally. As per the WHO, new cancer cases (14 million) are recorded each year. Over the next 20 years, that figure is expected to increase by over 70%. Despite substantial advances in the fight against cancer over the previous 30 years, with the survival rate has doubled, a universal cure remains elusive. A lack of useful prognostic data, the demand for tedious data input manually, a lack of full data, and, in some cases, full dependence on statistics all impede oncology risk categorization. Improved risk stratification models in cancer are driven by a need for more patient-centered care. Oncologists are being requested more and more to change treatment strategies according to patient's risk of confirmed results. Data on demographics, treatment episodes, and specific clinical issues must be collected and analyzed. According to estimates, the big data industry is predicted to reach $6.6 billion by 2021. By recognizing trends and increasing illness diagnosis, technology will be able to support traditional healthcare as it advances. In fact, it is already being used in several parts of the world to more accurately and early diagnose illnesses like cancer. Big data has a wide huge variety of skills and is best suited to dealing with massive large data. The volume, velocity, diversity, and authenticity of numerous, often complicated datasets determine the usefulness of big data gathering. Integration of various sources is critical and will benefit biological examination, treatment, and worth care monitoring.

In medical care, the goal of big data is to generate better health profiles and prediction models for specific people so that diseases may be detected and treated more effectively. The Cancer Genome Atlas (TCGA) and the Worldwide Cancer Genome Consortium (ICGC), which both make available investigators with the right of entry to 1000 patients with numerous diseases, are global models for sharing data with partners or the community at large. A large systematic review and machine learning-based tools have been used to incorporate different types of cancer as a result of the availability of this data, resulting in the discovery of unique malignancy candidate genes that are linked to certain processes and could be used as possible treatments therapeutic goals big data has already shown to be beneficial in the diagnosis of patients regularly. Big data analytics may be used to improve patient-centered services such as early detection of sickness outbreaks, new understandings into disease causes, inspecting medical and healthcare facilities for quality, and creating better treatment options.

Recent developments in genomics and associated information skills have sped up the spanning of methodical study and experimental presentation with the provision of publicly available figures, repositories, and investigative implements. Large companies are using cancer genome data from significant studies like TCGA and ICGC to discover novel targets and biomarkers. This chapter gives a quick overview of a few noteworthy web-based resources. CGHub, COSMIC, and cBioPortal are resources that assist as an encyclopedia of tumor and omics data types. Further sites, such as CPRG, GDAC, and canEvolve, provide data analysis and integration tools, as well as storing certain analysis results for querying. Visualization is mostly done with the UCSC cancer browser and CGWB. MethyCancer, SomamiR, and NONCODE databases are devoted to determining the relationship between particular biological characteristics and cancer. The GDSC and CanSAR databases assist the use of genomes in drug development. Simple clicks can perform comprehensive expression analysis. Data mining in research fields, scientific talks, and therapeutic development procedures would be greatly facilitated. Technologies can integrate and personalize prognostic information for individual patients, as well as give enhanced risk estimations for clinical scenarios with unknown outcomes. In the meanwhile, each database has its own set of advantages. Some databases, such as PROGgeneV2, PrognoScan, and TRGAted, focus on survival analysis by gathering records from diverse cancer types. Some databases, such as UALCAN and GEPIA, include additional features, such as top differential gene display, which allows doctors and examiners to identify potential objective genetic factors for analysis or therapy. Oncomine and TCPA both offer multi-variety data investigation and assessment. By examining the link between therapeutic targets and lncRNAs, GSCALite may be utilized for drug screening and therapy alternatives. Progresses in genetic expertise and digital biology have provided us with a once-in-a-lifetime opportunity to understand the molecular pathways that cause cancer and to precisely cure it. This review will be helpful to clinicians seeking extrapolative cancer features.

Acknowledgment

The writers are grateful to the management of Oriental University, Indore, for gracious assistance for providing the required amenities and giving inspiration to enable them to complete their work successfully.

Funding Information

No fund is received.

Conflicts of Interest

Nil

References

[1] Ferlay, J., Laversanne, M., Ervik, M., Lam, F., Colombet, M., Mery, L., Piñeros, M., Znaor, A., Soerjomataram, I.,and Bray, F.(2020) Global Cancer Observatory: Cancer Today. Lyon: International Agency for Research on Cancer.(https://gco.iarc.fr/today,accessedFebruary2021).

[2] de Martel, C., Georges, D., Bray, F., Ferlay, J.,and Clifford, G.M.(2020)Global burden of cancer attributable to infections in 2018: a worldwide incidence analysis. *Lancet Glob Health*. 8 , e180-e190. DOI: 10.1016/S2214-109X(19)30488-7.

[3] Assessing national capacity for the prevention and control of noncommunicable diseases: report of the 2019 global survey. Geneva, World Health Organization; 2020

[4] Wild, C. P., Weiderpass, E., Stewart, B. W.(2020)World Cancer Report: Cancer Research for Cancer Prevention. Lyon: International Agency for Research on Cancer.

[5] How Is Big Data Helping Fight Cancer. Available at: https://www.polestarllp.com/big-data-analytics-in-cancer.[accessedJune3,2020].

[6] Christakis, N. A., and Lamont, E. B.(2000) Extent and determinants of error in doctors' prognoses in terminally ill patients: prospective cohort study. *BMJ*.320, 469-472. DOI: 10.1136/bmj.320.7233.469.

[7] Sborov, K., Giaretta, S., Koong, A., Aggarwal, S., Aslakson, R., Gensheimer, M. F., Chang, D. T., & Pollom, E. L.(2019) Impact of accuracy of survival predictions on quality of end-of-life care among patients with metastatic cancer who receive radiation therapy. *J Oncol Pract*.15, e262-e270. DOI: 10.1200/JOP.18.00516.

[8] Fong, Y., Evans, J., Brook, D., Kenkre, J., Jarvis, P., & Gower-Thomas, K.(2015) The Nottingham prognostic index: five- and ten-year data for all-cause survival within a screened population. *Ann R Coll Surg Engl*. 97,137-139.DOI: 10.1308/003588414X14055925060514

[9] Alexander, M., Wolfe, R., Ball, D., Conron, M., Stirling, R. G., Solomon, B., MacManus, M., Officer, A., Karnam, S., Burbury, K., & Evans, S. M. (2017)Lung cancer prognostic index: a risk score to predict overall survival after the diagnosis of non-small-cell lung cancer. *Br J Cancer*.117,744-751. DOI: 10.1038/bjc.2017.232

[10] Lakin, J. R., Robinson, M. G., Bernacki, R. E.(2016) Estimating 1-year mortality for high-risk primary care patients using the "surprise" question. *JAMA Intern Med*. 176,1863-1865.DOI:10.1001/jamainternmed.2016.5928

[11] Morita, T., Tsunoda, J., Inoue, S., & Chihara, S. (1999)The Palliative Prognostic Index: a scoring system for survival prediction of terminally ill cancer patients. *Support. Care Cancer*.7, 128-133. DOI: 10.1007/s005200050242

[12] Chow, R., Chiu, N., Bruera, E., Krishnan, M., Chiu, L., Lam, H., DeAngelis, C., Pulenzas, N., Vuong, S.,and Chow, E.(2016) Inter-rater reliability in performance status assessment among health care professionals: a systematic review. *Ann Palliat Med*. 5, 83-92. DOI: 10.21037/apm.2016.03.02.

[13] Burwell, S. M.(2015) Setting value-based payment goals: HHS efforts to improve U.S. health care. *N Engl J Med*. 372,897-899. DOI: 10.1056/NEJMp1500445.

[14] Center for Medicare & Medicaid Innovation. Oncology care model. Available at: https://innovation.cms.gov/initiatives/oncology-care/. 2018.

[15] Kline, R., Adelson, K., Kirshner, J. J., Strawbridge, L. M., Devita, M., Sinanis, N., Conway, P. H.,and Basch, E.(2017)The Oncology Care Model: perspectives from the Centers for Medicare & Medicaid Services and participating oncology practices in academia and the community. *Am Soc Clin Oncol Educ Book.* 37, 460-466.DOI: 10.1200/EDBK_174909

[16] Kline, R. M., Bazell, C., Smith E., Schumacher, H., Rajkumar, R., and Patrick, H. C. (2015) Centers for Medicare and Medicaid Services: using an episode-based payment model to improve oncology care. *J Oncol Pract.* 11,114-116.DOI: 10.1200/JOP.2014.002337.

[17] Ostrovsky, A., O'Connor, L., Marshall, O., Angelo, A., Barrett, K., Majeski, E., Handrus, M., & Levy, J.(2016)Predicting 30- to 120-day readmission risk among Medicare fee-for-service patients using nonmedical workers and mobile technology. Perspect *Health Inf Manag.*13:1e.

[18] Conn, J.(2014) Predictive analytics tools help hospitals reduce preventable readmissions. *Mod Healthc.*44, 16-17.

[19] Shaikh, A. R., Butte, A. J., Schully, S. D., Dalton, W. S., Khoury, M. J., and Hesse, B. W., Collaborative biomedicine in the age of big data: the case of cancer. *J Med Internet Res.,*16 ,e101.2014. DOI: 10.2196/jmir.2496

[20] Roman-Belmonte, J. M., De la Corte-Rodriguez, H.,and Rodriguez-Merchan, E. C. (2018) How block chain technology can change medicine. *Postgrad Med.*130,420–7. DOI: 10.1080/00325481.2018.1472996.

[21] Bourne, P. E.(2014) What Big Data means to me. *J Am Med Inform Assoc.*21, 194.DOI: 10.1136/amiajnl-2014-002651.

[22] Anil J. The 5 Vs of Big data, Available at: https://www.ibm.com/blogs/watson-health/the-5-vs-of-big-data.[accessedSept17,2016].

[23] Zhang, C., Bijlard, J., Staiger, C., Scollen, S., van Enckevort, D., Hoogstrate, Y., Senf, A., Hiltemann, S., Repo, S., Pipping, W., Bierkens, M., Payralbe, S., Stringer, B., Heringa, J., Stubbs, A., Bonino Da Silva Santos, L. O., Belien, J., Weistra, W., Azevedo, R., van Bochove, and K.,Abeln, S.(2017)Systematically linking tranSMART, galaxy and EGA for reusing human translational research data. *F1000Res.*6, ELIXIR-1488. DOI: 10.12688/f1000research.12168.1

[24] Jameson, J. L.,and Longo, D. L.(2015) Precision medicine—personalized, problematic, and promising. *N Engl J Med.* 372, 2229–34. DOI: 10.1056/NEJMsb1503104.

[25] van Harten, W. H.(2014) Comprehensive cancer centres based on a network: the OECI point of view. *Ecancerms.* 8, ed43. DOI: 10.3332/ecancer.2014.ed43

[26] Joos, S., Nettelbeck, D. M., Reil-Held, A., Engelmann, K., Moosmann, A., Eggert, A., Hiddemann, W., Krause, M., Peters, C., Schuler, M., Schulze-Osthoff, K., Serve, H., Wick, W., Puchta, J., and Baumann, M.(2019) German Cancer Consortium (DKTK) - a national consortium for translational cancer research. *Mol Oncol.* 13, 535–42.

[27] Eggermont, A., Apolone, G., Baumann, M., Caldas, C., Celis, J. E., de Lorenzo, F., Ernberg, I., Ringborg, U., Rowell, J., Tabernero, J., Voest, E., and Calvo, F.(2019) Cancer core Europe: a translational research infrastructure for a European mission on cancer. *Mol Oncol.*, 13, 521–7. DOI: 10.1002/1878-0261.12447

[28] Pastorino, R., De Vito, C., Migliara, G., Glocker, K., Binenbaum, I., Ricciardi, W., and Boccia, S. (2019). Benefits and challenges of Big Data in healthcare: an overview of the European initiatives. *Eur. J. Public Health.* 29, 23–27. DOI : https://doi.org/10.1093/eurpub/ckz16

[29] Ooft, M. L.,Ipenburg, V. J.,Braunius, W. W., Stegeman, I., Wegner, I., Bree, R., Overbeek, L. I. H., Koljenović, S., Willems, S., and Stefan, W.(2016) Nation-wide epidemiological study on the risk of developing second malignancies in patients with different histological-subtypes of nasopharyngeal carcinoma. *Oral Oncol.*56, 40–46. DOI: 10.1016/j.oraloncology.2016.02.009.

[30] Datema, F. R., Ferrier, M. B., Vergouwe, Y., Moya, A., Molenaar, J., Piccirillo, J. F., and Baatenburg de Jong, R. J.(2013) Update and external validation of a head and neck cancer prognostic model. *Head Neck.* 35 ,1232–37. DOI: 10.1002/hed.23117.

[31] Datema, F. R., Moya, A., Krause, P., Bäck, T., Willmes, L., Langeveld, T., Baatenburg de Jong, R. J. and Blom, H. M.(2012) Novel head and neck cancer survival analysis approach: random survival forests versus Cox proportional hazards regression. *Head Neck.*34,50–8. DOI: 10.1002/hed.21698

[32] Brooks, G. A., Kansagra, A. J., Rao, S. R., Weitzman, J. I., Linden, E. A., and Jacobson, J. O.(2015) A clinical prediction model to assess

risk for chemotherapy-related hospitalization in patients initiating palliative chemotherapy. *JAMA Oncol.*1,441-447. DOI: 10.1001/jamaoncol.2015.0828.

[33] Yeo, H., Mao, J., Abelson, J. S., Lachs, M., Finlayson, E., Milsom, J., and Sedrakyan., A. (2016) Development of a nonparametric predictive model for readmission risk in elderly adults after colon and rectal cancer surgery. *J Am Geriatr Soc.*64, e125-e130. DOI: 10.1111/jgs.14448.

[34] Fieber, J. H., Sharoky, C. E., Collier, K. T., Hoffman, R. L., Wirtalla, C., Kelz, R. R., and Paulson, E. C.(2018) A preoperative prediction model for risk of multiple admissions after colon cancer surgery. *J Surg Res.*231, 380-386. DOI: 10.1016/j.jss.2018.05.079.

[35] Manning, A. M., Casper, K. A., Peter, K. S., Wilson, K. M., Mark, J. R., and Collar, R. M.(2018)Can predictive modeling identify head and neck oncology patients at risk for readmission? *Otolaryngol Head Neck Surg.* 159, 669-674.DOI: 10.1177/0194599818775938.

[36] Vogel, J., Evans, T. L., Braun, J.(2017) Development of a trigger tool for identifying emergency department visits in patients with lung cancer. *Int J Radiat Oncol Biol Phys.*99, S117.DOI: https://doi.org/10.1016/j.ijrobp.2017.06.276

[37] Furlow, B., Predictive analytics reduces chemotherapy-associated hospitalizations. Managed Healthcare Executive. Available at: https://www.managedhealthcareexecutive.com/mhe-articles/predictive-analytics-reduces-chemotherapy-associated-hospitalizations.[accessedMarch30,2017].

[38] Rajkomar, A., Oren, E., Chen, K., Dai, A. M., Hajaj, N., Hardt, M., Liu, P. J., Liu, X., Marcus, J., Sun, M., Sundberg, P., Yee, H., Zhang, K., Zhang, Y., Flores, G., Duggan, G. E., Irvine, J., Le, Q., Litsch, K., Mossin, A.,and Dean, J.(2018) Scalable and accurate deep learning with electronic health records. *NPJ Digit Med.*, 1. DOI: 10.1038/s41746-018-0029-1.

[39] Agarwal, M., Adhil, M., Talukder, A. K.,(2015) *International Conference on Big Data Analytics.* Cham, Switzerland: Springer International Publishing; Multi-omics multi-scale big data analytics for cancer genomics,228–43.

[40] He, K. Y., Ge, D., and He, M. M.(2017) Big data analytics for genomic medicine. *Int J Mol Sci.* 18, 412. DOI: 10.3390/ijms18020412.

[41] Tan, S. L., Gao, G., Koch, S.(2015) Big data and analytics in healthcare. *Methods Inf Med.* 54, 546–7. DOI: 10.3414/ME15-06-1001.

[42] Dinov, I. D., Heavner, B., Tang, M., Glusman, G., Chard, K., Darcy, M., Madduri, R., Pa, J., Spino, C., Kesselman, C., Foster, I., Deutsch, E. W., Price, N. D., Van Horn, J. D., Ames, J., Clark, K., Hood, L., Hampstead, B. M., Dauer, W., and Toga, A. W. (2016)Predictive big data analytics: a study of Parkinson's disease using large, complex, heterogeneous, incongruent, multi-source and incomplete observations. *PLoS One.* 11, e0157077. DOI: 10.1371/journal.pone.0157077.

[43] Archenaa, J., and Anita, E. M. (2015) A survey of big data analytics in healthcare and government. *Procedia Comput Sci.* 50, 408–13. DOI:10.1016/J.PROCS.2015.04.021

[44] Govers, T. M., Rovers, M. M., Brands, M. T., Dronkers, E., Baatenburg de Jong, R. J., Merkx, M., Takes, R. P., and Grutters, J. (2018) Integrated prediction and decision models are valuable in informing personalized decision making. *J Clin Epidemiol.* 104, 73-83. DOI: 10.1016/j.jclinepi.2018.08.016.

[45] Wong, A. J., Kanwa, A.,and Mohamed, A. S.(2016) Radiomics in head and neck cancer: from exploration to application. *Transl Cancer Res.*, 5, 371–82.DOI: 10.21037/tcr.2016.07.18.

[46] Parmar, C., Grossmann P., Rietveld, D., Rietbergen, M. M., Lambin, P.,and Aerts, H. J. W. L.(2015) Radiomic machine-learning classifiers for prognostic biomarkers of head and neck cancer. *Front Oncol.*3 , 272. DOI: 10.3389/fonc.2015.00272

[47] 7 Ways Cancer Research can Benefit from Big Data. Available at : https: //www.kolabtree.com/blog/7-ways-cancer-research-can-benefit-from-big-data/,[accessedFeb2,2018].

[48] van den Akker, J., Mishne, G., and Zimmer, A. D.(2018) A machine learning model to determine the accuracy of variant calls in capture-based next generation sequencing. *BMC Genom.*19,263. DOI: 10.1245/s10434-017-5959-3.

[49] Welsh, J. L., Hoskin, T. L., Day, C. N., Thomas, A. S., Cogswell, J. A., Couch, F. J.,and Boughey, J. C.(2017) Clinical decision-making in patients with variant of uncertain significance in BRCA1 or BRCA2 genes. *Ann Surg Oncol.* 24, 3067-3072.DOI: 10.1245/s10434-017-5959-3.

[50] Pashayan, N., Morris, S., Gilbert, F. J., and Pharoah, P.(2018) Cost-effectiveness and benefit-to-harm ratio of risk-stratified screening for breast cancer: a life-table model. *JAMA Oncol.* 4,1504-1510. DOI: 10.1001/jamaoncol.2018.1901.

[51] Hravnak, M., Devita, M. A., Clontz, A., Edwards, L., Valenta, C.,and Pinsky, M. R.(2011)Cardiorespiratory instability before and after implementing an integrated monitoring system. *Crit Care Med*. 39, 65-72. DOI: 10.1097/ccm.0b013e3181fb7b1c

[52] Raza, S. A., Barreira, C. M., Rodrigues, G. M., Frankel, M. R., Haussen, D. C., Nogueira, R. G.,and Rangaraju, S.(2019)Prognostic importance of CT ASPECTS and CT perfusion measures of infarction in anterior emergent large vessel occlusions. *J Neurointerv Surg*.11, 670–674. DOI:10.1136/neurintsurg-2018-014461.

[53] Parikh, R. B., Kakad, M., and Bates, D. W.(2016) Integrating predictive analytics into high-value care: the dawn of precision delivery. *JAMA*. 315,651-652. DOI: 10.1001/jama.2015.19417.

[54] Yu, K. H., Zhang, C., Berry, G. J., Altman, R. B., Ré, C., Rubin, D. L., and Snyder, M. (2016)Predicting non-small cell lung cancer prognosis by fully automated microscopic pathology image features. *Nat Commun*.7,12474.DOI: 10.1038/ncomms12474.DOI: 10.1038/ncomms12474.

[55] Sooriakumaran, P., Lovell, D. P., Henderson, A., Denham, P., Langley, S. E.,and Laing, R. W.(2005) Gleason scoring varies among pathologists and this affects clinical risk in patients with prostate cancer. *Clin Oncol*., 17:655-658.DOI: 10.1016/j.clon.2005.06.011.

[56] Ehteshami Bejnordi, B., Veta, M., Johannes van Diest, P., van Ginneken, B., Karssemeijer, N., Litjens, G., van der Laak, J., the CAMELYON16 Consortium, Hermsen, M., Manson, Q. F., Balkenhol, M., Geessink, O., Stathonikos, N., van Dijk, M. C., Bult, P., Beca, F., Beck, A. H., Wang, D., Khosla, A., Gargeya, R.,and Venâncio, R.(2017) Diagnostic assessment of deep learning algorithms for detection of lymph node metastases in women with breast cancer. *JAMA*. 318, 2199-2210.DOI: 10.1001/jama.2017.14585.

[57] Barlesi, F., Mazieres, J., Merlio, J. P., Debieuvre, D., Mosser, J., Lena, H., Ouafik, L., Besse, B., Rouquette, I., Westeel, V., Escande, F., Monnet, I., Lemoine, A., Veillon, R., Blons, H., Audigier-Valette, C., Bringuier, P. P., Lamy, R., Beau-Faller, M., and Pujol, J. L.(2016) Routine molecular profiling of patients with advanced non-small-cell lung cancer: results of a 1-year nationwide programme of the French Cooperative Thoracic Intergroup (IFCT). *Lancet* .387 ,1415–26.DOI: 10.1016/S0140-6736(16)00004-0. doi:10.1016/S0140-6736(16)00004-0.

[58] Petersen, J. F., Timmermans, A. J., and Van, D. B.(2018) Trends in treatment, incidence and survival of hypopharynx cancer: a 20-year population-based study in the Netherlands. *Eur. Arch. Otorhinolaryngol.* 275,181–9. DOI: 10.1007/s00405-017-4766-6.

[59] Timmermans, A. J., van Dijk, B. A., Overbeek, L. I., van Velthuysen, M. L., van Tinteren, H., Hilgers, F. J., and van den Brekel, M. W. (2016) Trends in treatment and survival for advanced laryngeal cancer: A 20-year population-based study in The Netherlands. *Head Neck* .38, E1247–55. DOI:10.1002/hed.24200.

[60] de Ridder, M., Balm, A. J., Smeele, L. E., Wouters, M. W.,and van Dijk, B. A.(2015) An epidemiological evaluation of salivary gland cancer in the Netherlands (1989–2010). *Cancer Epidemiol.* 39 ,14–20. DOI:10.1016/j.canep.2014.10.007.

[61] Wilkinson, M. D., Dumontier, M., Aalbersberg, I. J., Appleton, G., Axton, M., Baak, A., Blomberg, N., Boiten, J. W., da Silva Santos, L. B., Bourne, P. E., Bouwman, J., Brookes, A. J., Clark, T., Crosas, M., Dillo, I., Dumon, O., Edmunds, S., Evelo, C. T., Finkers, R., Gonzalez-Beltran, A., and Mons, B.(2016) The FAIR guiding principles for scientific data management and stewardship. *Sci. Data.* 15,3, DOI: http://dx.doi.org/10.1038/sdata.2016.18.

[62] Wilks, C., Cline, M. S., Weiler, E., Diehkans, M., Craft, B., Martin, C., Murphy, D., Pierce, H., Black, J., Nelson, D., Litzinger, B., Hatton, T., Maltbie, L., Ainsworth, M., Allen, P., Rosewood, L., Mitchell, E., Smith, B., Warner, J., Groboske, J.,and Maltbie, D.(2014) The Cancer Genomics Hub (CGHub): overcoming cancer through the power of torrential data. *Database.*, pii: bau093. DOI: 10.1093/database/bau093

[63] Forbes, S. A., Beare, D., Gunasekaran, P., Leung, K., Bindal, N., Boutselakis, H., Ding, M., Bamford, S., Cole, C., Ward, S., Kok, C. Y., Jia, M., De, T., Teague, J. W., Stratton, M. R., McDermott, U., and Campbell, P. J.(2015) COSMIC: exploring the world's knowledge of somatic mutations in human cancer. *Nucleic Acids Res.* 43, D805-11. DOI: 10.1093/nar/gku1075

[64] Marx, V.(2013) Drilling into big cancer-genome data. *Nat Methods*.10,293-7.DOI: 10.1038/nmeth.2410.

[65] Packer, B. R., Yeager, M., Burdett, L., Welch, R., Beerman, M., Qi, L., Sicotte, H., Staats, B., Acharya, M., Crenshaw, A., Eckert, A., Puri, V., Gerhard, D. S., and Chanock, S. J.(2006) SNP500 Cancer: a public resource for sequence validation, assay development, and frequency analysis for genetic variation in candidate genes. *Nucleic Acids Res*.34, D617-21. DOI: 10.1093/nar/gkj151

[66] Samur, M. K., Yan, Z., Wang, X., Cao, Q., Munshi, N. C., Li, C., and Shah, P. K.(2013) canEvolve: a web portal for integrative oncogenomics. *PLoS One*. 8: e56228.DOI: 10.1371/journal.pone.0056228.

[67] He, X., Chang, S., Zhang, J., Zhao, Q., Xiang, H., Kusonmano, K., Yang, L., Sun, Z. S., Yang, H.,and Wang, J.(2008) MethyCancer: the database of human DNA methylation and cancer. *Nucleic Acids Res*.36, D836-41. DOI: 10.1093/nar/gkm730.

[68] Bhattacharya, A., Ziebarth, J. D.,and Cui, Y.(2013) SomamiR: a database for somatic mutations impacting microRNA function in cancer. *Nucleic Acids Res*.41, D977-82. DOI : https://doi.org/10.1093/nar/gks1138

[69] Gao, J., Aksoy, B. A., Dogrusoz, U., Dresdner, G., Gross, B., Sumer, S. O., Sun, Y., Jacobsen, A., Sinha, R., Larsson, E., Cerami, E., Sander, C., and Schultz, N.(2013)Integrative analysis of complex cancer genomics and clinical profiles using the cBioPortal. *Sci Signal.* 6,pl1.DOI: 10.1126/scisignal.2004088

[70] Cerami, E., Gao, J., Dogrusoz, U., Gross, B. E., Sumer, S. O., Aksoy, B. A., Jacobsen, A., Byrne, C. J., Heuer, M. L., Larsson, E., Antipin, Y., Reva, B., Goldberg, A. P., Sander, C., and Schultz, N.(2012) The cBio cancer genomics portal: an open platform for exploring multidimensional cancer genomics data. *Cancer Discov* .2,401– 4. DOI: 10.1158/2159-8290.CD-12-0095.

[71] Tang, Z., Li, C., Kang, B., Gao, G., Li, C., and Zhang, Z.(2017) GEPIA: a web server for cancer and normal gene expression profiling and interactive analyses. *Nucleic Acids Res.* 45, W98–W102. DOI: 10.1093/nar/gkx247.

[72] Tang, Z., Kang, B., Li, C., Chen, T., and Zhang, Z. (2019). GEPIA2: an enhanced web server for large-scale expression profiling and interactive analysis. *Nucleic Acids Res*.47, W556–W560.DOI: https://doi.org/10.1093/nar/gkz430

[73] Yang, W., Soares, J., Greninger, P., Edelman, E. J., Lightfoot, H., Forbes, S., Bindal, N., Beare, D., Smith, J. A., Thompson, I. R., Ramaswamy, S., Futreal, P. A., Haber, D. A., Stratton, M. R., Benes, C., McDermott, U., and Garnett, M. J. (2013). Genomics of Drug Sensitivity in Cancer (GDSC): a resource for therapeutic biomarker discovery in cancer cells. *Nucleic Acids Res*, 41(Database issue), D955–D961. DOI: https://doi.org/10.1093/nar/gks1111

[74] Bulusu, K. C., Tym, J. E., Coker, E. A., Schierz, A. C., and Al-Lazikani, B.(2014) canSAR: updated cancer research and drug discovery knowledgebase. *Nucleic Acids Res.* 42, D1040-47. DOI: 10.1093/nar/gkt1182.

[75] Liu, C., Bai, B., Skogerbø, G., Cai, L., Deng, W., Zhang, Y., Bu, D., Zhao, Y., and Chen, R. (2005)NONCODE: an integrated knowledge database of non-coding RNAs. *Nucleic Acids Res.* 33,D112-115. DOI: 10.1093/nar/gki041

[76] Netwatch.(2005) Databases: decoding the noncode. Science, 307.

[77] Goldman, M., Craft, B., Swatloski, T., Ellrott, K., Cline, M., Diekhans, M., Ma, S., Wilks, C., Stuart, J., Haussler, D., and Zhu, J.(2013) The UCSC Cancer Genomics Browser: update 2013. *Nucleic Acids Res.* 41 ,D949-54. DOI: 10.1093/nar/gks1008

[78] Zhang, J., Finney, R. P., Rowe, W., Edmonson, M., Yang, S. H., Dracheva, T., Jen, J., Struewing, J. P., & Buetow, K. H.(2007) Systematic analysis of genetic alterations in tumors using Cancer Genome WorkBench (CGWB). *Genome Res.* 17, 1111-17.DOI: 10.1101/gr.5963407.

[79] Park, S. J., Yoon, B. H., Kim, S. K.,and Kim, S. Y.(2019) GENT2: an updated gene expression database for normal and tumor tissues. *BMC Med Genomics.* 12 ,101.DOI: 10.1186/s12920-019-0514-7.

[80] Goswami, C. P., and Nakshatri H. (2013) PROGgene: gene expression based survival analysis web application for multiple cancers. *J Clin Bioinform.*, 3,1-9. DOI: https://doi.org/10.1186/2043-9113-3-22

[81] Goswami, C. P., and Nakshatri, H.(2014) PROGgeneV2: enhancements on the existing database. *BioMed Central.* 14, 1-6.DOI: https://doi.org/10.1186/1471-2407-14-970

[82] Aguirre-Gamboa, R., Gomez-Rueda, H., Martínez-Ledesma, E., Martínez-Torteya, A., Chacolla-Huaringa, R., Rodriguez-Barrientos, A., Tamez-Peña, J. G.,and Treviño, V.(2013) SurvExpress: an online biomarker validation tool and database for cancer gene expression data using survival analysis. *PloS one.* 8, e74250, DOI: 10.1371/journal.pone.0074250

[83] Gentles, A. J., Newman, A. M., Liu, C. L., Bratman, S. V., Feng, W., Kim, D., Nair, V. S., Xu, Y., Khuong, A., Hoang, C. D., Diehn, M., West, R. B., Plevritis, S. K.,and Alizadeh, A. A.(2015)The prognostic landscape of genes and infiltrating immune cells across human cancers. *Nat med.* 21, 938–945. DOI: 10.1038/nm.3909.

[84] Rhodes, D. R., Yu, J., Shanker, K., Deshpande, N., Varambally,. R, and Ghosh, D.(2004) Oncomine: a cancer microarray database and integrated data-mining platform. *Neoplasia*. 6, 1–6. DOI: 10.1016/s1476-5586(4)80047-2.

[85] Rhodes, D..R, Kalyana, S..S, Mahavisno, V., Varambally,. R, Yu, J., and Briggs, B. B.(2007). Oncomine 3.0: genes, pathways, and networks in a collection of 18,000 cancer gene expression profiles. *Neoplasia*.,9, 166–80.2007. DOI: 10.1593/neo.07112.

[86] Mizuno, H., Kitada, K., Nakai, K., and Sarai, A.(2009) PrognoScan: a new database for meta-analysis of the prognostic value of genes. *BMC Med Genomics*. 2. DOI: 10.1186/1755-8794-2-18.

[87] Liu C. J., Hu, F. F., Xia, M. X., Han, L., Zhang, Q.,and Guo, A. Y. (2018) GSCALite: a web server for gene set cancer analysis. *Bioinformatics*. 34, 3771–2.DOI: 10.1093/bioinformatics/bty411.

[88] Chandrashekar, D. S., Bashel, B., Balasubramanya, S., Creighton, C. J., Ponce-Rodriguez, I., Chakravarthi, B., & Varambally, S.(2017) UALCAN: A Portal for Facilitating Tumor Subgroup Gene Expression and Survival Analyses. *Neoplasia*.19, 649–658. DOI: 10.1016/j.neo.2017.05.002

[89] Han, S., Kim, D., Kim, Y., Choi, K., Miller, J. E., Kim, D., and Lee, Y. (2018) CAS-viewer: web-based tool for splicing-guided integrative analysis of multi-omics cancer data. *BMC Med Genomics*.11. DOI :https://doi.org/10.1186/s12920-018-0348-8

[90] Alexander, K., Jana, J., Wim, V. C., Manon, V. E., and Tim, D. M. (2019) MEXPRESS update. *Nucleic Acids Res*. 47, W561–W565. DOI: 10.1093/nar/gkz445.

[91] Jang, Y., Seo, J., Jang, I., Lee, B., Kim, S., & Lee, S.(2019) CaPSSA: visual evaluation of cancer biomarker genes for patient stratification and survival analysis using mutation and expression data. *Bioinformatics*. 35, 5341–5343. DOI: 10.1093/bioinformatics/btz516.

[92] Li, J., Lu, Y., Akbani, R., Ju, Z., Roebuck, P. L., Liu, W., Yang, J. Y., Broom, B. M., Verhaak, R. G., Kane, D. W., Wakefield, C., Weinstein, J. N., Mills, G. B., and Liang, H.(2013)TCPA: a resource for cancer functional proteomics data. *Nat. methods*.10, 1046–1047. DOI: 10.1038/nmeth.2650.

[93] Chen, M. M., Li, J., Wang, Y., Akbani, R., Lu, Y., Mills, G. B.,and Liang, H. (2019). TCPA v3.0: An Integrative Platform to Explore the Pan-Cancer Analysis of Functional Proteomic Data. *Mol cell proteomics*. 18,S15–S25.DOI: 10.1074/mcp.RA118.001260

[94] Borcherding, N., Bormann, N. L., Voigt, A. P., and Zhang, W.(2018)TRGAted: A web tool for survival analysis using protein data in the Cancer Genome Atlas. *F1000Re.*7, 1235. DOI: 10.12688/f1000research.15789.2.

[95] Modhukur, V., Iljasenko, T., Metsalu, T., Lokk, K., Laisk-Podar, T., and Vilo, J.(2018) MethSurv: a web tool to perform multivariable survival analysis using DNA methylation data. *Epigenomics.* 10, 277–288. DOI: 10.2217/epi-2017-0118.

[96] Big Data Analytics Tool Could Help Guide Cancer Precision Medicine Available at :https://healthitanalytics.com/news/big-data-analytics-tool-could-help-guide-cancer-precision-medicine,[accessedMay20,2020].

Author Biography

Dr. Nitu Singh presently work as Professor, RGS College of Pharmacy, Itaunja, Lucknow, UP, India She has 12 years of wide experience in Pharmaceutical Education and Research. She has completed her B. Pharm, M. Pharm, and Ph. D from B.N University, Gyan Vihar University, and NIMS University, Rajasthan, India, respectively. She has to her credit more than 30 research papers, review publications, 5 books, and 2 patents. She has delivered more than 10 expert lectures as resource person in various national seminars and conferences. Her keen interest is in Standardization and Biological screening of Herbal drugs and Herbal formulations. She was also awarded the Woman Research Scientist award in 2021 at the National Conference on Recent Advancement in Pharmaceutical, Agriculture, Science, and Health Care Systems.

6

Big Data in Disease Diagnosis and Healthcare

Dhanalekshmi Unnikrishnan Meenakshi[1*], Alka Ahuja[1], Arul Prakash Francis[2], Narra Kishore[2], Pallavi Kurra[3], Shivkanya Fuloria[4], and Neeraj Kumar Fuloria[4]

[1]College of Pharmacy, National University of Science and Technology, Sultanate of Oman
[2]Centre for Molecular Medicine and Diagnostics (COMManD), Department of Biochemistry, Saveetha Dental College and Hospitals, Saveetha University, India
[3]Department of Pharmaceutics, Vignan Pharmacy College, India
[4]Faculty of Pharmacy, AIMST University, Malaysia
*Corresponding Author: College of Pharmacy, National University of Science and Technology, PO Box 620, PC 130 Muscat, Oman, EMail: dhanalekshmi@nu.edu.om, Phone: [+968] 24235000, Fax: [+968] 24504820, ORCID ID: 0000-0002-2689-4079.

Abstract

Optimal healthcare starts with ideal diagnostics. Accurate diagnosis for subsequent cancer care is one of the hurdles faced by oncologists while making clinical decisions. The success of treatment is entirely dependent on detecting cancer in its initial phase. This chapter spotlights the changing landscape of disease diagnostics and the importance of big data (BD) tools that are used to draw meaningful insights, especially in cancer diagnostics to ensure better patient outcomes. Data mining, optimal informatics, data analytics methods, and other resources that are helpful to realize the goal of successful diagnostics and treatment are briefed. Applications of new technology in clinical practice that support the development of vibrant, intelligent, and effective healthcare practitioners are highlighted. BD image analytics, which necessitates the processing of massive datasets employing dispersed memory and computational systems, is also emphasized. Current diagnostic challenges in cancer care and the advantages of using BD as a source for cancer diagnosis in the future are briefly deliberated.

Keywords: Diagnostics, Cancer, Big Data, Oncology, Image Analytics, Machine Learning, Tools, Healthcare, Patients.

6.1 Introduction

Cancer and its related mortality are on the rise all around the world. It has wreaked havoc on the population as the death rate continues to rise. Numerous scientific studies have been conducted in the previous few decades in an attempt to find a cure or manage cancer but with no success. Diagnostic uncertainty, diagnostic errors, inconsistencies in image analysis, limited communication, and lack of collaboration among oncologists, radiologists, and pathologists are the most important challenges in cancer care. The idea of individual doctors examining, diagnosing, and treating individual patients is as old as human history. In recent years, physicians and researchers are focused on the wealth of previously inaccessible data information of various types of cancers and bringing that data together to improve the care of millions. The rapidly changing landscape of diagnostics, precision oncology care, and bioinformatics are giving sureness for successful cancer care [1]. The increasing accessibility and growth rate of big data (BD) derived from numerous omics open a new platform to improve clinical diagnoses and treatment in cancer and hence it acts as a source of innovation for cancer diagnosis. To classify cancer, various machine learning (ML) approaches are developed, which will aid in the development of consistent curation and data integration into huge consolidated datasets. The phenotype procedure uses the natural language processing (NLP) technique to scrutinize and categorize data extracted from patients, including demographic data, gene mapping, images, medical and lab reports, disease details, and family history [2]. Advanced ML models can identify imaging biomarkers and analyze mountains of information that contribute greatly to understanding inner molecular signaling. It is possible to detect gene variations that led to tumor formation by analyzing data from multiple cancer types. Scientists have developed a novel tailored technique for detecting cancer and predicting patient survival and treatment response in immunotherapy 3–5].

In the realm of healthcare, artificial intelligence (AI) and computational algorithms were introduced to eliminate human error at every stage of disease diagnosis and treatment. Oncologists can utilize predictive analytics to recognize high-risk cancer patients. Tumor relapses are common in these people, especially following strong therapies like chemotherapy or surgery. Even the most competent medical practitioners may have difficulty diagnosing cancer patients at a preliminary phase, but ML and BD analytics can recognize particular patterns to predict the recurrence of cancer cells in them. The expenditures of expensive treatments can be saved if such people are identified early, and healthcare professionals can instead focus on adopting malignancy prevention strategies. Pathology and biopsy benefit from predictive analysis as well [6]. Using BD, it is possible to diagnose the disease and also to identify the precise therapeutic regimen for an individual patient without post-treatment complications. As we know, one of the most serious issues with cancer treatment is the possibility of over- or under-treating people which leads to a larger risk of the dead as a consequence of excessive therapy. Predictive analysis tools, such as Google's AI tool, help to improve cancer diagnosis accuracy, allowing clinicians to emphasis on variables other than potent medicines. Apart from this predictive analytics models also have a promising role in screening an entire population for cancer [7]. Clinical specialists increasingly use predictive analytics in the field of oncology due to its numerous advantages. Cancer therapy necessitates analyzing massive amounts of data that humans are incapable of handling but which can be successfully handled by computational algorithms. As a result, BD's function in oncology comes into its own. Health researchers can use BD analysis techniques to acquire new data directly. Oncologists can extract useful information from data using AI and predictive analysis, allowing them to make the best treatment decisions possible. Several renowned hospitals are using the BD tool to diagnose breast cancer in female patients and this tool is also helpful to predict the patient risk factors and the details of precise treatment [6]. To grasp the severity of malignancy present in a patient's body, AI-powered systems can be utilized to include significant patient data, such as familial history, genetic condition incidences, prior investigations, and concerns related to hormone production. Traditional cancer detection, treatment techniques, data collection, and analysis can be unreliable in several aspects. Incorporating AI and BD into oncology removes this step from the process [8]. Furthermore, the technologies allow for more precise, cost-effective, and time-efficient cancer therapies. To assure these things, BD with the support of other computational algorithms is employed to analyze massive volumes of patient data. BD is also useful to diagnose the variety of proteins expressed in benign and malignant cells and this will be greatly beneficial in drug discovery and development. As AI tools are completely dependent on BD it is obvious that one cannot work without the support of the other and this reflects the importance of BD in AI [6]. It is not hard to envision that, with these two technologies it is possible to overcome the long-standing health threat of different ailments, especially cancer. Based on the above-mentioned scenarios, this chapter addresses the evolving landscape of disease diagnostics and the value of BD techniques in gaining insightful information, particularly in disease diagnosis, to improve health outcomes.

6.2 Concepts of BD in Disease Diagnosis and Healthcare

6.2.1 BD and cancer diagnosis

Cancer is different from all of the terrible diseases in the past. Furthermore, as we all know, a variety of malignancies frequently reappear following treatment. Malignant cells mutate in bizarre and unanticipated ways and as a result, cancer continues to kill a significant number of people each year [9]. Early-stage cancer detection in a person can save their lives. Automated instruments to scrutinize molecular changes to determine whether tumors, malignant, can form in body tissue can be extremely beneficial to healthcare systems [10]. To identify a cancerous tumor, AI models are trained to categorize parameters such as the tumor's radius, area, density, proximity to organs, and others. Such criteria are frequently important in determining the mode of treatment that can be utilized to deal with aberrant cellular proliferation. It is conceivable to create several ways in which BD and predictive analytics influence oncology [5]. It is obvious that medical and pharmaceutical researchers are using data analytics and associated technologies to derive new meaning from massive diagnostic, medical and genomic information for therapeutic applications. There is so much raw clinical data collected from the pathology lab to the radiology lab to the surgical unit in the treatment and management of cancer. It is indeed rarely evident how to create and train algorithms to connect all the separate strings of data for patients and it remains a challenge as per several reports [11]. If an AI system can perform and guide the surgery, BD should be able to significantly improve cancer diagnosis and treatment.

6.2.2 BD platform in healthcare

Despite technical flaws, automated systems are already having an impact on cancer studies and treatment care, including early detection and prevention, medication development, patient screening for clinical trials, treatment, and healthcare decisions. Healthcare practitioners also have realized that using

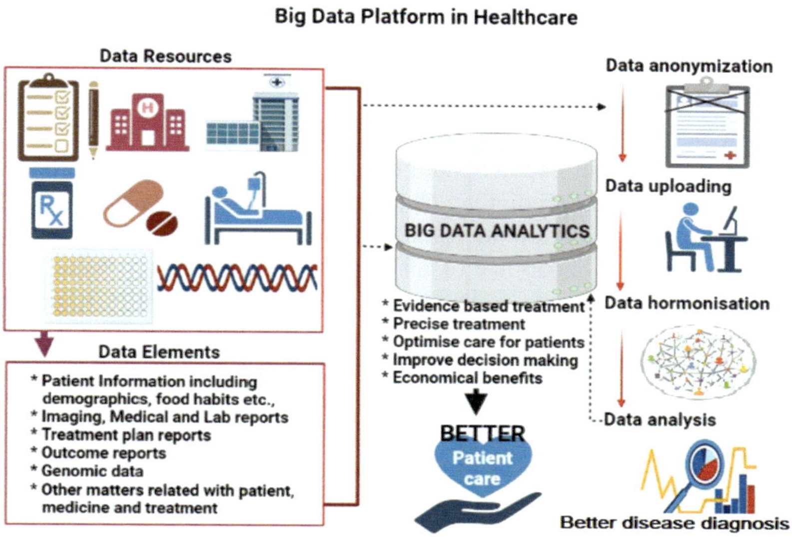

Figure 6.1 BD in healthcare involves the collection, examination, and use of customer, patient, physical, and medical evidence that is too large or complicated to be understood by conventional data processing techniques. Massive volumes of data produced from multiple sources are stored in data warehouses. To produce more intelligent and reasonably priced healthcare choices, this information is analyzed utilizing analytical algorithms.

automatic notifications and alerts for vaccines, abnormal test results, cancer screening, and other regular checkups can significantly improve their health services [5]. It is critical to know the differences in each patient in terms of tumor stage, comorbidity, (neo) adjuvant therapies, and so on to observe the excellence of treatment following precise surgical intrusions [12]. BD sources for healthcare are being developed and made available for integration into existing procedures worldwide. Clinical trial data, genomics and genetic mutation data, protein therapies data, and a slew of other emerging types of data can all be mined to improve day-to-day healthcare. The integration of BD into the healthcare management process will significantly impact the future efficacy of diagnosing and managing healthcare. The platform and the process created to integrate BD to identify (diagnose) and manage patient health are depicted in Figure 6.1. This operational plan can be utilized in a variety of fields but important issues like deep understanding, identifying the gaps, identifying the relevant BD sources, integration of data, data analytics, and proper decision-making should be dealt with [13].

6.3 Predictive Analysis, Quantum Computing, and BD

6.3.1 Predictive analytics

Predictive analysis can be a useful tool that helps healthcare workers to diagnose and treat cancer in patients that provide risk projections based on a participant's prognostic and predictive data. The predictive analysis supports reviewing historical information to forecast the events that will occur in the future. So, using this capability of an AI system can be promising for fighting several diseases in the future, especially cancer. It helps experts detect tumors and categorize them based on the level of danger they pose to patients and comply with global healthcare guidelines related to patient privacy [7, 8]. Predictive analytics encompasses the entire course of illness, including disease prevention, diagnosis, management, and outcome. For example, cigarette smoking is a significant risk factor for cancer especially lung, decreasing such factors can help to reduce the risk of lung cancer. If a patient is diagnosed with lung cancer, risk evaluation can help determine if surgery and/or chemo treatment should be employed. Lastly, precise forecasting is critical for communicating with family members and making medical decisions. BD analytics is increased significantly in clinical care [14]. Zhou et al. (2019) presented a full training about how to undertake predictive modeling. For simplicity of therapeutic applications, nearby sixteen sections address feature extraction, model calibration, usefulness, and nomogram. Researchers also highlighted several difficult circumstances, including the presence of conflicting risks and dimensional constraints. The R code for each phase of the modeling process is also delivered and thoroughly explicated. It is a useful guide for beginners who have little familiarity with R coding [15]. It has also been clarified about the difference between parametric and non-parametric models. Since numerous weights are associated with the nodes of the neural network (NN), it appears that NN must be characterized as a parametric modeling method. In reality, a single-layer neural network is just a linear regression model. Non-parametric models include k-nearest neighbors and decision trees, among other ML techniques. After the final model has been validated by means of a number of validation approaches, a nomogram, and/or risk ratings should be generated. The justifications in the training set and the external set are not conceptually equivalent [14].

6.3.2 Predictive analysis in health records and radiomics

Predictive analytics solutions use algorithms derived from historical patient data to forecast personal health outcomes in the future. The volume of electronic health records (EHR), radiological, genomic, and other data generated from cancer is potentially generalizable [16]. Predictive algorithms can categorize patients taking chemotherapy or needing acute care. This prediction can be helpful for physicians to monitor the cancer spectrum, including chemotherapy, surgery, or discharge planning [7]. Even though EHR data is often difficult to utilize, the Fast Healthcare Interoperability Resources (FHIR) helps to speed up the time-consuming progression of collecting data from EHRs [17]. Deep learning methods were built using over 46 billion data points in the FHIR setup to effectively predict several medical outcomes, including in-hospital death, period of stay, discharging conditions, and relapse. Predictive analytics models are incorporated in disease diagnosis and are extensively employed in oncology, as seen by the growing area of radiomics. Radiomics is a type of texture analysis that studies tumor properties using quantitative data from scans. These traits can help in solid tumor detection, characterization, and monitoring [18, 19]. Other disciplines which will profit from predictive analytics include pathology, which is crucial to cancer practice. There is a lot of variation among pathologists in detecting non–small cell carcinoma from bronchoscopic specimens. Biopsy results that are imprecise may result in a clinical judgment which is either undesirable or improper. AI procedures may accurately diagnose metastatic melanoma in scans of sensitive lymph node biopsies with a significant level of categorization that matches pathologists' assessments. These models improve the skill to scan vast cancer tissue and may assist pathologists to progress their workflow by enabling researchers to devote more time to other duties [7, 20].

6.3.3 Advances in quantum computing

AI as well as quantum computing are the technological areas that play a critical part in improving disease diagnosis. The principles of quantum physics support the design of scanners for the rapid and accurate diagnosis of diseases. Hence, it helps health professionals detect pathological changes resulting from an illness earlier to avoid more invasive treatment procedures. Oncologists can diagnose the disease condition using algorithms, increasing performance and reducing costs with greater precision. Advances in quantum computing made a group of researchers from Case Western Reserve University, together with Microsoft, develop a technique known as Magnetic Resonance Fingerprinting (MRF) that can detect the efficacy of chemotherapy after a single dose. MRF makes it possible to evaluate the tissues of magnetic resonance by comparing them with previously stored data and automatically estimates diagnostic results. The grouping of many patterns or stored images allows it to refine and speed up disease diagnosis without patients facing invasive procedures [21]. Conventional MRF lacks specificity in differentiating primary glioblastomas, lower-grade gliomas, and brain metastases [22]. Co-registered T1 and T2 mappings from MRF can benefit tumor cell characterization and treatment management. Brain water content used in assessing perilesional edema can be correlated with the T1 value [23]. While lesional T2 mapping related to the early detection of tumor progression under anti-angiogenic therapy [24] can differentiate molecular subtypes of grade II and III gliomas [25]. MR imaging scan values acquired in different tumor regions may reflect variations in the expression of tissue growth factors, indicating heterogeneity [26]. In contrast, the multiparametric output of MRF characterizes lesion heterogeneity by combining T1 and T2 measurements with ADC mapping or vp value. Studies have suggested that MRF with a multi-parametric approach can be a validated imaging tool for lesion characterization and hence lessens the encumbrance of biopsy [27].

6.4 Challenges in Early Disease Detection and Applications of BD in Disease Diagnosis

6.4.1 BD in cancer diagnosis

BD has a considerable impact on cancer diagnosis since there is trustworthy data across both malignant tumors and metastasis, as well as statistics on different diagnostic measures (also including mammography) for early sickness screening. For example, in case of breast cancer, the time it takes for a breast cancer tumor to double ranges between 30 days to further over a year, with an average of 150 days, while the time it takes for metastasis to double is approximately shorter than initial tumor [28]. Early diagnosis is very important at this juncture. To detect fast-growing cancers, screening would need to be done more regularly. However, a tumor with a doubling rate of 50 days would develop from the present mammography detection limit of 5 mm in diameter to a lesion massive enough to have been detected medically in just 6 months (2 cm) [29]. As a result, a successful yearly carcinoma diagnostic test would possible to detect increasingly smaller lesions. The identical tumor would take 16 months to develop from 3 mm to 3 cm, yet current technology is insufficiently effective in detecting relatively tiny tumors. Overdiagnosis or overprescribing may become more common, posing a risk of injury. The link between the likelihood of metastases and the size of the tumor holds crucial implications for early diagnosis [28]. Many new imaging technologies are indeed being researched; however, it looks unlikely that scanning will be possible to perceive incredibly tiny tumors; thus, the usefulness of scanning as a routine screening modality will be dictated by the likelihood of early metastasis for specific cancers. Those data can be extracted from historical research with prolonged follow-up of the incidence of lymph node metastasis in various cancer subjects. Individualized screening strategies could be aided by combining such information into parametric data mining and ML platforms that can simulate cancer growth and provide personalized predictions. [29]. In these aspects, AI, BD, and computational algorithms are playing a crucial role in disease diagnosis. Precision medicine has led to a significant volume of medical data which must be sorted and evaluated. The utilization of the scientific paradigm "Data-Intensive Scientific Discovery" is one solution for processing and analyzing large amounts of data for early disease detection (DISD) [30]. NOISeq R package, RNseqViewer, UCSC browser, and Genmoe Mapsor Savant, among others, are software tools for analyzing RN-seq data, expression analysis, alternative splicing research, as well as integration and visualization of diverse data formats. The Naive Bayes classifier is used to create the conceptual framework, and it is also used to conduct the experiments. Ivanova, 2017 uses the Wisconsin breast cancer database for diagnosis. The count of nucleotides in the sequences, the nucleotide bases content, and the guanine-cytosine composition are all integer numbers output by the software. The BLAST software, a database server search engine, can be used to double-check these results. This database associates both categorizations, and if any nucleotides are missing or incompatible, the program inserts gabs. The diagrammatic representation of the BD conceptual model is shown in Figure 6.2. Although many of the available datasets are somewhat difficult to read and have varying structures, several research teams around the world have used multiple techniques to analyze clinical data to identify cancer. It is crucial to categorize breast cancer anomalies and compile relevant datasets but major difficulties are tackled by the strength and unity of data experts and medical professionals. The establishment of benchmarking for cancer datasets seems to be a critical future issue. Mammography scans, 3D ultrasound, MRI, genomic data, and pathology data, among other things, were often used at an individual level over some time to create a diagnosis, a therapy regimen, track the disease's progression, and estimate a patient's prognosis. Furthermore, only organized data was used on a statistical scale. The rest was kept in data graveyards that were hardly visible to the medical professionals. Great data's big promise is that it will exploit all information, even unstructured ones like textual patient reports or photographs [30]. Some clinical investigations (ClinicalTrials.gov Identifier: NCT02810093) involve analyzing breast cancer patients' textual data. The study uses ML algorithms to extract and structure a wide range of data, including medical history, risk factors, tumor size, lymph node involvement, specific biomarkers, treatment, and patient progression. After this data is organized, the second round of statistical modeling is run, yielding many new insights into certain subpopulations, the value of certain biomarkers for prognosis, and the suitability of medical staff decision. Their findings will likely improve our knowledge of the numerous complex pathways that underpin cancer development and therapy resistance [31]. Based on population-based health databases, BD research could offer fresh light on the role of *H. pylori* extermination in stomach cancer. In a Swedish community-based study, H. pylori eradication therapy was linked to a decreased incidence of stomach cancer compared to the general population, however, this benefit did not show up until 5 years after treatment. BD analytics mentioned that diabetes type 2 was linked to a 1.4-fold increase in the incidence of colon cancer [32].

More broadly, all medical personnel must gradually secure ideas about the procedures to integrate the new possibilities given by the BD revolution in patient care. Machine intelligence along with BD is gaining popularity in oncologic endocrinology because of its potential to provide reliable non-invasive diagnoses. To assist doctors in their diagnostic interpretations, AI can be used to characterize tumors in pituitary, adrenal medulla, pancreas, and thyroid glands. In the context of endocrine cancer diagnosis, mitigation methods are necessary to overcome persistent issues with data availability and model interpretability [33]. In the field of neck and head malignancies, BD offers a lot of promise and it has the potential to alter how clinical and scientific data is exchanged. People and organizations will no longer be able to physically communicate datasets because actual data (streaming), combined with the vast volume of data, will make traditional data transfer unfeasible.

6.4.2 BD in the diagnosis of bipolar disorder

In addition to cancer diagnosis, ML delivered increased expertise and methodologies intended for an enhanced diagnosis of bipolar disorder similar to the field of oncology [34, 35]. The international society for bipolar disorders gathered renowned experts in the fields of bipolar disorder, ML, and

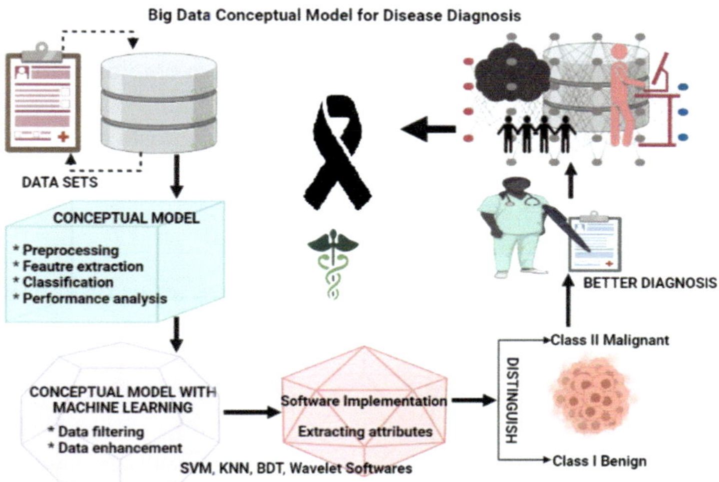

Figure 6.2 The effectiveness of diagnosing and managing healthcare in the future will be greatly improved by the integration of BD. Gadgets with a range of names depending on their use in diagnosis, detection, and other medical terminology have started to appear. The diagnosis of diseases is greatly aided by conceptual models, conceptual models with ML, and various software programs. To the advantage of the population, BD can help define needs, provide necessary services, and forecast and avert future crunches. A large amount of data can help identify demands and provide necessary services in the field of diagnosis.

BD analytics to assess the justification for ML and BD analytics strategies for this disorder [34, 36]. It has been documented that BD can provide vulnerability assessments to help with therapeutic interventions and estimate disease outcomes, such as suicide ideation, for the individual patient [36]. The BD method can help with diagnosis by allowing more relevant data-driven phenotypes and forecasting the information regarding the health condition of the patients who are at high risk. BD analytics applications' most common problems include heterogeneity, deficiency of external authentication, data budget, and lack of sufficient funding. Integrating international BD approaches with real-world healthcare therapies can be successful with large-scale coordinated activity, which would bring together government, business, and philanthropy to work toward a shared goal [34]. To diagnose bipolar disorder, another study discovered various ML models, including classification models, regression models, model-based clustering systems, natural language processing, clustering procedures, and DL-based models. When compared to other groups, magnetic resonance imaging data were commonly utilized

(11% and 34%), while microarray expression and genetic datasets were the least employed for diagnosis [35]. ML algorithms are also employed to create pre-diagnosis strategies for signaling a patient's susceptibility or danger for a psychopathological medical condition, forecasting a diagnostic pathway for identifying mental illnesses [4].

6.4.3 BD in orthodontics

Orthodontics is a notably good fit for BD technology in terms of improving clinical decision-making. Orthodontics has been portrayed as a discipline capable of providing customized and accurate orthodontic therapy owing to the accessibility of enormous genomic datasets, and the possibility of enormous medical information archives. Hence, the application of BD and related cloud computing technologies are magnificently welcomed in this arena [37]. The major purpose of data retention in the healthcare industry is to secure health records. BD and cloud solutions provide compliant data storage that is adaptable and responsive to memory storage requirements. Data storage not only makes orthodontists' jobs easier, but it also makes patient treatment more reliable and effective. It has been documented that patients have reacted favorably to the use of these digital technologies for their information.

The importance of BD analytics in cardiology cannot be overstated. One study looked into the use of ML approaches for predicting cardiac arrest [38]. Different factors, such as electrocardiographic parameters, heart rate variability, echocardiogram, etc., are continuously utilized as predictors in different investigations. The most often used supervised ML techniques to predict cardiac arrest episodes were regression techniques and SVM algorithms, with clear evidence of both. For risk score development and efficiency evaluation, the authors reported an AUC of 0.76. Correspondingly, research reports observed the application of intelligent systems in diagnosing acute coronary syndrome and heart failure, and found that numerous methods, including SVM, feature selection, and neural networks, had excellent accuracy levels. These investigations also identified clinical features that can be used to develop prediction and diagnosis models [38, 39].

6.4.4 BD in diabetes care

AI technologies connected with BD analytics in the care of patients with diabetes mellitus (DM) were investigated. Researchers looked at AI's accuracy, sensitivity, and specificity in screening and diagnosing type 1 and type 2 diabetes. Systolic blood pressure, body mass index, lipid levels, and other factors were among the variables [40, 41]. The best outcomes for assessing complications were allegedly supervised ML approaches, decision trees, deep neural networks, random forests (RF) learning, and support vector machine (SVM) [42, 43].

6.4.5 BD role in infectious diseases

Communicable diseases investigation spans dimension a decade ago, examining everything from pathogenic resistant strains, virulence factors, and reproduction to human, wildlife, and infectious disease migration over the globe. It is evident that communicable disease research areas have been highly influenced by the present development of large-scale data sources, BD, cloud computing , and data processing. In recent years, BD analytics is increasingly being a vital element in disease prevention and control policies as well as disaster management evaluations amid domestic or international outbreaks. The usage of BD in contagious diseases, on the other hand, has a variety of ethical implications. Study reports are focused on four main areas (i) enhanced monitoring and public intervention capacities (ii) the influence of BD profiling on explicit consent (iii) the impact of tracking on personal and social identities, and (iv) the effect of stereotyping on judicial accessibility. Recent expansions in the extent of communicable disease datasets have enabled the use of expressive predictive methods and in turn, leads to deep insights into scientific knowledge. Population sample density is particularly pertinent for contagious diseases because of two variables: transmissibility and immune or drug selection. Researchers examined huge computations using huge simulated databases, massive perturbation investigations, vast medical information, and community, web, and tracking devices connected to communicable diseases for data science [44, 45].

BD information systems are crucial in the COVID-19 period for acquiring the information necessary to develop decisions and preventative measures. Due to the existence of large amounts of COVID-19 relevant data across multiple sources, studies related to the function of BD analysis in COVID-19 spread control are warranted. The main obstacles and possibilities of COVID-19 data analysis give a foundation for associated information systems and studies, enabling new COVID-19 analysis exploration to be more efficient. DL detects the features and complications of COVID-19 disease in the infected person's chest in the CT+AI system (Computerized Tomography and Artificial Intelligence). This taught system could then do scanning of the COVID-19 diseased lung as well as an experienced clinician, but considerably faster. This strategy has already been utilized in over 100 medical centers in China, as well as in several other countries [46, 47]. Despite the lack of reproducibility and uniformity in the outcomes on the application of BD and AI tools in regard to abrupt public health disasters, a current study in China suggests that employing BD and AI in relation to the COVID-19 outbreak is feasible. These research reports concluded that the use of BD and AI technologies can aid in the prevention, diagnosis, treatment, and management of unexpected public health catastrophes like COVID-19 [48, 49].

6.4.6 BD analytics in healthcare

BD analytics technologies improve disease diagnosis and event prediction, including a dramatic improvement in sepsis prediction using ML approaches [50]. Another study found proof of the ML model's influence to achieve extraordinary concert levels in recognizing diseases linked to wellbeing [51]. AI investigative accurateness to evaluate radiogram pictures for lung-related tuberculosis was the subject of one of the studies, which largely focused on development rather than clinical evaluation [52]. Multiple sclerosis diagnosis was also evaluated in one study. Based on increased accuracy and positive predictive value, rule-based and NLP methods were found to have superior diagnostic performance among detection methodologies. This research suggests that these strategies could help with early disease detection, improved quality of life, and prompt pharmaceutical and nonpharmacological treatments [53]. In one review, incidences of asthmatic aggravation and prognostic techniques for primary identification are assessed, with a collective analytical frequency of 77% (95% CI 73–80%) [54]. On the other hand, most of the reports advocated for the creation of models based on huge datasets in order to improve forecast accuracy. One study looked at gastric tissue disease and the usefulness of DL approaches [55]. The most widely utilized approaches found to be logistic and Cox proportional hazard regression techniques. The most common model for detecting or classifying gastrointestinal problems was CNN. Additionally, residual NN and CNN have been providing satisfactory models for disease generation, classification, and segmentation.

6.5 Data Mining in Clinical Big Data

Data mining has introduced a variety of approaches and tools that may be used on a complicated therapeutic databank to mine useful buried or unrecognizable data that can be utilized to accurately diagnose a patient's ailment. Disease diagnostics are among the medical corporate enterprise whereby data

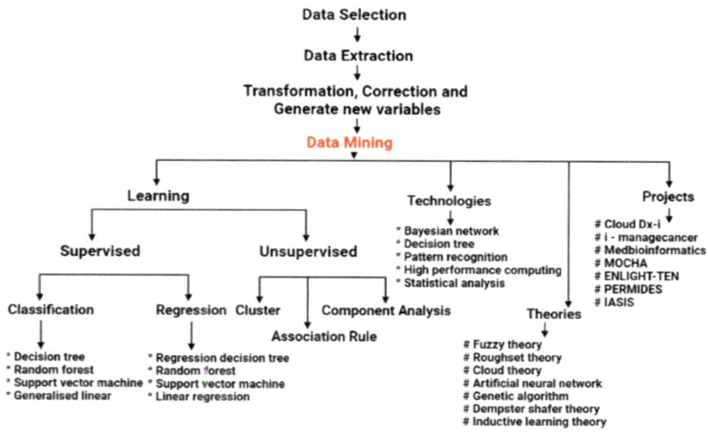

Figure 6.3 Data Mining in Clinical Big Data: To extract hidden or unrecognizable data from a complex therapeutic dataset, data mining can be used to precisely assess a patient's condition. Data mining has introduced a variety of methods and tools. This figure depicts an interdisciplinary program that combines data science,technologies, analytics, machine learning, algorithms and pattern matching to profit from each of these fields.

mining algorithms have shown to be effective, particularly in the treatment of heart, cancer, and thyroid disorders. Data-mining approach has already been a cutting-edge topic in clinical investigations, with outstanding results in assessing the risk factors and supporting clinical decision-making in the development of disease-prediction models. It is an interdisciplinary program that integrates data science technology, analytics, ML, and pattern recognition to reap the benefits of all of these disciplines and the preliminary details are presented in Figure 6.3 Regardless of the reality, this strategy has not yet been commonly employed in clinical research. Some investigations have shown the potential of data mining in creating disease-prediction models, evaluating risk in patients, and supporting doctors in developing treatment strategies [56, 57]. As a result, data mining offers various benefits in BD investigations, particularly in large-scale public clinical datasets. Common databases for cancer include Surveillance, Epidemiology, and End Results (SEER), The Cancer Genome Atlas (TCGA), International Cancer Genome Consortium (ICGC), eICU Collaborative Research Database, Health and Retirement Study (HRS), Medical Information Mart for Intensive Care (MIMIC), for children health includes National Health and Nutrition Examination Survey (NHANES), Paediatric Intensive Care (PIC), etc., Data mining are applied by the use of public data bases as mentioned. The medical sector contains a large amount of heterogeneous data, which is turned into meaningful information using data mining techniques, and then used by clinicians to diagnose different ailments with high precision. Data mining models are both descriptive and predictive. The examination of cancer patients' predicted risk factors is aided by data mining [58].

A multi-layer perceptron neural network trained with the back propagation algorithm was used to diagnose heart problems, with a 94 percent accuracy rate. Medical domain specialists deliver better and earlier diagnosis findings for patients since it functions realistically well even without retraining [59]. A back-propagation algorithm was also utilized to diagnose heart illness, with an average accuracy of 98.65%, outperforming earlier research [59]. For breast cancer diagnosis, a comparison of multi-layer perceptron (MLP) and statistical neural network was conducted. The statistical neural networks used are radial basis function (RBF), probabilistic neural network (PNN), and general regression neural network (GRNN), and when used for classification, they performed well with 96.18 percent for RBF, 97 percent for PNN, 98.8 percent for GRNN, and 95.74 percent for MLP. As a result, it is obvious that these statistical neural network topologies can be used for breast cancer diagnosis. The RBF and PNN outperformed the training set for breast cancer diagnosis. RapidMiner data mining tool uses neural nets, vote ensemble methods, and stacking ensemble methods to diagnose thyroid illnesses [58].

Data mining technologies have given disease detection a new lease on life. Using BD mining technology, the prior diagnosis of a patient can be retrieved, along with laboratory results and clinical complaints, to assist in the investigation of related manifestations. The benefits of data mining and the experiences of doctors can be combined to provide the highest possibility for accurate disease detection. Armstrong et al, 2011 conducted a detailed examination of 240 microcalcifications in mammograms across 220 cases for typical indexes using neural network technology [50]. Data mining was eventually shown to accurately identify whether microcalcifications photographs of suspected breast cancer patients were benign or malignant. A novel kernel-based fuzzy level set for medical image categorization and recognition was used in disease diagnosis [61]. They chose CT scans of blood arteries and the heart, MRIs of the brain and breast, and silver microscopic images of nuclei as image examples. Matlab R2008 software was used to process and analyze the picture data, and the processed image was then associated with the original. The image boundaries based on the new fuzzy algorithm were clearer and this strategy gave clinicians more solid evidence to back up their diagnoses [62].

Data mining aids in the development of early illness prediction models. Sepsis was revealed to be the primary reason for mortality in intensive care unit patients in a previous study [63]. Wu et al., 2021 pointed out that the previous prediction model had a small number of variables and that model performance needed to be improved [62]. In the future, the medical industry should adopt BD with more dimensionality and larger capacities. The goals and objectives of data analysis will be more demanding, with more levels of visualization, more accurate outcomes, and better real-time performance. As a result, the technologies for mining and processing big amounts of information will continue to advance. In terms of application, the generation of information analytics and disease-screening solutions for a wide range of community, such as the military, would aid in the identification of the most appropriate treatments and the formation of supplementary standards that will benefit both cost-efficiency and personnel.

6.6 BD and mHealth in Healthcare

mHealth has taken new dimensions worldwide through the development of smart devices/gadgets like smart phones, desktops, laptops, iPad's, wearables, digital cameras, and sensors. Modern mHealth applications rely heavily on smart phone, wireless high-speed communication systems, and Bluetooth for relatively brief telecommunications to connect healthcare providers and patients. On the other hand, mHealth could be seen as an alternative to current healthcare systems which have discovered benefits, especially in developing nations where the majority of health services are located in urban areas, which are too far away for those living in rural areas [64, 65]. Instead of relying on face-to-face encounters, mHealth applications helped customers and hospital workers to get access to medical help via technologies [66]. mHealth application plays a leading role in healthcare assistance programs, infant birth assistance, diagnostic systems, and other areas 67–69]. Patients may prefer to never see health expert people in person, according to Ippoliti and L'Engle, who note an increasing interest and attention on sexual and reproductive health (SRH) mHealth applications in both developed and developing countries [70]. The rise of Apps that can be downloaded from App stores and run on smartphones is a new component of mHealth. According to a 2017 survey, there are 318,000 health-related apps are available to download in app stores [71]. However, prominent health Apps are healthy life Applications, such as the iPhone's Health App, which cater to elite urban customers [72, 73].

Healthcare sectors are predicted as significant beneficiaries of BD, with the origin since collecting data into electronic records [74]. Studies have found that over 90% of basic healthcare professionals from developed countries, including the UK, Netherlands, etc., used electronic health records (EHRs) for more than a decade [75]. In contrast, research reveals that healthcare facilities in developed countries are lagging behind in developing EHRs and implementing BD [76]; however, some countries, such as the US, have made significant advancements, particularly over the last four years. The data from the US Department of Health and Human Services' health-related IT information website reports showing a vast increase in EHR version among US hospitals from 25% of the hospitals to almost 100% from 2012 to 2015 [77]. In all domains, particularly healthcare, modern technical advancements in data processing and analysis have created an opportunity to access massive data resources. Reports published by consulting firm Mc Kinsey in 2013 titled "The 'big data' transformation in healthcare" piqued the curiosity of the IT startup community [78–80]. Although human healthcare grew exponentially in the 20th century, the 21st era is shaping up to be a time when people and machines collaborate for greater health, integrating human logic and ML. Recognition of BD activities, particularly mHealth software solutions related to rural healthcare, has been steadily increasing.

The main challenge concerning BD in healthcare is data collection in rural areas. BD collection in rural areas faces significant issues due to insufficient facilities and a lack of healthcare personnel. According to WHO statistics, severe scarcity of doctors resulted in a lack of medical care in rural areas reported in several countries worldwide. The technology infrastructures, including data entry procedures that are well-organized, computer facilities, and channels of information exchange required for data transmission and storage, do not exist in many countries. On the other hand, developed countries also faced problems with data collection methods. A study reported by Kim et al. (2014), point out the issues created by government regulations in the US and Europe, associated with health-related data collection problems [81]. Another study on BD in the health sector indicated a lack of data quality, i.e., the data obtained could imply things unrelated to the goal for which the data have been gathered [82]. Furthermore, due to various data-gathering methods and the aggregation of data from multiple sources, 90 percent of global data remains unstructured [83]. As mentioned earlier, in developing countries, many people still live in rural areas where medical care remains a significant problem [84, 85]. The majority of frontline health workers lack the necessary skill in executing their ideas and they also lack advanced medical equipment to obtain reliable medical data.

6.7 Utility of BD

The future potential of BD in biomedical research is still unknown. It has been assured that BD would add value to daily diagnoses, treatment quality, and biomedical research [86]. BD has been shown to be beneficial in the treatment of a variety of ailments. Every qualified pathologist and molecular biologist gets direct access to a database that tracks every patient's pathology instantaneously. It also aids in recognizing a pertinent patient's oncological history, including ruling out a previous cancerous growth in situations of a suspicious lesion or offering pathophysiological documentary evidence on previous relevant pathological features (such as excision margins, lymph node cancer) in cases where pathology was performed in some other laboratory facility. This database can equally be used to investigate the concomitant incidence of various diseases or undiscovered relationships in less prevalent disorders that generally give an unrelated impression at first glance [87]. Computerized patient data generate a huge set of health data that is used to predict prognosis. Models can be automatically updated as new information is collected when statistical prognostication methods are automated. Such statistics could also be utilized to develop therapeutic judgment platforms for more effective patient guidance and improved patient assessments [12].

Genetic modifications are the prognostic or predictive markers as they are the precursors in many different endpoints. Next-generation sequencing (NGS), an automated, high-throughput sequencing technique, provides greater accuracy and efficiency at a reduced price. NGS could be used to do an extensive assessment of the genomes at various scales, such as particular genes, exomes, as well as whole genomes. With the help of sophisticated computational facilities and algorithms, NGS processes enormous amounts of information. Depending on the breadth and depth of the sequencing, a FASTQ document could be several hundred terabytes in size. The sequenced files should then be transmitted to a computer and kept in a specialized database system following assessment. The genomic data pipeline includes assessing the quality of sequencing, assembly of the reads, and variant detection [88]. SeqAcademy developed a user-friendly pipeline that highlights the use of widely available RNA-Seq and ChIP-Seq data and integrates widely used strategies for creating a connection between basic genotyping and physiological understanding [89]. Proteomics gives you a broad picture of how proteins are expressed at any particular time. Proteomes can be explored using different technologies like mass spectrometry [46], chemical labeling [90], and specific affinity chromatography [91]. Metabolites are estimated using mass spectrometry combined with chromatography or spectroscopy [92]. An enormous amount of data generated through spectroscopic techniques were analyzed interactively with genomics and are applied in the field of medical diagnoses and treatment.

The quantitative features of images mined from various imaging systems like magnetic resonance imaging, computed tomography, or positron emission tomography-computed tomography could be quantified using radiomics [93]. These features were interrelated using cancer characteristics in patients and as well as clinical outcomes. In contrast to an independent assessment of cancer appearance, radiomics render a mineable feature space with the help of BD characterization techniques, which are autonomously retrieved from a demarcated targeted area. The image features include first-order topographies relating to the dissemination of voxel intensities, profile, texture, and wavelet features. The same patient should be retested with the same volume to access the stability of the feature values and compared with that of different people to verify the uncertainties for better reproducibility of the resulting model. A predictive model has been developed by combining the patient reports' outstanding features and then assessing the model on an external validation dataset. Furthermore, imaging techniques play an important role in the model's repeatability [94]. Radiomics is a modern technique to extract data from imaging that provides specific information on tumoral heterogeneity, lymphocytic infiltration, etc. Several microarray characteristics could be damaged or non-replicable [95]. However, using standardized imaging acquisition techniques, data quality can be enhanced. Clinical informatics develop the prediction model using the medical data gathered from individual patients to personalize the treatments for patients. Many data on patient characteristics, treatment features, adverse events, and follow-up can be digitally captured from EHR. Moreover, various studies of surveillance, epidemiology, and end results (SEER) produced fast results based on essential queries [96]. Existing models, such as translational research platforms, may combine massive databases of health-related data with genomic information and facilitate the accessibility of patient information [97]. Harvard Medical School for clinical data warehouses (CDW) developed the Informatics for Integrating Biology and the Bedside – i2b2, a data source accepted by most academic hospitals worldwide [98, 99].

6.8 Challenges and Future Perspectives

Managing BD with unknown medical impact is one of the most significant setbacks that confines the usage of BD and creates the opportunity for further research in data analytics and the cloud computing arena. [100]. One example is employing genome sequencing to detect specific conformational changes which emphasize independent or undisclosed mutations. Furthermore, limited genetic changes may be causing further ethical concerns. This is exacerbated by limitations in various analytical techniques, data processing science, standardization, and technological inexperience or insufficient libraries, all of which necessitated improvisation [101]. Furthermore, the EHR-connected procedure for voluntary participation and privacy must be recognized as the foundation for revolutionary treatment/clinical procedures. A solid relationship between healthcare practitioner and patients have been developed through transparency, i.e., conveying the treatment possibilities, creating awareness of the risk–benefit ratio, and informing about medical options [102]. Only if the patient is well-educated about the advantages and disadvantages of various diagnoses and therapies will they be able to make fully informed judgments. Alternatively, the rapid growth of innovative technologies and the large volume of BD generation continuously raise challenges [103]. Indeed, BD is defined by a vast data volume that is always expanding due to meaningful critical research and various data acquisition as well as its analysis. The development of cost-effective BD, information storage and analysis, data transformation and analysis, and personal and international economic relevancy are all part of this problem. We found that the distinctive storage of data in different hospitals and EHRs do not provide the formats needed to record genetic results. These difficulties are intertwined with data standardization, analysis, and evaluation, as well as their application. Collecting genetic data into electronic medical records (EMR) contributes considerably to data processing and uncovering the genetic component of disease risk and treatment outcomes. As a result, using emerging innovations, it is required to enhance data collecting and storage in the data bank. Only through increased investments in genomics, biomathematics, and statistics will this be attainable [100]. Rapid advancements in EMR systems used by patient care providers reveal a huge opportunity to improve the efficiency of gathering, amending, and offering proper access to personalized patient records in a variety of settings like emergency rooms, travel circumstances, etc. Real-time medical record platforms would provide up-to-date information, such as human response to treatments in particular disease conditions, and allow for cost-effective therapies and tailored therapy [101]. Finally, it was clear that the most significant encounters are commercial, not technical.

Despite the fact that BD is already being used in clinical practice and holds immense promise for the future, producers of BD confront obstacles in making it optimally relevant. To begin with, the number of data continues to grow exponentially as new technologies emerge, like next-generation sequencing and radiomics. The intricacy of these huge datasets further adds to the difficulty of data interpretation. The increase in data (volume and velocity) along with an increased data heterogeneity, comprising of variations in treatment protocol, study design, analytical procedures, outcomes, and data interpretation pipelines, make it challenging to reach firm conclusions from the data [12]. Correct data administration, especially when linked from diverse sources, is a second difficulty. Who owns the data, and how and which data are made available? Is the patient still in charge of their data? Is it the researcher who governs it or some authorized person? [4, 104]. BD investigations revealed medium to high reliability in diagnosing and controlling serious illnesses, as well as assistance for quick and real-time analysis of huge amounts of heterogeneous input data in order to forecast treatment outcomes [4]. Another key issue is maintaining patient personal data, like a genetic sequence, in a manner that allows for data reuse while respecting patient privacy. Researchers confront an additional barrier when they want to explore generated biologically relevant datasets instead of being able to link precise markers regressively. There are various systems that use fine-graining access control to represent the datasets without displaying specific indicators [105]. Finally, combining disparate data repositories in a confidentiality manner while still being able to track the actual computational processing performed on the data remains a challenge [106].

6.9 Conclusion

Throughout the last decades, tremendous progress has been made in the treatment of numerous diseases, and technological advancements have enabled the integration of data from the myriad of public data sources for better patient care. BD, which generates comprehensive reports from both internal and public data, has the ability to provide detailed information regarding disease diagnosis, genetic relationships, drug and treatment impacts, and so on. The relevance of BD as a source of information for the diagnosis of diseases and the challenges associated with it was briefly discussed in this chapter. With its technical infrastructure, BD made it feasible to curate, enrich, and integrate large, messy, and even incomplete datasets in less time and at a lower cost. By focusing more narrowly on specific applications that may address tedious challenges with the help of BD, healthcare providers and researchers can avoid several risks and ensure pragmatic success in therapy, diagnosis, and drug discovery. Extraordinary changes in computational approaches, as well as a new competitive landscape for disease detection and treatment, will benefit patients and society as a whole. Critical research connected to the issues of adopting BD in disease diagnosis should be established in order to make deliberate decisions, and this demands full human-level intelligence and joint collaborations.

Acknowledgment

The authors are thankful to their respective Universities for supporting the successful completion of this chapter.

References

[1] Elemento O, Leslie C, Lundin J and Tourassi G 2021 Artificial intelligence in cancer research, diagnosis and therapy *Nat Rev Cancer* 21 747–52

[2] Sivakumar K, Nithya N S and Revathy O 2019 Phenotype algorithm based big data analytics for cancer diagnose *J Med Syst* 43 264

[3] Pastorino R, de Vito C, Migliara G, Glocker K, Binenbaum I, Ricciardi W and Boccia S 2019 Benefits and challenges of big data in healthcare: an overview of the European initiatives *Eur J Public Health* 29 23–7

[4] do Nascimento I J B, Marcolino M S, Abdulazeem H M, Weerasekara I, Azzopardi-Muscat N, Goncalves M A and Novillo-Ortiz D 2021 Impact of big data analytics on people's health: Overview of systematic reviews and recommendations for future studies *J Med Internet Res* 23 e27275

[5] Dash S, Shakyawar S K, Sharma M and Kaushik S 2019 Big data in healthcare: management, analysis and future prospects *J Big Data* 6 54

[6] Joshi N 2021 Catalyzing cancer treatment with big data and predictive analytics

[7] Parikh R B, Gdowski A, Patt D A, Hertler A, Mermel C and Bekelman J E 2019 Using big data and predictive analytics to determine patient risk in oncology. *Am Soc Clin Oncol Educ Book* 39 e53–8

[8] van Calster B, Wynants L, Timmerman D, Steyerberg E W and Collins G S 2019 Predictive analytics in health care: how can we know it

works? *J Am Med Inform Assoc* 26 1651–4

[9] Bray F, Ferlay J, Soerjomataram I, Siegel R L, Torre L A and Jemal A 2018 Global cancer statistics 2018: GLOBOCAN estimates of incidence and mortality worldwide for 36 cancers in 185 countries *CA Cancer J Clin* 68 394–424

[10] Shameer K, Badgeley M A, Miotto R, Glicksberg B S, Morgan J W and Dudley J T 2017 Translational bioinformatics in the era of real-time biomedical, health care and wellness data streams *Brief Bioinform* 18 105–24

[11] Lehman C D, Yala A, Schuster T, Dontchos B, Bahl M, Swanson K and Barzilay R 2019 Mammographic breast density assessment using deep learning: Clinical implementation *Radiology* 290 52–8

[12] Willems S M, Abeln S, Feenstra K A, de Bree R, van der Poel E F, Baatenburg de Jong R J, Heringa J and van den Brekel M W M 2019 The potential use of big data in oncology *Oral Oncol* 98 8–12

[13] Hurwitz J S, Nugent A, Halper F and Kaufman M 2016 How to incorporate big Data into the diagnosis of diseases *Big Data For Dummies* ed J S Hurwitz and A Nugent (John Wiley & Sons, Inc.)

[14] Zhang Z 2020 Predictive analytics in the era of big data: opportunities and challenges *Ann Transl Med* 8 68

[15] Zhou Z-R, Wang W-W, Li Y, Jin K-R, Wang X-Y, Wang Z-W, Chen Y-S, Wang S-J, Hu J, Zhang H-N, Huang P, Zhao G-Z, Chen X-X, Li B and Zhang T-S 2019 In-depth mining of clinical data: the construction of clinical prediction model with R *Ann Transl Med* 7 796

[16] Elfiky A A, Pany M J, Parikh R B and Obermeyer Z 2018 Development and application of a machine learning approach to assess short-term mortality risk among patients with cancer starting chemotherapy *JAMA Netw Open* 1 e180926–e180926

[17] Holtedahl K 2020 Challenges in early diagnosis of cancer: the fast track *Scand J Prim Health Care* 38 251–2

[18] Bi W L, Hosny A, Schabath M B, Giger M L, Birkbak N J, Mehrtash A, Allison T, Arnaout O, Abbosh C, Dunn I F, Mak R H, Tamimi R M, Tempany C M, Swanton C, Hoffmann U, Schwartz L H, Gillies R J, Huang R Y and Aerts H J W L 2019 Artificial intelligence in cancer imaging: Clinical challenges and applications *CA Cancer J Clin* 69 127–57

[19] Sorace A G, Wu C, Barnes S L, Jarrett A M, Avery S, Patt D, Goodgame B, Luci J J, Kang H, Abramson R G, Yankeelov T E and Virostko J 2018 Repeatability, reproducibility, and accuracy of quantitative MRI of the breast in the community radiology setting *J Magn Reson Imaging* 48 695–707

[20] Bejnordi B E, Veta M, van Diest P J, van Ginneken B, Karssemeijer N, Litjens G, van der Laak J A W M, Hermsen M, Manson Q F, Balkenhol M, Geessink O, Stathonikos N, van Dijk M C R F, Bult P, Beca F, Beck A H, Wang D, Khosla A, Gargeya R, Irshad H, Zhong A, Dou Q, Li Q, Chen H, Lin H J, Heng P A, Haß C, Bruni E, Wong Q, Halici U, Öner M Ü, Cetin-Atalay R, Berseth M, Khvatkov V, Vylegzhanin A, Kraus O, Shaban M, Rajpoot N, Awan R, Sirinukunwattana K, Qaiser T, Tsang Y W, Tellez D, Annuscheit J, Hufnagl P, Valkonen M, Kartasalo K, Latonen L, Ruusuvuori P, Liimatainen K, Albarqouni S, Mungal B, George A, Demirci S, Navab N, Watanabe S, Seno S, Takenaka Y, Matsuda H, Phoulady H A, Kovalev V, Kalinovsky A, Liauchuk V, Bueno G, Fernandez-Carrobles M M, Serrano I, Deniz O, Racoceanu D and Venâncio R 2017 Diagnostic assessment of deep learning algorithms for detection of lymph node metastases in women with breast cancer *JAMA* 318 2199–210

[21] Nayak L, Lee E Q and Wen P Y 2012 Epidemiology of brain metastases *Curr Oncol Rep* 14 48–54

[22] Cha S 2009 Neuroimaging in neuro-oncology *Neurotherapeutics* 6 465–77

[23] Schwarcz A, Berente Z, Ösz E and Dóczi T 2002 Fast in vivo water quantification in rat brain oedema based on T(1) measurement at high magnetic field *Acta Neurochir* 144 811–6

[24] Hattingen E, Jurcoane A, Daneshvar K, Pilatus U, Mittelbronn M, Steinbach J P and Bähr O 2013 Quantitative T2 mapping of recurrent glioblastoma under bevacizumab improves monitoring for non-enhancing tumor progression and predicts overall survival *Neuro Oncol* 15 1395–404

[25] Kern M, Auer T A, Picht T, Misch M and Wiener E 2020 T2 mapping of molecular subtypes of WHO grade II/III gliomas *BMC Neurol* 20 1–9

[26] Jackson A, O'Connor J P B, Parker G J M and Jayson G C 2007 Imaging tumor vascular heterogeneity and angiogenesis using dynamic contrast-enhanced magnetic resonance imaging *Clin Cancer Res* 13 3449–59

[27] Ahmed H U, El-Shater Bosaily A, Brown L C, Gabe R, Kaplan R, Parmar M K, Collaco-Moraes Y, Ward K, Hindley R G, Freeman A, Kirkham A P, Oldroyd R, Parker C and Emberton M 2017 Diagnostic accuracy of multi-parametric MRI and TRUS biopsy in prostate cancer (PROMIS): a paired validating confirmatory study *Lancet* 389 815–22

[28] Avanzini S, Kurtz D M, Chabon J J, Moding E J, Hori S S, Gambhir S S, Alizadeh A A, Diehn M and Reiter J G 2020 A mathematical model

of ctDNA shedding predicts tumor detection size *Sci Adv* 6 eabc4308
[29] Pashayan N and Pharoah P D P 2020 The challenge of early detection in cancer *Science* 368 589–90
[30] Ibnouhsein I, Jankowski S, Neuberger K and Mathelin C 2018 The big data revolution for breast cancer patients *Eur J Breast Health* 14 61–2
[31] Dheeba J, Albert Singh N and Tamil Selvi S 2014 Computer-aided detection of breast cancer on mammograms: A swarm intelligence optimized wavelet neural network approach *J Biomed Inform* 49 45–52
[32] Cheung K S, Leung W K and Seto W K 2019 Application of big data analysis in gastrointestinal research *World J Gastroenterol* 25 3008
[33] Thomasian N M, Kamel I R and Bai H X 2021 Machine intelligence in non-invasive endocrine cancer diagnostics *Nat Rev Endocrinol* 18 81–95
[34] Manchia M, Vieta E, Smeland O B, Altimus C, Bechdolf A, Bellivier F, Bergink V, Fagiolini A, Geddes J R, Hajek T, Henry C, Kupka R, Lagerberg T v., Licht R W, Martinez-Cengotitabengoa M, Morken G, Nielsen R E, Pinto A G, Reif A, Rietschel M, Ritter P, Schulze T G, Scott J, Severus E, Yildiz A, Kessing L V, Bauer M, Goodwin G M and Andreassen O A 2020 Translating big data to better treatment in bipolar disorder - a manifesto for coordinated action *Eur Neuropsychopharmacol* 36 121–36
[35] Jan Z, AI-Ansari N, Mousa O, Abd-Alrazaq A, Ahmed A, Alam T and Househ M 2021 The role of machine learning in diagnosing bipolar disorder: Scoping review *J Med Internet Res* 23 e29749
[36] Passos I C, Ballester P L, Barros R C, Librenza-Garcia D, Mwangi B, Birmaher B, Brietzke E, Hajek T, Lopez Jaramillo C, Mansur R B, Alda M, Haarman B C M, Isometsa E, Lam R W, McIntyre R S, Minuzzi L, Kessing L v., Yatham L N, Duffy A and Kapczinski F 2019 Machine learning and big data analytics in bipolar disorder: A position paper from the International Society for Bipolar Disorders Big Data Task Force *Bipolar Disord* 21 582–94
[37] Allareddy V, Rengasamy Venugopalan S, Nalliah R P, Caplin J L, Lee M K and Allareddy V 2019 Orthodontics in the era of big data analytics *Orthod Craniofac Res* 22 8–13
[38] Layeghian Javan S, Sepehri M M and Aghajani H 2018 Toward analyzing and synthesizing previous research in early prediction of cardiac arrest using machine learning based on a multi-layered integrative framework *J Biomed Inform* 88 70–89

[39] Wang W, Kiik M, Peek N, Curcin V, Marshall I J, Rudd A G, Wang Y, Douiri A, Wolfe C D and Bray B 2020 A systematic review of machine learning models for predicting outcomes of stroke with structured data *PloS one* 15 e0234722

[40] Woldaregay A Z, Årsand E, Walderhaug S, Albers D, Mamykina L, Botsis T and Hartvigsen G 2019 Data-driven modeling and prediction of blood glucose dynamics: Machine learning applications in type 1 diabetes *Artif Intell Med* 98 109–34

[41] Abhari S, Kalhori S R N, Ebrahimi M, Hasannejadasl H and Garavand A 2019 Artificial intelligence applications in Type 2 diabetes mellitus care: Focus on machine learning methods *Healthc Inform Res* 25248–61

[42] el Idrissi T, Idri A and Bakkoury Z 2019 Systematic map and review of predictive techniques in diabetes self-management *Int J Inf Manage* 46263–77

[43] Nielsen K B, Lautrup M L, Andersen J K H, Savarimuthu T R and Grauslund J 2019 Deep learning-based algorithms in screening of diabetic retinopathy: A systematic review of diagnostic performance *Ophthalmol Retina* 3 294–304

[44] Garattini C, Raffle J, Aisyah D N, Sartain F and Kozlakidis Z 2017 Big data analytics, infectious diseases and associated ethical impacts *Philos Technol* 32 69–85

[45] Kasson P M 2020 Infectious disease research in the era of big data *Annu Rev Biomed Data Sci* 3 43–59

[46] Alsunaidi S J, Almuhaideb A M, Ibrahim N M, Shaikh F S, Alqudaihi K S, Alhaidari F A, Khan I U, Aslam N and Alshahrani M S 2021 Applications of big data analytics to control COVID-19 pandemic *Sensors* 21 2282

[47] Dong J, Wu H, Zhou D, Li K, Zhang Y, Ji H, Tong Z, Lou S and Liu Z 2021 Application of big data and artificial intelligence in COVID-19 prevention, diagnosis, treatment and Management decisions in China *J Med Syst* 45 84

[48] Wu J, Wang J, Nicholas S, Maitland E and Fan Q 2020 Application of big data technology for COVID-19 prevention and control in china: Lessons and recommendations *J Med Internet Res* 22 e21980

[49] Vaishya R, Javaid M, Khan I H and Haleem A 2020 Artificial Intelligence (AI) applications for COVID-19 pandemic *Diabetes Metab Syndr* 14 337–9

[50] Scardoni A, Balzarini F, Signorelli C, Cabitza F and Odone A 2020 Artificial intelligence-based tools to control healthcare associated infections: A systematic review of the literature *J Infect Public Health*

13 1061–77

[51] Fleuren L M, Klausch T L T, Zwager C L, Schoonmade L J, Guo T, Roggeveen L F, Swart E L, Girbes A R J, Thoral P, Ercole A, Hoogendoorn M and Elbers P W G 2020 Machine learning for the prediction of sepsis: a systematic review and meta-analysis of diagnostic test accuracy *Intensive Care Med* 46 383–400

[52] Harris M, Qi A, Jeagal L, Torabi N, Menzies D, Korobitsyn A, Pai M, Nathavitharana R R and Khan F A 2019 A systematic review of the diagnostic accuracy of artificial intelligence-based computer programs to analyze chest x-rays for pulmonary tuberculosis *PloS one* 14 e0221339

[53] Arani L A, Hosseini A, Asadi F, Masoud S A and Nazemi E 2018 Intelligent computer systems for multiple sclerosis diagnosis: a systematic review of reasoning techniques and methods *Acta Inform Med* 26 258–64

[54] Bridge J, Blakey J D and Bonnett L J 2020 A systematic review of methodology used in the development of prediction models for future asthma exacerbation *BMC Med Res Methodol* 20 22

[55] Gonçalves W G E, Santos M H D P dos, Lobato F M F, Ribeiro-Dos-Santos Â and Araújo G S de 2020 Deep learning in gastric tissue diseases: a systematic review *BMJ Open Gastroenterol* 7 e000371

[56] Ngiam K Y and Khor I W 2019 Big data and machine learning algorithms for health-care delivery *Lancet Oncol* 20 e262–73

[57] Huang C, Murugiah K, Mahajan S, Li S X, Dhruva S S, Haimovich J S, Wang Y, Schulz W L, Testani J M, Wilson F P, Mena C I, Masoudi F A, Rumsfeld J S, Spertus J A, Mortazavi B J and Krumholz H M 2018 Enhancing the prediction of acute kidney injury risk after percutaneous coronary intervention using machine learning techniques: A retrospective cohort study *PLoS Med* 15 e1002703

[58] Sidiq U and Mutahar Aaqib S 2019 Disease diagnosis through data mining techniques *International Conference on Intelligent Computing and Control Systems (ICCS)* (Institute of Electrical and Electronics Engineers Inc.) pp 275–80

[59] Sayad A T and Halkarnikar P P 2014 Diagnosis of heart disease using neural network approach *Int J Adv Sci Eng Tech* 2 88–92

[60] Armstrong A J, Marengo M S, Oltean S, Kemeny G, Bitting R L, Turnbull J D, Herold C I, Marcom P K, George D J and Garcia-Blanco M A 2011 Circulating tumor cells from patients with advanced prostate and breast cancer display both epithelial and mesenchymal markers *Mol Cancer Res* 9 997–1007

[61] Zhang Y, Guo S L, Han L N and Li T L 2016 Application and exploration of big data mining in clinical medicine *Chin Med J* 129 731–8
[62] Wu W T, Li Y J, Feng A Z, Li L, Huang T, Xu A D and Lyu J 2021 Data mining in clinical big data: the frequently used databases, steps, and methodological models *Mil Med Res* 8 44
[63] Layeghian Javan S, Sepehri M M, Layeghian Javan M and Khatibi T 2019 An intelligent warning model for early prediction of cardiac arrest in sepsis patients *Comput Methods Programs Biomed* 178 47–58
[64] Hampshire K, Porter G, Owusu S A, Mariwah S, Abane A, Robson E, Munthali A, DeLannoy A, Bango A, Gunguluza N and Milner J 2015 Informal m-health: How are young people using mobile phones to bridge healthcare gaps in Sub-Saharan Africa? *Soc Sci Med* 142 90–9
[65] Labrique A B, Vasudevan L, Kochi E, Fabricant R and Mehl G 2013 mHealth innovations as health system strengthening tools: 12 common applications and a visual framework *Glob Health Sci Pract* 1 160–71
[66] Ganapathy K and Ravindra A 2008 mHealth: A potential tool for health care delivery in india *Making the ehealth Connection: Global Partnerships, , Local Solutions* (Bellagio, Italy)
[67] Agarwal S and Labrique A 2014 Newborn health on the line: the potential mHealth applications *JAMA* 312 229–30
[68] Conway N, Campbell I, Forbes P, Cunningham S and Wake D 2016 mHealth applications for diabetes: User preference and implications for app development *Health Informatics J* 22 1111–20
[69] Miller A S, Cafazzo J A and Seto E 2016 A game plan: Gamification design principles in mHealth applications for chronic disease management *Health Informatics J* 22 184–93
[70] Ippoliti N B and L'Engle K 2017 Meet us on the phone: Mobile phone programs for adolescent sexual and reproductive health in low-to-middle income countries *Reprode Health* 14 11
[71] Byambasuren O, Sanders S, Beller E and Glasziou P 2018 Prescribable mHealth apps identified from an overview of systematic reviews *NPJ Digit Med* 1 12
[72] Agarwal S, Perry H B, Long L A and Labrique A B 2015 Evidence on feasibility and effective use of mHealth strategies by frontline health workers in developing countries: systematic review *Trop Med Int Health* 20 1003–14

[73] Diwan V, Agnihotri D and Hulth A 2015 Collecting syndromic surveillance data by mobile phone in rural India: implementation and feasibility *Glob Health Action* 8 26608
[74] Raghupathi W and Raghupathi V 2014 Big data analytics in healthcare: promise and potential *Health Inf Sci Syst* 2 3
[75] Jha A K, Doolan D, Grandt D, Scott T and Bates D W 2008 The use of health information technology in seven nations *Int J Med Inform* 77 848–54
[76] van Velthoven M H, Mastellos N, Majeed A, O'Donoghue J and Car J 2016 Feasibility of extracting data from electronic medical records for research: an international comparative study *BMC Med Inform Decis Mak* 16 90
[77] Health IT 2017 Non-federal acute care hospital electronic health record adoption *Office of the National Coordinator for Health Information Technology*
[78] Kaka N, Madgavkar A, Manyika J, Bughin J and Parameswaran P 2014 *India's technology opportunity: Transforming work, empowering people* (McKinsey Global Institute)
[79] Groves P, Kayyali B, Knott D and van Kuiken S 2013 *The big-data revolution in US health care: Accelerating value and innovation*
[80] Senthilkumar S, Rai B K, Meshram A A, Gunasekaran A and S C 2018 Big data in healthcare management: A review of literature *Am J Theor Appl Bus* 4 57–69
[81] Kim G H, Trimi S and Chung J H 2014 Big-data applications in the government sector *Commun ACM* 57 78–85
[82] Sukumar S R, Natarajan R and Ferrell R K 2015 Quality of Big Data in health care *Int J Health Care Qual Assur* 28 621–34
[83] Mathew P S and Pillai A S 2015 Big data solutions in healthcare: Problems and perspectives *2015 IEEE International Conference on Innovations in Information, Embedded and Communication Systems (ICIIECS)* (Institute of Electrical and Electronics Engineers Inc.) pp 1–6
[84] Strasser R, Kam S M and Regalado S M 2016 Rural health care access and policy in developing countries *Annu Rev Public Health* 37 395–412
[85] Mills A 2014 Health care systems in low- and middle-income countries *N Engl J Med* 370 552–7
[86] Roman-Belmonte J M, de la Corte-Rodriguez H and Rodriguez-Merchan E C 2018 How blockchain technology can change medicine *Postgrad Med* 130 420–7
[87] Ooft M L, Vanpenburg J, Braunius W W, Stegeman I, Wegner I, de Bree R, Overbeek L I H, Koljenović S and Willems S M 2016

A nation-wide epidemiological study on the risk of developing second malignancies in patients with different histological subtypes of nasopharyngeal carcinoma *Oral Oncol* 56 40–6

[88] Gogol-Döring A and Chen W 2012 An overview of the analysis of next generation sequencing data *Methods Mol Biol* 802 249–57

[89] Ather S H, Awe O I, Butler T J, Denka T, Semick S A, Tang W and Busby B 2020 SeqAcademy: An educational pipeline for RNA-Seq and ChIP-Seq analysis *F1000Research* 7 628

[90] Vidova V and Spacil Z 2017 A review on mass spectrometry-based quantitative proteomics: Targeted and data independent acquisition *Anal Chim Acta* 964 7–23

[91] Cifani P and Kentsis A 2017 Towards comprehensive and quantitative proteomics for diagnosis and therapy of human disease *Proteomics* 17 1600079

[92] Fuhrer T and Zamboni N 2015 High-throughput discovery metabolomics *Curr Opin Biotechnol* 31 73–8

[93] Gillies R J, Kinahan P E and Hricak H 2016 Radiomics: Images are more than pictures, they are data *Radiology* 278 563–77

[94] Ger R B, Zhou S, Chi P C M, Lee H J, Layman R R, Jones A K, Goff D L, Fuller C D, Howell R M, Li H, Stafford R J, Court L E and Mackin D S 2018 Comprehensive investigation on controlling for CT imaging variabilities in radiomics studies *Sci Rep* 8 13047

[95] Berenguer R, del Rosario Pastor-Juan M, Canales-Vázquez J, Castro-García M, Villas M V, Legorburo F M and Sabater S 2018 Radiomics of CT features may be nonreproducible and redundant: Influence of CT acquisition parameters *Radiology* 288 407–15

[96] Song Y, Wang W, Tao G, Zhu W, Zhou X and Pan P 2016 Survival benefit of radiotherapy to patients with small cell esophagus carcinoma: an analysis of Surveillance Epidemiology and End Results (SEER) data *Oncotarget* 7 15474–80

[97] Canuel V, Rance B, Avillach P, Degoulet P and Burgun A 2015 Translational research platforms integrating clinical and omics data: a review of publicly available solutions *Brief Bioinform* 16 280–90

[98] Rance B, Canuel V, Countouris H, Laurent-Puig P and Burgun A 2016 Integrating heterogeneous biomedical data for cancer research: the CARPEM infrastructure *Appl Clin Inform* 7 260–74

[99] Zapletal E, Rodon N, Grabar N and Degoulet P 2010 Methodology of integration of a clinical data warehouse with a clinical information system: the HEGP case *Stud Health Technol Inform* 160 193–7

[100] Ersek J L, Black L J, Thompson M A and Kim E S 2018 Implementing precision medicine programs and clinical trials in the community-based oncology practice: Barriers and best practices *Am Soc Clin Oncol Educ Book* 38 188–96

[101] Garralda E, Dienstmann R, Piris-Giménez A, Braña I, Rodon J and Tabernero J 2019 New clinical trial designs in the era of precision medicine *Mol Oncol* 13 549–57

[102] Salgado R, Moore H, Martens J W M, Lively T, Malik S, McDermott U, Michiels S, Moscow J A, Tejpar S, McKee T, Lacombe D, Becker R, Beer P, Bergh J, Bogaerts J, Dovedi S, Fojo A T, Gerstung M, Golfinopoulos V, Hewitt S, Hochhauser D, Juhl H, Kinders R, Lillie T, Herbert K L, Maheswaran S, Mesri M, Nagai S, Norstedt I, O'Connor D, Oliver K, Oyen W J G, Pignatti F, Polley E, Rosenfeld N, Schellens J, Schilsky R, Schneider E, Senderowicz A, Tenhunen O, van Dongen A, Vietz C and Wilking N 2017 Societal challenges of precision medicine: Bringing order to chaos *Eur J Cancer* 84 325–34

[103] Joshi Y B and Light G A 2018 Using EEG-guided basket and umbrella trials in psychiatry: A precision medicine approach for cognitive impairment in schizophrenia *Front Psychiatry* 9 554

[104] Wang X, Williams C, Liu Z H and Croghan J 2019 Big data management challenges in health research-a literature review *Brief Bioinform* 20 156–67

[105] Christoph J, Knell C, Bosserhoff A, Naschberger E, Stürzl M, Rübner M, Seuss H, Ruh M, Prokosch H U and Sedlmayr B 2017 Usability and suitability of the omics-integrating analysis platform tranSMART for translational research and education *Appl Clin Inform* 8 1173–83

[106] He S, Yong M, Matthews P M and Guo Y 2017 tranSMART-XNAT Connector tranSMART-XNAT connector—image selection based on clinical phenotypes and genetic profiles *Bioinformatics* 33 787–8

AUTHOR BIOGRAPHY

Dhanalekshmi Unnikrishnan Meenakshi holds a Doctorate in Pharmacology (specialization in Nanomedicine) from the Council of Scientific and Industrial Research (CSIR) – CLRI, India. She worked as a Scientist in the Council of Scientific and Industrial Research – NEIST, India, and was involved in government-funded projects for the North East Exploration of Pharmaceuticals from natural sources. She has over 10 years of research and teaching experience with leading National and International Organizations. She has been working on an array of projects relating to cancer and gene therapy, nanotechnology, and pharmacology. Her research vicinity focuses on preclinical and clinical trials, mechanisms of intoxication, etc. She uses state-of-the-art technology, for systematic evaluation of the efficiency of novel polymeric nanoparticles encapsulated with biologically active agents. Currently, she is a faculty member at the College of Pharmacy, National University of Science and Technology, Muscat, Sultanate of Oman. She published extensively in nanomedicine, drug delivery, and formulation technology in peer-reviewed reputed journals and books. She bagged various awards and fellowships for her excellent contributions in her respective fields.

7

Big Data as a Source of Innovation for Disease Diagnosis

Deepika Bairagee[1*], Nitu Singh[2], Neetesh Kumar Jain[2], Neelam Jain[2], Sumeet Dwivedi[3], and Javed Ahamad[4]

[1]Pacific College of Pharmacy, Pacific University, India
[2]Oriental College of Pharmacy and Research, Oriental University, India
[3]University Institute of Pharmacy, Oriental University, India
[4]Department of Pharmacognosy, Faculty of Pharmacy, Tishk International University, Iraq
*Corresponding Author.

Abstract

"Big data" denotes the massive quantities of data that can be used to accomplish incredible feats. It has piqued people's imagination over the past two decades because of the enormous potential it has. Big data is generated, stored, and analyzed by a variety of public and commercial sector organizations in order to improve the services they provide. Clinic records, patient clinical records, clinical assessment results, and web of things gadgets are all examples of big data hotspots in the medical industry. Biomedical research creates a lot of big data that may be utilized in public health. In order to make meaningful judgments, this information must be correctly administered and analyzed. In any case, sifting through large amounts of data for patterns quickly becomes akin to searching for a needle in a haystack. The implementation of high-quality registration solutions for big data analysis must overcome a variety of problems related to each level of big data management. As a result, medical care providers must be fully equipped with a proper framework for rapidly creating and analyzing bid data in order to provide relevant answers for working on overall health. A successful administration, inspection, and comprehension of big data can alter the game by allowing existing medical services to branch out in new directions. As a result, many businesses, including the medical care industry, are working hard to turn this potential into better administration and medical service information. Clinical treatments and personalized medication may be reformable by current medical service organizations.

Keywords: Big Data, Disease Diagnosis, Innovation; Bioinformatics, Internet of Things.

7.1 Introduction

7.1.1 Data

Data has been the key to a better relationship and unexpected consequences. We can better organize our efforts to attain the best results if we have more data. As a result, obtaining data is a crucial part of every company's operations. We may also utilize this data to anticipate recent changes in specified boundaries as well as future occurrences. As we have gotten more conscious of this, we have started to develop and gather more information about almost everything by providing inventive advancements along the way. We now live in a world where we are bombarded with large volumes of data from every part of our life, including social activities, research, job, health, and so on. The current scenario may be compared to an information storm. Innovative developments have aided us in creating an ever-increasing volume of data, to the point where it is now unmanageable with existing accessible advancements. As a result, the phrase "big data" was invented to characterize large and unmanageable amounts of data. We wish to promote innovative ways of gathering this data and inferring crucial facts in order to suit our existing and forthcoming social demands. Medical services are not a frequent societal requirement. Medical services associations, like any other business, produce information at breakneck speed, bringing a plethora of benefits and problems at the same time.

Individuals working for numerous organizations all around the world create large amounts of data on a regular basis. The term "computerized universe" refers to the enormous volumes of data that are created, recreated, and consumed in a single year. In 2005, the International Data Corporation (IDC) estimated the advanced universe's size to be 130 exabytes (EB). In 2017, the computerized globe surpassed 16,000 EB (16 zettabytes) in size (ZB). According to IDC, by 2020, the computerized universe will have grown to 40,000 EB. To get a sense of the scale, we would need to provide everyone with around 5100 gigabytes (GB) of data. This demonstrates the remarkable rate at which the computerized cosmos is growing. Google and Facebook, for example, have been assembling and storing vast quantities of data. Google may retain a range of data, including client location, commercial preferences, a list of used programs, web browsing history, associates, bookmarks, messages, and other vital data relating to the client, depending on our settings. Facebook also saves and decodes more than 30 petabytes (PB) of data created by its users. The term "enormous information" refers to a big volume of data. The IT industry has successfully used big data to generate basic data that can generate critical revenue over the last decade.

These views have grown so widespread that a new field of research called "Data Science" has been created to address them. Data Science manages different angles including data from the board and investigation, to separate further bits of knowledge for working on the usefulness or administrations of a framework (for instance, medical services and transport framework). Furthermore, with the availability of some of the most inventive and significant methods of visualizing big data post-investigation, the operation of any complex framework has become more clear. It is vital to clarify what big data is when a huge portion of society becomes aware of and involved with its development. As a result, we want to give insights into the influence of big data on the expansion of the international medical services sector and its influence on our everyday routines [1] in this audit.

7.1.2 Big data

Big data, as the name indicates, deals with massive volumes of data that are hard to manage using standard programming or web-based systems. It outperforms the widely used capacity, handling, and logical power metrics. Despite the fact that there are several definitions for big data, Douglas Laney's is the most well-known and commonly accepted. Laney recognized that (big) data was filling in three separate characteristics, namely volume, speed, and variety (often referred to as the three Vs) [2]. The "large" in big data alludes to its enormous size. Apart from volume, the representation of big data also involves speed and diversity. Speed relates to the pace at which data is collected and made accessible for further analysis, whereas assortment refers to the many sorts of coordinated and random data that any company or framework may gather, such as exchange level data, video, sound, and text or log documents. These three Vs have become the industry standard for defining huge datasets. Despite the addition of a few more Vs to this description by others, the most well-known fourth V remains "veracity" [3].

Big data is a concept that has lately received a lot of traction throughout the world. Almost every field of study, regardless of whether it is associated with industry or academia, is producing and analyzing huge quantities of data for a diversity of objectives. The most challenging aspect is managing this vast volume of data, which may be both organized and disorganized. We need improved applications and programming that can employ a speedy and cost-effective top-of-the-line processing capacity for such jobs since huge data is unmanageable with traditional programming. To seem legitimate from this massive amount of data, it would be necessary to carry out artificial intelligence (AI) calculations and innovative combination calculations. Without a doubt, achieving mechanical decision-making via the use of AI (ML) tactics such as neural networks and other AI methods would be a tremendous achievement. Regardless, without the right programming and equipment, massive amounts of data may become quite hazy. We would want to see better techniques for dealing with this "infinite ocean" of data, as well as clever online apps for effective inspection and the acquisition of important bits of knowledge. The facts and experiences gained from enormous data can make fundamental social framework parts and administrations (like medical services, wellness, or transportation) more mindful, intelligent, and effective [4] with genuine capability and logical gadgets close by. Furthermore, a crucial component of cultural shifts will be the perception of big data in an easy-to-use manner.

Big data has changed the manner in which we make due, dissect, and influence data across ventures. Quite possibly the most striking area where data examination is rolling out enormous improvements is medical services.

In reality, medical care examination can potentially lower therapy costs, anticipate pandemic outbreaks, avoid preventable infections, and improve overall personal satisfaction. The average human life expectancy is rising across the board, bringing significant problems for current treatment options. Experts in well-being, like business visionaries, are skilled in accumulating large volumes of data and determining the best approaches for interpreting them [5].

7.2 Electronic Health Records

Note that the National Institutes of Health (NIH) as of late declared the "We all" drive (https://allofus.nih.gov/) that expects to gather at least 1,000,000 patients' data like EHR, including clinical imaging, socio-conduct, and natural information throughout the following not many years. EHRs enjoy presented many benefits for dealing with current medical care-related data. Beneath, we portray a portion of the trademark benefits of utilizing EHRs. The main benefit of EHRs is that medical care experts have a further developed admittance to the whole clinical history of a patient. The data incorporates clinical analyses, remedies, data identified with known sensitivities, socioeconomics, clinical accounts, and the outcomes got from different lab tests. The acknowledgment and therapy of ailments subsequently is time proficient because of a decrease in the slack season of past test outcomes. With time we have noticed a huge abatement in the excess and extra assessments, lost requests and uncertainties brought about by obscured penmanship, and a further developed consideration coordination between numerous medical services suppliers. Conquering such calculated mistakes has prompted a decrease in the number of medication sensitivities by lessening blunders in medicine portion and recurrence. Medical care experts have additionally observed admittance over online and electronic stages to further develop their clinical practices essentially utilizing programmed updates and prompts in regard to inoculations, unusual research facility results, malignant growth screening, and other intermittent exams. There would be a more noteworthy progression of care and convenient mediation by working with correspondence among numerous medical care suppliers and patients. They may be linked to electronic clearance and faster protection endorsements due to reduced desk work. EHRs empower quicker data recovery and work by revealing key medical care quality markers to the associations, and further develop general well-being reconnaissance by prompt detailing of illness episodes. EHRs likewise give important information in regard to the nature of care for the recipients of worker medical coverage programs and can assist with controlling the expanding expenses of healthcare coverage benefits. At last, EHRs can decrease or dispense with postponements and disarray in the charging and claims of the executive's region. The EHRs and the web together assist with giving admittance to a great many wellbeings-related clinical data basic for patient life.

7.3 Digitization of Healthcare and Big Data

Standard clinical and clinical information acquired from patients is stored in an electronic medical record (EMR), which is similar to an electronic health record (EHR). EHRs, EMRs, personal health records (PHR), medical practice management software (MPM), and a number of other medical information components have the potential to enhance the excellence, competence, and

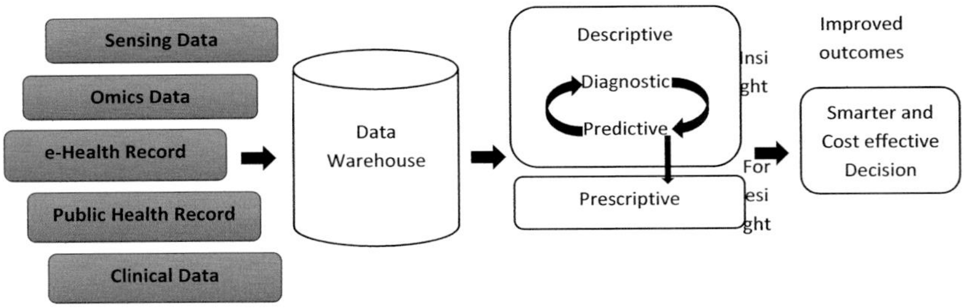

Figure 7.1 Big data management, analysis, and future possibilities in healthcare.

affordability of medical treatment while also decreasing clinical mistakes. Medical payer-supplier data (e.g., EMRs, drug store treatment, and protection records) are included in big data in medical services, as are genomics-driven analyses (e.g., genotyping, quality articulation data) and other data gathered through the smart trap of the Internet of things (IoT) (Figure 7.1). EHRs were not widely accepted until the early twenty-first century, but after 2009, it has become more liberal [6, 7]. The management and utilization of such medical information have been altered by data innovation over time. The evolution of events has gained traction, particularly in the establishment of a continuous biomedical and health-monitoring framework, and the use of health-monitoring gadgets and related programming that can produce signals and share a patient's health-related information with specific medical service providers. These devices provide a large quantity of data, which may be analyzed to offer ongoing clinical or clinical consideration [8]. The utilization of vast volumes of data from medical treatment has the potential to improve health outcomes while also lowering costs.

7.4 Healthcare and Big Data

Big data in medical care refers to vast amounts of data created by modern medical improvements that collect patient records and aid in clinic execution, but which are too large and complex for traditional medical breakthroughs.

The application of big data analysis in medical treatment has yielded a slew of beneficial and even life-saving results. Fundamentally, immense style data refers to the massive volumes of data generated by the digitalization of everything that is cemented and broken down by explicit advances. When used for medical services, it will make use of particular health data from a population (or a single individual) to help prevent epidemics, treat illnesses, reduce expenditures, and so on.

Because we are living longer, treatment paradigms have altered, and a huge percentage of these progressions are now data-driven. Specialists need to learn as much as they can about a patient and as early in their life as possible in order to recognize signs of true illness as they arise treating any infection at any stage is indisputably simpler and less expensive. By using key execution markers in medical care and medical services information examination, anticipation is superior to fixing, and figuring out how to draw an exhaustive image of a patient will allow protection to give a custom-made bundle. This is the business' endeavor to handle the siloes issues a patient's information has: wherever are gathered pieces and chomps of it and filed in medical clinics, facilities, medical procedures, and so forth, with the difficulty to convey appropriately [9].

For a long time, obtaining enormous volumes of data for clinical purposes has been expensive and time-consuming. With today's ever-evolving technology, it is getting simpler to not only gather such information but also to compile full medical care reports and transform them into crucial fundamental bits of knowledge that may be utilized to deliver better treatment. This is the motivation behind medical care data investigation: utilizing data-driven discoveries to foresee and take care of an issue before it is past the point of no return, yet additionally survey strategies and therapies quicker, monitor stock, include patients more in their own wellbeing, and enable them with the devices to do as such.

These days, medical care associations are under expanding strain to use their assets all the more productively while ensuring a high understanding of quality consideration [10]. To address these issues, different scientists and specialists prescribe medical services associations to apply enormous data examination to guarantee viable clinical and authoritative independent direction [11], work on tolerant consideration [12, 13], and clinical expense decrease [14]. Lately, the medical services industry has encountered a remarkable development of well-being data [15]. Exclusively in the U.S. medical services framework, well-being data has arrived at the size of zettabytes (1021) of which 80% is unstructured [16]. Instances of unstructured well-being data are doctors' clinical notes and clinical pictures. In 2015 alone, 60 billion pictures were produced in the United States that could be utilized for more exact clinical conclusions and worked on persistent consideration.

Despite the expanding volume and intricacy of clinical image data, the improvement of huge scope examination to overcome any issues among pictures and conclusions in close to constant could not stay up with the requirements of PC helped analysis (CAD) or choice emotionally supportive networks in present-day medical care associations [17].

7.4.1 Descriptive analytics

Descriptive analytics is a technique for recounting a series of prior occurrences. Data mining is a technique for extracting information from aggregated data in order to gain an improved understanding of a previous occurrence. In a nutshell, the descriptive analysis may be summarized as "What happened?" (a subgroup of TF7). In healthcare, descriptive analytics may reveal the source and causation of illness transmission, as well as the length of quarantine periods that may be necessary. It can also tell you which drugs did not work and which drug combinations did. It can help pinpoint the disease's etiology and treatment options.

7.4.2 Diagnostic analytics

A diagnostic analysis is an analytic approach that aims to discover the multiple aspects that led to the occurrence of a given event in the past using expressive analytics data. In a nutshell, diagnostic analytics is a sort of enquiry that asks, "Why did it happen?" The year 2016 is upon us (TF7 Healthcare subgroup). Diagnostic analytics in healthcare can reveal the origins of numerous disease outbursts, as well as managerial and patient care problems. It is also useful for figuring out why some illnesses are latent and when they become active, that is the circumstances that allow infections to germinate. The diagnostic analysis is concerned with using data mining, correlation, and other techniques to determine the reason for a prior occurrence.

7.4.3 Predictive Analytics

Predictive analytics is a progressive analytics method that forecasts upcoming occurrences using data mining, machine learning, artificial intelligence, and statistical modeling. In healthcare, predictive analytics enables us to put the phrase "prevention is better than cure" into practice. Using predictive analytics in healthcare, we can predict diseases that will affect a person based on their behaviors, genetic makeup, and history, as well as their work or home environment. On a bigger scale, this sort of research can aid us in anticipating population health in the future, allowing us to take necessary preventative actions. When utilized effectively, predictive analytics has the ability to keep a big number of people out of hospitals in the future.

7.4.4 Prescriptive analytics

Prescriptive analytics is concerned with determining the best course of action for a future event based on predictive analysis. The prescriptive analysis is a type of analysis that combines predictive and descriptive analysis to determine what to do in the event of an anticipated event. The predictive analysis addresses the question "What should we do?" based on the available forecasts (TF7 Healthcare subgroup, 2016). It has the potential to give risk-mitigation decision-making options. Prescriptive analytics in healthcare has the potential to revolutionize current medicine by providing not only speedy diagnosis but also therapy based on a patient's medical history. In terms of pharmaceutical incompatibilities and adverse effects, this will result in better treatment.

7.5 Big Data Analytics in Healthcare

Research characterizes the idea of Big information by the three V's of information: volume, velocity, and variety [18, 19]. In the medical care setting, data is spread among various substances, like clinics, well-being frameworks, specialists, and legislatures put away in storehouses that need worldwide straightforwardness and access [11]. Inside this domain, different analysts see bid data investigation as an empowering agent to get to, store, break down, and imagine a lot of data in a coordinated, interoperable, and constant methodology that will ultimately uphold numerous medical care associations in navigation and activity taking [20]. The three unique sorts of data sources that can be found in clinics are (1) clinical data sources like electronic health records (EHR), research center outcomes, and clinical pictures, (2) regulatory data sources like faculty and monetary data, and (3) outside data sources like measurable or web-based media data [21]. To use big data investigation and benefit from this massive volume of well-being data, medical services associations should execute present-day big data answers for data stockpiling, examination, and visualization.

Raghupathi and Raghupathi [20] have proposed an engineering system for big data examination in medical care that features various advances (e.g., Hadoop, Pig, and Hive). Wang et al. [22] present a comparative structure that adds data administration across the data catching, changes, and utilization

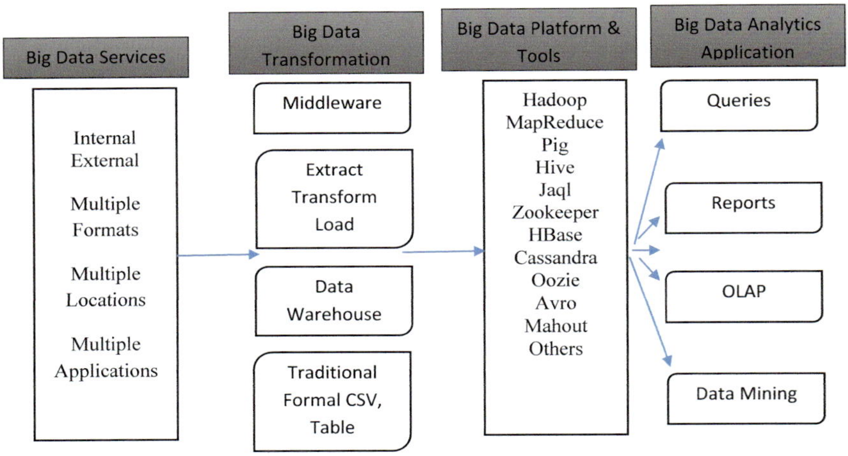

Figure 7.2 Conceptual architecture of big data analytics.

layer. The system of Mettler and Vimarlund [21] incorporates medical care partners and cycles inside the innovation domain. They contend that big information examination in medical care should uphold the board in understanding accessible inward and outer capacities and work with clinical and managerial decision-making by coordinating a wide range of measurements about an assortment of entertainers. This system joins the center methodologies of the previously mentioned specialists and takes into account an organized investigation dependent on the normal transaction of individuals, cycles, and innovation in an organization (Figure 7.2).

However, execution of this coordinated system includes different difficulties, for example, restricted information sharing because of data security guidelines; frustrated information incorporation because of required principles, merchant lock-in, and data design assortment; complicated, ideal, and precise examination of huge scope data; and frail data representation arrangements [12].

7.5.1 Image analytics in healthcare

According to Siuly and Zhang [23] pictures are a significant hotspot for clinical analysis, treatment evaluation, and arranging. Notable imaging procedures are processed tomography (CT), X-beam, magnetic resonance imaging (MRI), and mammography. Produced pictures are shared utilizing standard conventions like digital imaging and communication in medicine (DICOM) and put away in picture archiving and communication system (PACS) [24]. The data size of clinical pictures can go from a couple of megabytes to many megabytes per study [11]. The developing measure of clinical pictures created consistently in current clinics requires a shift from customary clinical picture examination towards generally adaptable arrangements offering openings for more prominent utilization of PC-helped analysis (CAD) and choice of emotionally supportive networks [22, 25]. The volume, speed, and assortment of clinical image data require big data stockpiling limits just as quick and precise calculations. Numerous earlier investigations have tried various techniques for picture examination in medical care [23]. Among the old-style AI techniques for data mining, like directed learning (e.g., grouping) and unaided learning (e.g., bunching), further developed techniques like support vector machines (SVM), neuronal organizations, and computerized reasoning (AI) are frequently applied in this domain [11, 25]. For instance, grouping and division comprise relegating a mark (e.g., solid or ailing) to a given picture, which is addressed in a component space that portrays the picture (e.g., shading and surface). Subsequent to having been prepared, directed AI calculations are utilized to anticipate test picture classes dependent on input visual highlights [25]. To empower these techniques, late examinations propose answers for huge scope clinical picture investigation dependent on equal figuring and calculation improvement [22, 24]. Apache Hadoop is an open-source system that considers the dispersed handling of big datasets across PC bunches utilizing straightforward but incredible programming models, like MapReduce and Spark [11].

7.6 Big Data in Diseases Diagnosis

Across the world, big data hotspots for medical services are being made and made accessible for reconciliation into existing cycles. Clinical trial data, hereditary qualities and hereditary change information, protein therapeutics data, and numerous other new wellsprings of data can be reaped to further develop everyday medical care processes

Web-based media can and will be utilized to expand existing data and cycles to give more customized perspectives on treatment and therapies. New clinical gadgets will control medicines and send telemetry information for constant and different sorts of examination. The errand ahead is to comprehend these new wellsprings of data and supplement the current data and cycles with the new big data types.

Anyway, what might the medical services process resemble with the presentation of big data into the functional course of recognizing and overseeing patient wellbeing? Here is an illustration of what the future may resemble:

7.6.1 Comprehend the issue we are attempting to settle (need to treat a patient with a specific type of cancer)

Specialists are anticipating that big data investigation can speed up malignant growth treatment and follow through on the guarantee of more customized medication and treatment.

Data is being used to help oncologists in offering tailor-made medicines based on – biopsy examples, patient history, and other related information. Foundations all over the planet are occupied with examining a bunch of types of malignant growth-related data from patient case narratives and worldwide exploration and reviews.

Immense amounts of this information are currently being investigated and analyzed to spot discoveries that can be utilized to make new treatments that can prompt more customized medicines. For example, scientists in the United States are currently filtering through many gigabytes of picture data from a huge number of patients to attempt to find contrasts that recognize the distinctive subtypes of bosom tumors from one another. Considering that no disease is something similar, this could be the sacred goal of malignant growth treatment. For patients who have non-repetitive malignant growth, this could prompt milder treatments and medicines. The inverse may be appropriate for those with more forceful or intermittent strains of the infection.

In the momentum situation, scientists are utilizing hazard definition modules to slither through the huge number of well-being records in a populace to get examples, patterns, and patient comparability measurements utilizing natural language processing (NLP) frameworks. This additionally takes into setting unstructured information, for example, release archives and doctor notices that are not regularly considered for examination, building up the significance of big data investigation around here of treatment.

Big data additionally helps experts and specialists with explicit subtleties of how the huge number of medications connect with the human body, making ideas of those that may communicate with malignant growth-plagued cells. Curiously, it was as of late reported that 14 malignant growth organizations across the United States and Canada would utilize investigation and artificial intelligence (AI) to coordinate with disease patients with the medicines probably going to help them. As well as prescribing applicable medications to treat specific diseases, Artificial Intelligence could likewise recommend treatments not attempted previously.

Facilities and medical clinics should take on web-based clinic data frameworks (HIS) and electronic health records (EHR) that can additionally incorporate with the National Cancer Registry Program (NCRP). The assortment and the board of this information should be orderly and legitimate records ought to be kept up with. This examination of data, alongside demographical insights, should be concurrent with existing data collections for more compelling results around disease and treatment [26].

7.6.2 Distinguish the cycles in question

7.6.2.1 Determination and testing (identify genetic mutation)

Enormous information is thought for putting away huge assortments of informational collections and further investigating, envisioning, and moving them. Gathering information from a wide range of gadgets prompts putting away a great deal of information yet, what is noteworthy with regards to it is not the amount of data yet how we can manage it. We have gone from a few terabytes of acquired data in 2012 to full petabytes now, and it is motionlessly rising.

Enormous information can be depicted as huge pools of information that are caught and aggregated with benefits that lead to expanding present-day financial matters, medical care, transportation, and numerous different enterprises. NoSQL implies Not Only SQL, suggesting that when planning a product arrangement or item, there is also more than one stockpiling system that could be utilized depending on the necessities. There are roughly 150 NoSQL dataset administration frameworks, and they are classed according to their information model: key-esteem (Dynamo, Riak), graph (Allegro, Infinite Graph), multi-model (OrientDB, FoundationDB), document (MongoDB, Couchbase), and column (MongoDB, Couchbase) (Accumulo, Cassandra). One of the most distinguishing features of NoSQL databases is that they are composition agnostic, which means that no simple mapping strategy is required for data storage. Another key factor to consider is robust consistency, which means that all clients should view the same version of data. Finally, segment resistance is a need for NoSQL databases: the entire structure must preserve its attributes even when transferred across many servers.

Big data is the data identified with quality and all of its possible inconsistencies that must be stored in order to carry out a meaningful examination in the case of hereditary alterations. The ability to record diverse attributes without being bound by an existing inflexible schema is crucial, which is why NoSQL and Genomics go hand in hand [27].

7.6.2.2 Results investigation including exploring treatment choices, clinical preliminary examination, hereditary examination, and protein investigation

With such an immense measure of clinical information, the following inquiry is the way to consider clinical examination thoughts. Right off the bat, I might want to list a few kinds of clinical exploration by utilizing huge information (Table 7.1). The first comprises looking at risk variables, which often necessitates high-level data for puzzling controls. This sort of research can benefit from multivariable models, delineation, and inclination score analysis. The effectiveness of mediation is assessed in the second stage. Furthermore, this sort of research demands high-level information in order to adjust for confusing factors. Inclination to a certain therapy can have a big impact and should be factored into the focus strategy. The third component is the expectation model structure, which seeks to fit a model for future forecasting. A cost-effectiveness examination of a healthcare approach is the first category.

The approach for conceptualizing research topics that can be addressed using data collection is another key subject of interest. There are two options: do information mining based on your research query or customize your investigation question to the data collection. These two procedures may be used in tandem in rare circumstances. We planned to investigate lactate measured in ICU transit when we started looking into the relationship between lactate and clinical outcome. In any event, after doing information mining, we observed that lactate was frequently approximated, so we decided to look into the pattern of estimation (lactate freedom). The review procedure was tweaked as needed.

Another option is to explore research that is based on your data. Using traditional focal propensity (mean and middle) and discrete inclination, one may conduct a factual depiction for a dataset (full reach, 95% certainty span). Graphical exhibit might be especially useful. For instance, a form plot might assist with investigating the relationship between three factors; a histogram can be utilized to investigate the dispersion of a variable. Nonetheless, somebody might fight that a look at the dataset prior to drafting a review convention might present a predisposition (e.g., the issue of numerous testing and particular detailing are of this sort). In other words, multiple times measurable testing for affiliation will bring about one with $P < 0.05$ among

Table 7.1 Categories of clinical research by using big data.

Categories of studies	Examples	Requirement of clinical data	Note
Evaluation of risk factor	Is urine production on ICU admission related to mortality outcome?	High resolution	It is possible to use a multivariable model, stratified analysis, and propensity score analysis.
Effectiveness of intervention	Will PiCCO monitoring improve the outcome of septic shock patients?	High resolution	Intervention may be given to patients suffering from a variety of conditions. These conditions must be managed in order to avoid "selective treatment."
Prediction model	ICU prediction model	Reasonable resolution	Rather than focusing on a single risk factor, the predictive value of the entire model is emphasized.
Epidemiological study	The prevalence and incidence of catheter-related bloodstream infection in intensive care units	Little resolution	There is no need to change risk factors because a simple explanation suffices.
Implementation and efficacy of healthcare	Is the policy of hypertension screening and control effective in lowering the rate of cardiovascular events?	Low resolution	There is no need for complex clinical data.

autonomously created arbitrary factors. I recognize this restriction, yet such review can in any case be utilized as speculation creating and give reasoning to additional investigations.

One more kind of study is to examine basic and effective possible boundaries. By utilizing such boundaries, your review and thoughts can be tended to by utilizing assortments of information base. Our review bunch has recently performed an affiliation study including pee results and mortality. Since pee yield is a fundamental yet effectively possible variable, it ought to be recorded in a wide range of ICUs. There is not any justification to preclude this recording, actually like any remaining indispensable signs. We were sure at the beginning that pee yield should be painstakingly recorded in MIMIC-II, and the review was relied upon to be direct without a hitch. Such straightforward factors included temperature, electrolytes, and pulse. In one more review, we examined the relationship between ionized calcium and mortality. In any case, studies including basic factors are normally condemned as being the absence of curiosity, and this might be the main motivation to dismiss our paper [25, 26].

7.7 Meaning of Treatment Convention, Perhaps Including Quality or Protein Treatment

It would be extremely difficult to predict protein structures just based on genomic sequence data without the use of computer modeling. Analyzing such data could be just as beneficial as reading tea leaves. While computers' number-crunching methods may appear as mysterious as divination, they can produce valuable findings if enough data from a large enough amount of tea leaves! is available.

A new study shows that metagenomics, or the sequencing of DNA from environmental samples, can provide enough data for protein structure prediction. Metagenomics has been utilized to characterize the genetic diversity of microbial communities on several occasions. The approach, on the other hand, is helping to map the protein universe, which is still mostly unknown.

7.8 Screen Patients and Change Treatment Depending on the Situation Utilizing the New Remote Gadget for Customized Treatment Conveyance and Observing

7.8.1 The patient uses web-based media to archive general insight

Ten years' worth of hospital admissions information was crunched by data scientists using "time series analysis" techniques. As a consequence of their investigation, the researchers were able to identify key trends in admission rates. They could then use machine learning to figure out which algorithms are the most accurate in predicting future admissions patterns.

To summarize the findings of this research, the data science team developed a web-based user interface that assesses patient loads and aids in resource allocation planning via online data visualization, intending to improve overall patient care [28].

7.8.2 Recognize the data needed to tackle the issue

7.8.2.1 Patient history

The electronic health record is the most common application of big data in medicine. Every patient is given their digital record, which includes demographics, medical history, allergies, and laboratory test results. Secure information systems are used to share records, and both public and private sector vendors have access to them. Doctors may make adjustments over time without having to deal with paperwork or the danger of data replication because each record is made up of a single editable file.

When a patient requires a new lab test, EHRs can send alerts and reminders, as well as track prescriptions to see if they have been followed.

Despite the fact that electronic health records are a fantastic idea, many nations are still having difficulty implementing them fully. According to the HITECH research, the US has made great progress, with 94% of hospitals implementing electronic health records (EHRs), while the EU is still lagging. However, it is anticipated that comprehensive legislation drafted by the European Commission would reverse this trend.

Kaiser Permanente is leading the way in the United States, and it might serve as a model for the European Union. They have finished implementing a system called Health Connect, which exchanges data across all of their sites and streamlines the use of electronic health records. According to a McKinsey report on big data healthcare, "the integrated system has improved outcomes in cardiovascular disease and produced an estimated $1 billion in savings from decreased doctor visits and lab testing."

7.8.2.2 Blood, tissue, test results, etc.

Smart data is data that provides accurate information based on the context. It enables doctors to make informed decisions, such as coordinating a patient's care for a variety of illnesses, improving disease diagnosis, and gaining a comprehensive picture of the patient based on family history. Smart data precisely advises the right treatment for a particular patient's disease and predicts any potential consequences based on symptoms. Smart data is useful information collected from healthcare big data based on the kind, volume, and authenticity of the data.

Precision preparation and execution are necessary to draw logical conclusions from big data. The focus is on what the organization will require following the implementation of big data. Depending on the system's requirements, it is critical to populate big data with the right kind of legitimate data in terms of quality and volume. Scanning endless PDFs, fracture X-rays, and blood test results, for example, may not always help a practitioner identify a patient with gastrointestinal sickness or figure out why a patient is having a negative reaction to a medicine. To generate smart data, big data should be supplemented with relevant and related data. It is critical to keep in mind that what is gathered is also delivered [29].

7.8.2.3 Measurable consequences of treatment choices

Patient outcome databases combined with information on patient characteristics and therapy have the potential to give unrivaled feedback on care quality and efficacy. The landscape of molecular testing for targeted therapy in non-small cell lung cancer (NSCLC), as well as treatment regimens based on it, was described in a recent French study. This gives you fast feedback on which test-treatment correlations are the most beneficial. More importantly, it could be a strong incentive for failed laboratories to change their protocol/workflow in order to improve overall care quality. To highlight variation in clinical care in the Netherlands, data from the national cancer registry (which included clinical stage, treatment, and outcome data) were combined with the aforementioned PALGA database. Transparency of such data is the only way to enhance the quality of care, but laboratories and hospitals may be concerned about reputational damage, name, and shaming, thus such data, particularly findings and standards, are often withheld. With the utmost caution, feedback should be delivered. Several hospitals are happy to participate in a such duplicate review if it is anonymous, public, and shared only on an individual basis. As a result, the following algorithms have been developed in the Netherlands. B. The Dutch Institute for Clinical Auditing provides regular, automated feedback on pathological and treatment-related factors (www.dica.nl). If the mirror data show a higher recurrence rate than peer hospitals, there may be an inducement to focus on the underlying chain to recognize (and correct) potential errors [30].

7.8.2.4 Clinical preliminary information

The next challenge is how to produce clinical research concepts with such a big number of clinical data. To begin, I would want to describe the several sorts of clinical research that employ big data (Table 7.1). To prevent confusing controls, the initial phase is examining risk variables, which needs the use of high-resolution data. For this sort of study, multivariable models, stratification, and tendency score investigation are all relevant methods. The effectiveness of interventions is assessed in the second stage. To account for confounding effects, high-resolution data is also essential for this sort of study. Treatment selection bias may have a big influence on study design, thus it is important to think about it.

Conceiving research topics that can be addressed using a database is another key difficulty. Data mining based on your research question or adjusting your research question to the database are the two approaches. These two approaches may be used in tandem at times. We planned to look at lactate measured on ICU admission as the initial stage of a study looking at the association between lactate and clinical outcome. However, after doing some data mining, we realized that lactate was frequently recorded, so we decided to look at the lactate trend (lactate clearance). As a result, the study protocol was altered.

Another approach is to develop study ideas based on your data. Traditional central tendency (mean and median) and discrete tendency can be utilized to statistically describe a dataset (full range, 95% confidence interval). A visual demonstration could be particularly helpful. For example, a contour plot can be used to investigate correlations between three variables, whereas a histogram can be used to investigate the distribution of a variable. However, some would argue that looking at a dataset before creating a study process could contribute to bias (e.g., the problem of multiple testing and selective reporting are of this kind). That is, if you test for association between randomly produced random variables twenty times, you will get one with a $P = 0.05$. Despite this limitation, I believe that such research can still be beneficial. Although I recognize this limitation, such research can still be used to generate hypotheses and provide justification for further investigation.

Another form of research is to look at parameters that are basic and easy to collect. Your study and thoughts can be addressed by using such parameters and a variety of databases. Our research team has previously conducted a study that looked at the relationship between urine production and mortality. Urine output should be documented in all types of ICUs since it is an important but easily accessible variable. This, like all other essential indications, has no reason to be overlooked. We were convinced from the start that urine output would need to be meticulously recorded in

MIMIC-II, and the study would go well. Temperature, electrolytes, and heart rate were examples of basic variables. Another study looked into the link between ionized calcium and death. However, research with modest variables is frequently criticized for being unoriginal, and this may be the most crucial reason for our paper's rejection [31, 32].

7.8.2.5 Hereditary qualities information

The examination of the germline for DNA mutations that carry down from generation to generation and cause negative phenotypic features has changed substantially throughout time. This is primarily due to advancements in technologies that enable comprehensive genome testing, such as microarrays and, in particular, next-generation sequencing. The following stage of the pipeline, bioinformatic analysis, and statistical validation, has major hurdles as a result of these technologies, which create huge datasets. When these laboratory procedures are paired with future bioinformatic studies, they establish a platform for the development of "big data," which is subsequently filtered down into therapeutically interpretable genetic information or immediately relevant information for the patient.

The technical hurdles of genetic data analysis for knowledge discovery begin in the lab with sample preparation and laboratory analysis processes that are appropriate for the samples. The problems continue in the first stages of the bioinformatic pipeline for finding clinically relevant variations. Among the early problems are selecting appropriate algorithms for optimal genetic variant filtering, as well as selecting relevant algorithms at all future informatic phases required to uncover significant variations. Furthermore, the huge number of loci explored in such discovery research is individual testing for clinically relevant genetic changes, posing a statistical difficulty as well as concerns about false positives. The data created at the conclusion of this genetic process, i.e. the data utilized to draw clinical links, must be assessed regularly. Bias and uneven criteria for determining whether or not certain genetic variations are clinically significant might lead to inaccurate data findings. To maximize the chance of finding true positive variations and eliminate missed false negatives, it is necessary to evaluate the data at each stage of the pipeline. Furthermore, not only do technological decisions cause delays in the development of trustworthy, relevant data, but so do misunderstandings among researchers at various phases of the genome process. If users of genetic discoveries in the healthcare industry are to have trust in the quality of the underlying data, careful approaches are necessary.

7.8.2.6 Protein information

The Protein Data Bank (PDB) is a three-dimensional structural data collection for proteins and nucleic acids, among other biological elements. The data, which is frequently obtained using X-ray crystallography, NMR spectroscopy, or, increasingly, cryo-electron microscopy and donated by biologists and biochemists from around the world, is made freely available on the Internet through the websites of its member organizations (PDBe, PDBj RCSB, and BMRB). The PDB is managed by the Worldwide Protein Data Bank, or wwPDB.

The PDB is essential in structural biology fields like structural genomics. Most major scientific publications and funding sources now ask scientists to give their structural data to the PDB. Many other databases use protein structures that have been deposited in the PDB. Protein structures are classified using SCOP and CATH, for example, whereas PDBsum visualizes PDB entries using data from other sources, such as Gene Ontology [33].

7.8.2.7 Online media information

Social media has become an increasingly important aspect of people's daily lives as the internet has grown in popularity. Not only has social media shown to be a useful tool for connecting with individuals, but it has also proven to be a useful tool for businesses to reach out to their target market. Since the advent of big data, social media marketing has taken on an entirely new meaning. The total volume of big data is expected to hit 44 trillion gigabytes by 2020, according to forecasts. Marketers may use the massive amount of data accessible to create effective social media marketing campaigns.

Users' social media grade updates, photographs, and videos reveal vital information about their demographics, interests, and dislikes, among other things. In order to obtain a competitive edge, companies are managing and analyzing this data in a number of ways. Marketers utilize big data to plan future social media initiatives by knowing as much as they can about their target clientele before approaching them. The usage of big data in community media marketing will be discussed in this article, as well as its present and future consequences.

Personalization: Big data allows companies to engage customers in more personalized ways depending on their preferences and interests. It provides extensive information and a complete picture of the target audience, allowing firms to adjust communications to increase retention and trust.

Using big data, marketers will be able to show only the adverts that customers are interested in, making marketing a non-intrusive experience.

Advertisements will be customized based on what people post on social media, what they watch and share on YouTube and other variables. Marketers will be able to interact with social media users and convert them into consumers through targeted marketing after determining the most successful platform, timing, and format for their advertising.

Marketers may utilize big data to uncover social media trends and get insights on which individuals to contact and which groups of users to send marketing emails to, among other things. It also makes it easy to keep track of demographics and choose which social media site to target.

Big data allows businesses to analyze market moods fast and develop winning strategies. Big data assists in making intelligent decisions to better match future customer desires and expectations, rather than depending just on past performance to understand what improvements are required.

Big data is essential for analyzing the efficacy of social media campaigns and identifying how ROI evolves. Marketers may also use it to test campaigns before launching them, analyze the results, make adjustments, and retest them. Predictive analytics may be used by businesses to determine when a campaign should be discontinued to avoid losses.

Businesses may use actionable insights from big data to better identify peak customer timings, preferences, behavior, and other aspects, resulting in more successful social media campaigns. From the early phases of the purchasing cycle through post-purchase contacts, marketers may receive vital knowledge about their clients' buying processes, allowing them to fine-tune their campaigns at each level.

Product Insights: Big data may be used efficiently by social media marketers to estimate future purchasing behavior and trends. Consumers are more definite about what they want when they want it, and how they want it thanks to big data. This provides information to firms on how these new items should be developed. Businesses may use big data to look at people's preferences, complaints, what products are missing, and product flaws, among other things. They will be able to improve the current product while also developing new, inventive products as a result of this.

7.9 Accumulate the Information, Process It, and Examine the Outcomes

7.9.1 Begin treatment

In addition to detecting illness states, medical imaging provides critical information about anatomy and organ function. Only a few of the applications include organ delineation, lung tumor identification, spinal deformity diagnosis, artery stenosis detection, and aneurysm detection. In these applications, image processing methods such as enhancement, segmentation, and denoizing, as well as machine learning algorithms, are used. To grasp data linkages and create efficient, accurate, and computationally effective methods, new computer-aided approaches and platforms are necessary as data amount and dimensions rise. Computer-aided medical diagnostics and decision support systems are being employed increasingly often in clinical settings as a result of the fast increase of healthcare organizations and patients. Computational intelligence has the potential to enhance several aspects of healthcare, including diagnosis, prognosis, and screening. Computer analysis, when combined with adequate care, has the potential to aid doctors in improving diagnosis accuracy. Integrating medical pictures with other forms of EHR data and genetic data can also upgrade diagnostic exactness and speed up the process.

7.9.2 Screen patients and change treatment on a case-by-case basis

Healthcare organizations in several countries have presented numerous types of healthcare information systems in order to deliver the greatest services and care to patients. These personalized, predictive, participatory, and preventative medicine models make use of electronic health records (EHRs), as well as large amounts of complex biological data and high-quality omics data.

Genomic and postgenomics technologies are already producing massive amounts of raw data about complex biochemical and regulatory processes in living organisms. These -omics data are diverse, and they are frequently stored in several formats. Electronic health record data, such as these omics data, is available in a variety of formats. Electronic health records contain organized, semi-structured, and unstructured data, as well as discrete and continuous data (EHRs).

In healthcare and medicine, big data refers to huge and complex datasets that are difficult to analyze and handle with standard software or technology. Big data analytics include heterogeneous data integration, data quality management, data analysis, modeling, interpretation, and validation. Big data analytics allows for the extraction of comprehensive information from large amounts of data.

Big data analytics, particularly in medicine and healthcare, allows for the investigation of massive datasets from thousands of patients, the finding

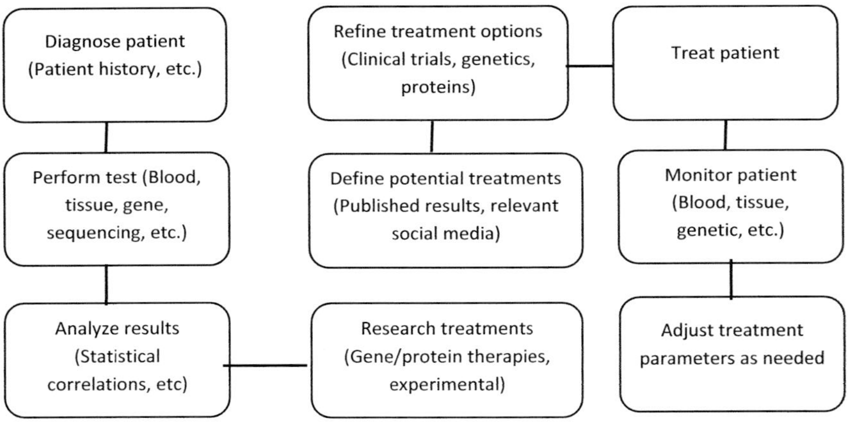

Figure 7.3 Process of disease diagnosis.

of clusters and linkages among datasets, and the construction of prediction models using data mining techniques (Figure 7.3).

Big data analytics in medicine and healthcare encompasses bioinformatics, medical imaging, sensor informatics, medical informatics, and other scientific subjects.

This satisfies the ideal scenario, in which no additional cycles are required to facilitate huge data assimilation. The hidden advancements include applications that must be adjusted to accommodate the impact of large amounts of information, such as the volume of info, the variety of info sources, and the speed or speed required to deal with that information, while the cycles remain largely unchanged.

The incorporation of massive amounts of data into the most common method of overseeing medical services will have a significant impact on the adequacy of diagnosing and overseeing medical services in the future. This equivalent functional methodology interaction can be used in a variety of businesses. What are the keys to effectively applying massive amounts of data to functional cycles? The following are the absolute most important issues to consider:

- Understand the current interaction completely.
- Completely understand where data gaps exist.
- Distinguish between relevant large information sources.
- Create an interaction that will flawlessly coordinate the information both now and as it changes [34].

7.10 Conclusion

The gap between structured and unstructured data sources is exploited by big data analytics. The transition to a unified data environment is a well-known challenge. Surprisingly, the premise of big data is based on the assumption that the more data there is, the more insights one may get from it, and the more forecasts one can make about future occurrences. Various reputable consulting organizations and healthcare corporations have correctly predicted that the big data healthcare sector will develop at an exponential rate. However, in a short period, we have seen a wide range of analytics in use that has had major implications on healthcare decision-making and performance. The exponential increase of medical data from diverse fields has compelled computational professionals to devise novel ways of analyzing and interpreting such vast amounts of data in a short amount of time. The use of computer systems for signal processing by both researchers and practicing doctors has increased. As a result, merging physiological data with "-omics" approaches to create a comprehensive model of the human body might be the next great goal. This revolutionary concept has the potential to improve our understanding of illness states and aid in the creation of new diagnostic tools. The growing amount of available genetic data, as well as inherent hidden mistakes from experiments and analysis procedures, need more research. However, there are chances to implement systemic changes in healthcare research at every stage of this lengthy process.

Big data's greatest advantage is its inexhaustible potential. Within the last several years, the emergence and integration of big data have resulted in significant advances in the healthcare industry, spanning from medical data management to drug discovery programs for complicated human diseases such as cancer and neurological disorders. To give a basic example, since the late 2000s, developments in the EHR system in terms of data gathering, administration, and usability have been seen in the healthcare business. We think that, rather than replacing qualified people, subject knowledge specialists, and intellectuals, big data will complement and boost the existing pipeline of healthcare improvements, as many have stated. The shifts in the healthcare business from a broad volume basis to a tailored or individual-specific domain are seen. As a result, technicians and experts must be aware of this changing circumstance. Big data analytics is expected to progress toward a predictive system in the following year. This would imply predicting future results in a person's health based on present or historical data (such as EHR-based and Omics-based). Similarly, structured data gathered from a specific location may lead to the development of population health data. Big data will enhance healthcare by allowing for the prediction of epidemics (in connection to population health), early warnings of illness conditions, and the identification of novel biomarkers and intelligent therapeutic intervention tactics for a better quality of life.

Acknowledgment

The authors are grateful to the administration of Oriental University, Indore, for their assistance.

Conflict of Interest

There is no potential for a conflict of interest.

Funding

There is no funding issued.

References

[1] Dash, S., Shakyawar, S. K., Sharma, M., and Kaushik, S. (2019). Big data in healthcare: management, analysis and future prospects. *J. Big Data, 6*. 1-25. DOI: 10.1186/s40537-019-0217-0.

[2] Laney, D. (2001). 3D data management: Controlling data volume, velocity and variety. *META group research note*, *6*, 1. DOI: 10.4236/vp.2021.74013.

[3] De Mauro, A., Greco, M., and Grimaldi, M. (2016). A formal definition of Big Data based on its essential features. *Libr. Rev*. DOI: 10.1108/LR-06-2015-0061

[4] Gubbi, J., Buyya, R., Marusic, S., and Palaniswami, M. (2013). Internet of Things (IoT): A vision, architectural elements, and future directions. *Future Gener. Comput. Syst. 29*, 1645-1660. DOI: 10.1016/j.future.2013.01.010.

[5] Reiser, S. J. (1991). The clinical record in medicine part 1: learning from cases. *Ann. Intern. Med. 114*, 902-907. DOI: 10.7326/0003-4819-114-10-902

[6] Reisman, M. (2017). EHRs: the challenge of making electronic data usable and interoperable. *Pharm. Ther. 42*, 572. PMID: 28890644; PMCID: PMC5565131.

[7] Murphy, G. F., Hanken, A. M., and Waters, K. A. (Eds.). (1999). *Electronic health records: changing the vision*. WB Saunders Company.

[8] Shameer, K., Badgeley, M. A., Miotto, R., Glicksberg, B. S., Morgan, J. W., and Dudley, J. T. (2017). Translational bioinformatics in the era of real-time biomedical, health care and wellness data streams. *Brief. Bioinformatics. 18*, 105-124. DOI: 10.1093/bib/bbv118

[9] Pastorino, R., De Vito, C., Migliara, G., Glocker, K., Binenbaum, I., Ricciardi, W., and Boccia, S. (2019). Benefits and challenges of Big Data in healthcare: an overview of the European initiatives. *Eur. J. Public Health. 29*, 23-27. DOI: 10.1093/eurpub/ckz168

[10] Foshay, N., and Kuziemsky, C. (2014). Towards an implementation framework for business intelligence in healthcare. *Int. J. Inf. Manage. 34*, 20-27. DOI: 10.1016/j.ijinfomgt.2013.09.003 .

[11] Belle, A., Thiagarajan, R., Soroushmehr, S. M., Navidi, F., Beard, D. A., and Najarian, K. (2015). Big data analytics in healthcare. *BioMed Res. Int. 2015*. DOI: 10.1155/2015/370194

[12] Wang, Y., and Hajli, N. (2017). Exploring the path to big data analytics success in healthcare. *J. Bus. Res. 70*, 287-299. DOI: 10.1016/j.jbusres.2016.08.002

[13] Foshay, N., and Kuziemsky, C. (2014). Towards an implementation framework for business intelligence in healthcare. *Int. J. Inf. Manage. 34*, 20-27. DOI: 10.1016/j.ijinfomgt.2013.09.003

[14] Bates, D. W., Saria, S., Ohno-Machado, L., Shah, A., and Escobar, G. (2014). Big data in health care: using analytics to identify and manage high-risk and high-cost patients. *Health Aff. 33*, 1123-1131. DOI: 10.1377/hlthaff.2014.0041

[15] Fang, C., Weng, H., Dai, X., and Fang, Z. (2016). Topological nodal line semimetals. *Chin. Phys. B. 25*, 117106. DOI: 10.1088/1674-1056/25/11/117106

[16] Wang, G., Wei, Y., Qiao, S., Lin, P., and Chen, Y. (2018). *Generalized inverses: theory and computations* (Vol. 53). Singapore: Springer. DOI: 10.1007/978-981-13-0146-9_2.

[17] Zhang, H., Xu, T., Elhoseiny, M., Huang, X., Zhang, S., Elgammal, A., and Metaxas, D. (2016). Spda-cnn: Unifying semantic part detection and abstraction for fine-grained recognition. In *Proceedings of the IEEE conference on computer vision and pattern recognition* (pp. 1143-1152),

[18] Phillips-Wren, G., and Hoskisson, A. (2015). An analytical journey towards big data. *J. Decis. Syst. 24*, 87-102. DOI:10.1080/12460125.2015.994333.
[19] Gandomi, A., and Haider, M. (2015). Beyond the hype: Big data concepts, methods, and analytics. *Int. J. Inf. Manage. 35*, 137-144. DOI: 10.1016/j.ijinfomgt.2014.10.007
[20] Raghupathi, W., and Raghupathi, V. (2014). Big data analytics in healthcare: promise and potential. *Health Inf. Sci. Syst. 2*, 1-10. DOI: 10.1186/2047-2501-2-3
[21] Mettler, T., and Vimarlund, V. (2009). Understanding business intelligence in the context of healthcare. *Health Inform. J. 15*, 254-264. DOI: 10.1177/1460458209337446
[22] Wang, G., Wei, Y., Qiao, S., Lin, P., and Chen, Y. (2018). *Generalized inverses: theory and computations* (Vol. 53). Singapore: Springer. DOI: 10.1007/978-981-13-0146-9
[23] Siuly, S., and Zhang, Y. (2016). Medical big data: neurological diseases diagnosis through medical data analysis. *Data Science and Engineering. 1*, 54-64. DOI: 10.1007/s41019-016-0011-3
[24] Luo, Y., Benali, A., Shulenburger, L., Krogel, J. T., Heinonen, O., and Kent, P. R. (2016). Phase stability of TiO2 polymorphs from diffusion Quantum Monte Carlo. *New J. Phys. 18*, 113049. DOI: 10.1088/1367-2630/18/11/113049
[25] Markonis, D., Holzer, M., Baroz, F., De Castaneda, R. L. R., Boyer, C., Langs, G., and Müller, H. (2015). User-oriented evaluation of a medical image retrieval system for radiologists. *Int. J. Med. Inform. 84*, 774-783. DOI: 10.1016/j.ijmedinf.2015.04.003
[26] Hurwitz, J. S., Nugent, A., Halper, F., and Kaufman, M. (2013). *Big data for dummies*. John Wiley & Sons.
[27] Sinclair, A. (2002). Genetics 101: detecting mutations in human genes. *Cmaj. 167*, 275-279. PMID: 12186176; PMCID: PMC117476.
[28] Kunin, V., Copeland, A., Lapidus, A., Mavromatis, K., and Hugenholtz, P. (2008). A bioinformatician's guide to metagenomics. *Microbiol. Mol. Biol.Rev. 72*, 557-578. DOI:10.1128/MMBR.00009-08.
[29] Bresnick, S. D. (2016). Management of a common breast augmentation complication: treatment of the double-bubble deformity with fat grafting. *Ann. Plast. Surg. 76*, 18.
[30] Oliver, A., and Greenberg, C. C. (2009). Measuring outcomes in oncology treatment: the importance of patient-centered outcomes. *Surg. Clin. North Am.*, *89*, 17-25. DOI:10.1016/j.suc.2008.09.015.

[31] Harcus, J. W., and Stevens, B. J. (2021). What information is required in a preliminary clinical evaluation? A service evaluation. *Radiography. 27*, 1033-1037. DOI: 10.1016/j.radi.2021.04.001.
[32] Paterson, A. (2013). Preliminary clinical evaluation and clinical reporting by radiographers: policy and practice guidance.
[33] "Protein Data Bank: the single global archive for 3D macromolecular structure data." Nucleic acids research 47, no. D1 (2019): D520-D528.
[34] The Diagnostic Process: Rediscovering the Basic Steps (thesullivan-group.com)

Author Biography

Deepika Bairagee

Ms. Deepika Bairagee, B. Pharm, M. Pharm (Quality Assurance), is an Assistant Professor at the Oriental College of Pharmacy and Research, Oriental University in Indore (India). She has five years of teaching experience and two years of research experience. She has spoken at more than 20 national and international conferences and seminars, presenting over 20 research articles. She has over 20 publications in international and national journals. She is the author of over 18 books. She has over 50 abstracts that have been published at national and international conferences. Young researchers, young achievers, and excellent researchers were among the honors bestowed upon her. Proteomics and metabolomics are two areas of research that she is currently interested in.

8

Various Cancer Analysis Using Big Data Analysis Technology – An Advanced Approach

Ayush Chandra Mishra[1], Ratnesh Chaubey[1], Smriti Ojha[2], Sudhanshu Mishra[2*], Mahendran Sekar[3], and Swati Verma[4]

[1]Dr MC Saxena College of Pharmacy, India
[2]Department of Pharmaceutical Science & Technology, Madan Mohan Malaviya University of Technology, India
[3]Department of Pharmaceutical Chemistry, Faculty of Pharmacy and Health Sciences, Royal College of Medicine Perak, Universiti Kuala Lumpur, Malaysia
[4]Department of Pharmacy, School of Medical and Allied Sciences, Galgotias University, Greater Noida, India
*Corresponding Author: Department of Pharmacy, Birla Institute of Technology and Science, Pilani, EMail: msudhanshu22@gmail.com.

Abstract

In this modern era, big data provides an enormous amount of collected data about various diseases in India and throughout the world, which is created by data available in vast diversity, high speed, and volume. Because such data is becoming more prevalent, methods for handling and extracting insights from it must be studied and provided. Furthermore, decision-makers should be able to retrieve important from such a distinct, diverse, and frequently changing dataset, which encompasses everything from ordinary transactions to customer/patient interactions, as well as social media sites and other digital platforms. Big data is viewed as a revolution with the potential to change the way businesses function in a variety of industries. This study investigates the research data available from existing challenges and cases and applies big data analytics to enhance cancer risk categorization. We have divided the input data into three categories: pathology, radiomics, and population health management. Big data provides computational methodologies which will improve clinical risk stratification for cancer patients in a minimum period.

Keywords: Big Data, Disease, Computational Approach, Predictive Analysis, Cancer, Datasets, Clinical Rill Stratification.

8.1 Introduction

The key to greater organization and new advances has been informed. We can better arrange ourselves to offer the greatest results if we have more information. As a result, data collecting is an essential component of any company. We may also utilize this information to forecast present trends in specific metrics as well as future occurrences. As we become more conscious of this, we have begun to produce and gather more data on nearly everything by using technical advancements in this regard. We are now in a scenario where we are flooded with data gathered from multiple sources which include health, job, social activities, and so on. In some ways, the current situation resembles a data flood. Technological advancements have aided us in creating an increasing amount of data, to the point that it is now unmanageable with currently available technology. As a result, the phrase "big data" was coined to characterize data that is enormous and unmanageable. We need to find innovative techniques to organize this data and derive relevant information to fulfill our current and future societal demands [1].

Predictive analytics is a subset of advanced analytics that makes assumptions about events or actions that have not been determined yet to help decision-makers to make better judgments. It is a scientific discipline that analyses historic and legit data and produces predictions using several methodologies like artificial intelligence, data mining, machine learning, modeling, and statistics (Figure 8.1). A few times in your life, you have the chance to investigate the future and predict future trends by applying these projections in patient care, both on a personal level and as a group. In computing, the term "big data" refers to a collection of information that is both large and constantly growing in scope and volume. The five Vs that represent it are veracity, velocity, volume, variability, and variety. Big data is capable of handling all types of data, including structured, unstructured, and semi-structured. Since structured data is composed of words and numbers, it is simple to recognize and analyze. Structured data provides complex information that cannot be easily reviewed since the data is presented in the form of visuals. A well-rounded implementation of these technologies should

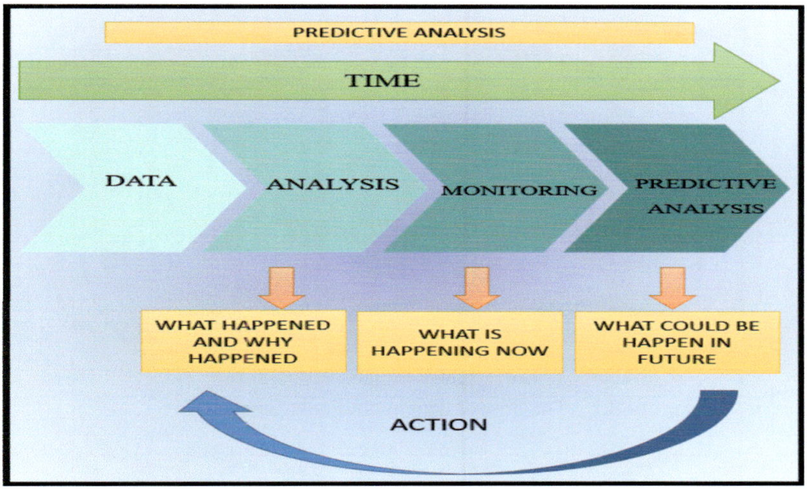

Figure 8.1 Implementation of big data and predictive analysis.

include the use of both supervised and unsupervised predictive modeling as legitimate analytical methods [3, 4]. It is possible to utilize predictive analytics to make predictions based on either a human's hypothesis or by using an algorithm devoid of any guiding hypothesis but instead employing an algorithm to analyze information [4]. Predictive analytics is becoming increasingly popular, and it is quite beneficial in a variety of areas including criminal justice, marketing & management, fraud detection, healthcare, and law. Predictive analytics has the potential to be a significant boon for the healthcare industry, which has a wide range of stakeholders and is increasingly recognized as a vital facet. This article will examine the distinct ethical pitfalls necessary to be prevented by doctors, individuals working in government agencies, and many more while using the available predictive analytics. New technologies bring new threats to them. This article will look at how predictive analytics can be used in a variety of situations, with a focus on how healthcare services are delivered [5].

8.2 Predictive Analysis and Big Data in Oncology

Predictive Analysis, as the name implies, examines previous data to foretell future events. As a result, leveraging an AI system's power to combat cancer in the future might be promising (Figure 8.2).

Predictive analysis can be a beneficial technique for healthcare professionals who are diagnosing and treating cancer patients. It enables such

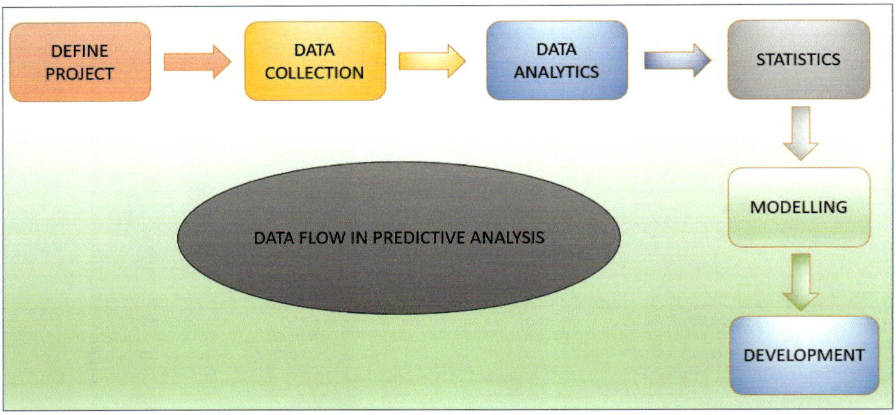

Figure 8.2 The flow of data in predictive analysis.

professionals to not only detect cancers and categorize them according to the amount of risk they represent to patients but also to adhere to worldwide healthcare privacy rules.

As we all know, artificial intelligence (AI) was brought into the healthcare profession to eliminate human error at every stage of illness diagnosis and treatment. As a result, incorporating cancer-related treatment procedures into an AI system helps clinicians to minimize the number of errors associated with utilizing the wrong therapies for certain types of tumors [6].

Oncologists can utilize predictive analytics to identify high-risk cancer patients. Cancer relapses are common in these people, especially after high-intensity therapies like chemotherapy or various types of surgery. Even the most experienced health professionals may have difficulty identifying such individuals at an early stage, but machine learning can recognize particular patterns to predict the return of cancer cells in them. The expenditures of expensive treatments can be saved if such people are identified early, and health professionals can instead focus on adopting malignancy prevention strategies (medicines and lifestyle guidelines) to a larger extent [7].

Second, AI-based predictive analysis might allow doctors to investigate tumor features in specific patients in greater depth. This research can help workers identify people whose bodies can handle chemotherapy without causing too much harm afterward. The bodies of certain individuals may respond to some therapies more favorably than those of others [8].

Finally, pathology and biopsy can benefit from predictive analysis. One of the most serious issues with cancer treatment is the possibility of over-or under-treating people who do not require it. As a result, compared to cancer itself, such individuals have a larger risk of dying as a result of excessive therapy. As a result, predictive analytic tools, such as Google's AI tool, improve cancer diagnosis accuracy, allowing clinicians to focus on measures other than potent medicines [9].

Predictive analytics' advantages

1. Improving the efficiency of healthcare company operations' operational management.
2. In personal medicine, the accuracy of diagnosis and treatment is crucial.
3. Increased understanding to improve cohort therapy [10].

8.3 Application of Big Data Analytics in Biomedical Science

A biological system is characterized by the complex interaction of molecular and physical processes that occur inside it. A biomedical or biological experiment is frequently designed to collect data received from multiple components to better comprehend the interconnectedness of several mechanisms and events in a complicated system. Therefore, a large number of reduced trials are required to construct a comprehensive plot of a genetic phenomenon of concern. Hence, the greater the extent of data variability more comprehending the biological processes will be. As an example, consider the massive amount of data that has been created since efficient technologies such as GWAS (genome-wide association studies) and NGS (next-generation sequencing) were used to decode human genetics. To view or record biological processes linked with certain disorders in real-time at a higher resolution is possible all because of advanced technology. This new age in science began with the realization that vast amounts of data might supply us with a significant quantity of information that was previously unnoticed or buried in more traditional experimental approaches. Instead of analyzing a single "gene," scientists may now examine an organism's whole "genome" in "genomics" research, which can be completed in a very short period. The same is true for gene expression research. Instead of looking at the transcription of a single gene, we may now use "transcriptomics" research to look at the expression of all genes or the expression of the whole "transcriptome" of an organism. Each of these tests generates a lot of data that is more detailed than anything else we have seen before.

While this level of accuracy and clarity is admirable, it is not always sufficient for understanding a certain process or occurrence. Because of this, it is common to find oneself faced with the task of assessing a huge quantity of data gathered from diverse sources to uncover new insights. This is backed by the fact that the number of papers on big data in healthcare has been steadily increasing. The analysis of such massive volumes of data from medical and healthcare systems might be immensely valuable in the development of new healthcare policies, among other things. A customized medical revolution is expected to occur shortly, thanks to recent technological advancements in data collection, collection, and analysis [11].

8.4 Data Analysis from Omics Research

Next-Generation Sequencing has provided ease in the process of obtaining whole-genome sequence data much simpler and less expensive than it was before. From millions to a few thousand dollars, the cost of whole-genome sequencing has decreased dramatically over time (Figure 8.3).

As a result of the advancement of NGS technology, the volume of biomedical data generated by genomic and transcriptome research has expanded significantly. By 2025, sequencing of 100 million to 200 billion (approximately) human genomes are expected to be sequenced, according to one estimate [12].

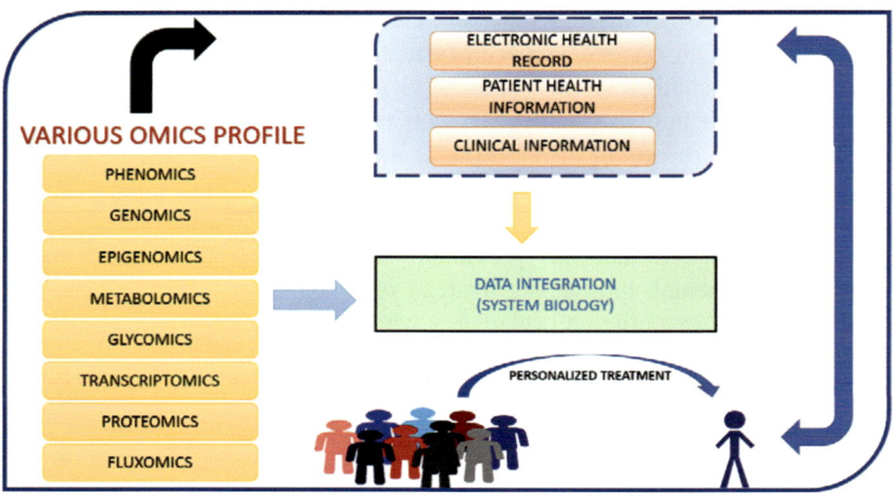

Figure 8.3 Omics data integration and healthcare analytics framework.

Combining genetic and transcriptome data with proteome and metabolome data may significantly improve our understanding of a patient's unique profile, a process known as "individualized, tailored, or precision healthcare," according to the National Institutes of Health. Systematic and comprehensive omics data analysis, when combined with healthcare analytics, can help in the development of improved treatment solutions. In medicine, the goal is to achieve more precision and customization. Big data sources in biological healthcare include genomic-driven research such as NGS, gene expression, electronic health records, genotyping-based studies, and many more. The healthcare industry requires sophisticated integration of biological data from a variety of sources to deliver well patient care. Even though genetic data from patients would have a large number of factors to contend with when dealing with big data, these possibilities are so enticing that they are worth exploring further [13].

8.5 Oncology Predictive Data Analysis: Recent Application and Case Studies

Forecasting future health outcomes for individuals or communities has been made possible via the use of predictive analytics solutions that are generated using previous patient data (Figure 8.4). A variety of potentially generalizable cases in cancer have emerged with the increase in the quantity of data obtained from studies like genomic, radiological, and EHR [14].

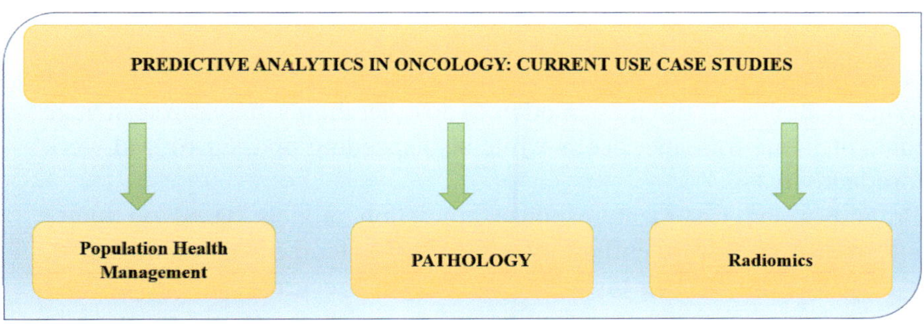

Figure 8.4 Image depicting current use case studies of predictive analysis in oncology.

8.5.1 Population health management

The delivery of therapies to high-risk patients to minimize negative effects is a critical component of community health management strategies. Predictive algorithms can be used to locate chemotherapy patients who are at the urge of death or who require immediate medical attention. As a result of this prediction, physician actions across the cancer spectrum could be influenced [15, 16].

Dealing with these high-risk individuals may be able to help you save money on medical expenses. Predictive algorithms are used by organizations to detect cancer patients who are urged of being admitted to the hospital or visit the emergency department and then target care management solutions for them through pre-emptive phone calls or visits [17]. Take, for example, Google employees [18]. The quick healthcare interoperability resources format, even though EHR data is often difficult to use, has been suggested as a means of speeding up the time-consuming process of acquiring data from EHRs [19].

8.5.2 Pathology

In addition to pathology, which is critical in cancer therapy, predictive analytics will benefit other fields of medicine as well as pathology. For example, when it comes to the interpretation of the Gleason score and identification of non-small cell lung cancer [20]. As a result of incorrect biopsy findings, patients may be offered treatment options that are either unnecessary or inappropriate for them. If you have sentinel lymph node biopsies, artificial intelligence systems can tell if you have metastatic or nonmetastatic breast cancer with a lot of accuracy. This is similar to how pathologists make their decisions [21].

When pathologists use these models to do other things, they can scan huge amounts of tissue for cancer cells while also speeding up their overall work and productivity.

Prognosis and treatment response prediction models based on tumor pathologic traits are frequently used to forecast outcomes and predict treatment responses in specific diseases. The 21-gene recurrence score and the 70-gene recurrence score are examples of scores that are widely used in clinical practice to assess the efficacy of chemotherapy in early-stage breast cancer patients [22].

8.5.3 Radiomics

A developing field of radiomics, which is expanding at a rapid pace, demonstrates predictive analytics as an emerging field in cancer. Textural analysis using quantitative data from scans is known as radiomics, and it may be used to learn more about the properties of tumors. Solid tumors contain several characteristics that may be utilized to aid in the identification, characterization, and monitoring of tumors. Solid tumors have the following characteristics: In addition to using computer-aided detection (CAD) to detect malignant lung nodules and prostate lesions on magnetic resonance imaging (MRI) scans, computer-aided detection may also be used to automate tumor staging [23, 24].

Another encouraging breakthrough is that AI can be used for the analysis of lung cancer CT images and to predict crucial outcomes like the presence or absence of mutations and the likelihood of distant metastasis. It is possible to detect early responses to therapy with dynamic magnetic resonance imaging (MRI) and to warn physicians of tumor responses before known predictors of response are employed. This is because radiologic data systems might assist in the decision-making process for care delivery [25, 26].

8.6 Utilizing Cases for the Future

By analyzing large amounts of data, we can gain new insights into the factors that contribute to disease. Patient data from mobile health applications and connected devices can be imported, allowing us to interact more closely

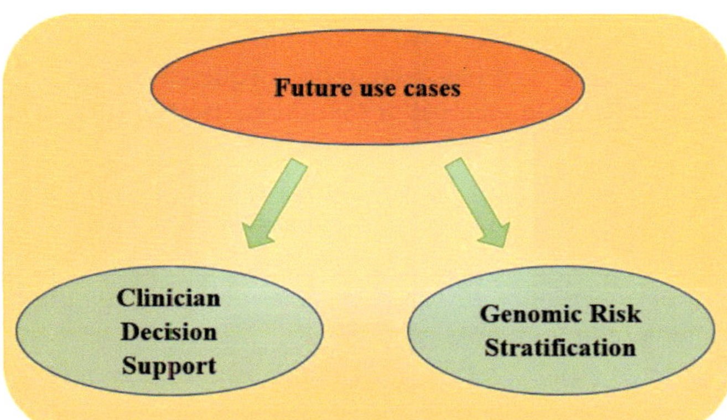

Figure 8.5 Future use cases in predictive analysis.

with individual patients. It is possible to look at and use this information right now, to help people, change their behavior and lower their health risks, eliminate harmful environmental exposures, or improve their health outcomes (Figure 8.5).

8.6.1 Decision-making assistance for clinicians

When predictive analytics tools reach a certain level of competence, oncologists may depend heavily on them to influence regular components of treatment, such as chemotherapy regimens. In prospective trials, the use of predictive algorithms was shown to allow for more expeditious treatment of stroke victims and minimize the time required to respond to patients with sepsis. Analytics can assist clinicians in making treatment decisions by forecasting the side effects of chemotherapy, how long it will take for chemotherapy to work, how likely cancer will recur, and how long the treatment will take to work [27, 28].

As a working prototype, real-time EHR-based algorithms were developed to evaluate short-term death in people before chemotherapy treatment. These algorithms, which are based on both structured and unstructured EHR data, may potentially be applied to any EHR system. Even though the future of these technologies is unknown, oncologists may find precise mortality predictions very useful during treatment [29].

8.6.2 Genomic risk stratification

With the growing popularity of germline testing and next-generation tumor sequencing in the cancer community, sophisticated algorithms capable of estimating risk based on hundreds of newly identified genes are needed. Because of the high expense of next-generation sequencing, it is unfeasible to use it as a random screening approach for a large sample. Instead, genetic testing may be tailored to patients using prediction algorithms based on the history and clinical features. Researchers were effective in separating real differences from artifacts after applying machine learning algorithms to selected NGS panels. Because variations with unknown values can generate a significant deal of uncertainty among physicians and patients [30, 31], this could be a valuable prediction tool.

Furthermore, genetic risk stratification can be used to determine whether or not a person will be benefitted from breast cancer screening in the first place. In research conducted in the United Kingdom, the delivery of breast cancer mammography to women with a high hereditary risk of breast cancer, as opposed to the existing age-based breast cancer screening paradigm, was found to minimize overdiagnosis and enhance cost-effectiveness [46]. Because of the advancements in genetics, risk prediction will continue to progress, and it will most likely be most useful when used in a multimodal setting that combines other clinical data points, a paraphrase that has been formalized [32].

8.7 Challenges for Big Data Analysis and Storages

Despite the alluring promise of big data analytics, it is necessary to analyze the "state of the science" and accept that, for the time being, big data analytics applications are largely speculative, according to the National Institute of Standards and Technology. Because of this, it is vital to understand some of the issues that big data applications in healthcare are confronted with. Starting with the lack of proof that big data analytics has any practical benefits, let us examine this issue more closely. Next, there are several methodological concerns to take into account, including data quality, inconsistency and instability in data, the limitations of observational research, validation, analytical hurdles, and legal difficulties, to name just a few of the issues. Improving data quality in electronic health records is vital, and attempts to do so must be made immediately [33]. The fact that chronic kidney disease is a significant issue in nephrology is obscured by the fact that its codes are not assigned in many administrative claim databases and the fact that the vast majority of cases of acute renal injury that do not necessitate dialysis are not documented in claim databases. Because of this, it is necessary to change these behaviors. Many of these technical concerns are being solved at present. Finally, there are concerns about clinical integration and usefulness. The benefits of big data analytics must be incorporated into clinical practice to be realized. However, before this can happen, the clinical value of big data analytics must be proven in the clinical setting. The challenges of clinical integration and usefulness have mostly gone unnoticed until now [34].

Identifying and removing these hurdles is crucial to accelerate the implementation of big data technologies in the medical sector and, as a result, improve patient outcomes and minimize healthcare waste, which should be the true value of big data research.

8.7.1 Current challenges

(1) Lack of awareness: Healthcare professionals are unaware of the possibilities of big data analytics.
(2) Data security: Difficulties include privacy concerns, a lack of transparency, a lack of integrity, and an inherent data structure.
(3) Data structure: Issues with incompatible or heterogeneous data and fragmented data formats
(4) Data standards: restricted interoperability, data gathering, mining, and sharing, as well as linguistic barriers, must all be addressed.
(5) Expensive storage equipment and expansions.
(6) Inconsistency, correctness, and data timeliness are all issues.
(7) Legal criteria and regularity authority of various countries and political organizations.
(8) Inadequate skills and training: Professionals lack the essential competencies to collect, process, and extract data.

8.7.2 Perspectives and challenges for the future

(1) To aid in disease detection and prognostication by anticipating epidemics and pandemics, boosting sickness surveillance, developing and tracking healthy behaviors, and forecasting patients' vulnerabilities.
(2) To enhance data quality, structure, and accessibility by simplifying the transparent collection of massive amounts and types of data, as well as the detection of data mistakes.
(3) Improve your decision-making process by using real-time data.
(4) Improving patient-centered care and personalization of medicine.
(5) Improving the detection of fraud [35].

8.7.3 Data acquisition

The four Vs that regulate the most important aspects of big data collection are diversity, value, velocity, and volume. Data acquisition scenarios include a large volume of data collected at a fast rate with a wide variety of data types, but the low value is common. Following data collection, data analysis, categorization, and packing of extremely large data volumes are all critical for firms dealing with large amounts of data. Flexible and time-efficient data collection and screening are required to ensure that only high-value data bits are processed by data-warehouse analysis. However, for some businesses, any data is potentially valuable since it may assist them in attracting new clients [36].

8.7.4 Prospective validation of the algorithm

Another issue that has been hotly debated is algorithmic bias. Because contemporary artificial intelligence algorithms are heavily reliant on data, their performance is entirely determined by the data. Algorithms take into account the intrinsic qualitative and quantitative characteristics of the data they process. Suppose the data is unequally distributed and contains disproportionately more information about general population students than it does about minority students. In this case, the algorithms may cause systematic and repetitive mistakes that penalize minorities. Because every kid is significant, it is necessary to solve these challenging concerns before they can be widely implemented in educational practice at scale. Even though studies in this field are just getting started, more in-depth evaluations and validation in real-world learning settings are necessary [37].

8.7.5 Interstation

When people are unable to comprehend the findings of big data analysis, the capacity to analyze big data is rendered worthless. The final step is for the decision-maker to evaluate the results of the analysis once he or she has received them. The completion of this task will not be possible in a vacuum. It is common for it to need to go over all of the assumptions again and redo the analytical steps. As previously noted, there are other potential sources of error: computer systems may have defects, models nearly always contain assumptions, and conclusions may be based on erroneous information. All of these considerations lead to the conclusion that no responsible user will delegate authority to the computer system. Instead, she will make an effort to comprehend and assess the computer's results. She will likely find it simple to do the task thanks to the computer system. Because of its complexity, big data presents a unique set of challenges. In most cases, critical assumptions are incorporated into the data that is given. In analytical pipelines, steps are distinguished from one another by the assumptions that they make. Rather than adopting a financial institution's declared soundness at face value, decision-makers must thoroughly scrutinize the various assumptions at several levels of analysis, as was demonstrated by the recent mortgage-related financial system shock [38].

8.8 Big Data in Cancer Treatment

The term "big data" has become a popular catchphrase across a wide range of industries. Each person's genome sequencing, clinical records linked to medical imaging, the mRNA expression landscape of healthy and unhealthy tissues, insurance claims data, and biobank tissue-derived data are examples of big data in medicine [39, 40].

8.8.1 The cancer genome atlas (TCGA) research network

The cancer genome atlas (TCGA) is a cancer genomics research network. The TCGA dataset, which includes 2.5 petabytes of data from tumors and matched normal tissues from more than 11,000 patients and is free to the public, has now been made public. In collaboration with the National Cancer Institute (NCI) and the National Human Genome Research Institute (NHGRI), the TCGA has developed precise, multi-dimensional maps of the key genetic changes in 33 types of cancer [40].

8.8.2 The International Cancer Genome Consortium (ICGC)

Tumor genome research is the focus of the International Cancer Genome Consortium (ICGC), a non-profit organization dedicated to this field. The International Cancer Genome Consortium (ICGC) data (which was published on August 22, 2016) includes information from over 19,290 cancer donors from 70 research and 21 cancer locations. Currently, the complete dataset is securely stored on Amazon Web Services (AWS) cloud, where cancer researchers from across the world can have access to it at their convenience [41].

8.8.3 The cancer genome hub

The University of California, Santa Cruz has built a cancer genome center to facilitate the study of cancer. The cancer genome atlas (TCGA) was established in August 2011 to serve as a repository for cancer genomics research. After years of hard work, the cancer genome hub has grown to become the world's largest cancer genome library. It now has over 2.5 petabytes of data, with roughly three petabytes of data downloaded each month [42].

8.8.4 The cosmic database

When it comes to determining the effects of genetic changes on tumor growth, a catalog of somatic mutations in cancer (cosmos library) is among the greatest thorough resource available anywhere in the world. The most recent version (v70; August 2014) contains information on 2,002, 811 coding point mutations that have been detected in over one million tumor samples as well as in the vast majority of human gene sequences [43]. Another source of cancer information is the oncomine® genomes [44], which includes information from more than 500 sources and includes the PCOR portal for cancer genomics. A huge number of cancer genomic datasets are available for viewing, analysis, and download. The Mini-Sentinel Program and the patient-centric clinical research network of the United States Food and Drug Administration are two of the most important programs in the country [45]. An amazon data management client and cloud-based analytical in both open source and web-based are built using the Galaxy platform on top of the globus genomics systems, a collection of companies that are only focused on genomics [46]. The American Society of Clinical Oncology in addition to the cancer LINQ also provides claims data. Advancements in cloud-based computing have considerably increased the scope of data mining in the field of cancer. A 200 TB Amazon cloud-based data repository system is being used to categorize human sequence variants for the 1000 genomes project that has been identified by deep sequencing of 1000 genomes from around the world [47].

8.9 Conclusion

As the world's population grows and ages, cancer is expected to become more prevalent. Cancer is recognized as one of the major causes of premature death, as well as a major contributor to job-related productivity and participation in the labor market. According to a plethora of reports and studies, a major fraction of premature deaths is due to cancer. The advent of big data analytics has provided cancer researchers with powerful new tools to extract information from diverse sources. As big data analytics technologies continue to grow from research labs to clinical areas, organizations are increasingly utilizing these tools and techniques to design more comprehensive cancer treatments.

References

[1] Abbasi, Ahmed, et al. "Enhancing predictive analytics for anti-phishing by exploiting website genre information." Journal of Management Information Systems 31.4 (2015): 109-157.

[2] Loon, Lee KhAI, and Lim Cean Peing. "Big Data and Predictive Analytics Capabilities: A Review of Literature on Its Impact on Firm's Financials Performance." *KnE Social Sciences* (2019): 1057-1073.

[3] Bender, Ralf, and Ulrich Grouven. "Ordinal logistic regression in medical research." *Journal of the Royal College of physicians of London* 31.5 (1997): 546.

[4] ArunJalanila and Nirmal Subramanian. Comparing SAS® Text Miner, Python, R: Analysis on Random Forest and SVM Models for Text Mining. 2016 IEEE International Conference on Healthcare Informatics (ICHI), pages 316–316, 2016

[5] Harshawardhan S Bhosale and Devendra P Gadekar. A Review Paper on Big Data and Hadoop. International Journal of Scientific and Research Publications, 4(1):2250–3153, 2014

[6] Arora, M., Som, S., & Rana, A. (2020, June). Predictive Analysis of Machine Learning Algorithms for Breast Cancer Diagnosis. In *2020 8th International Conference on Reliability, Infocom Technologies and Optimization (Trends and Future Directions) (ICRITO)* (pp. 835-839). IEEE.

[7] Paesmans, M. (2012). Prognostic and predictive factors for lung cancer. *Breathe*, *9*(2), 112-121.

[8] Steward, L., Conant, L., Gao, F., & Margenthaler, J. A. (2014). Predictive factors and patterns of recurrence in patients with triple negative breast cancer. *Annals of surgical oncology*, *21*(7), 2165-2171.

[9] Degnim, A. C., Reynolds, C., Pant Vaidya, G., Zakaria, S., Hoskin, T., Barnes, S., & Newman, L. A. (2005). Nonsentinel node metastasis in breast cancer patients: assessment of an existing and a new predictive nomogram. *The American journal of surgery*, *190*(4), 543-550.

[10] Steward, L., Conant, L., Gao, F., & Margenthaler, J. A. (2014). Predictive factors and patterns of recurrence in patients with triple negative breast cancer. *Annals of surgical oncology*, *21*(7), 2165-2171.

[11] Dash, S., Shakyawar, S. K., Sharma, M., & Kaushik, S. (2019). Big Data in healthcare: management, analysis and future prospects. *Journal of Big Data*, *6*(1), 1-25.

[12] Service, R. F. (2006). The race for the $1000 genome.
[13] Stephens, Z. D., Lee, S. Y., Faghri, F., Campbell, R. H., ZhAI, C., Efron, M. J., ... & Robinson, G. E. (2015). Big Data: astronomical or genomical?. PLoS biology, 13(7), e1002195.
[14] Elfiky, A. A., Pany, M. J., Parikh, R. B., & Obermeyer, Z. (2018). Development and application of a machine learning approach to assess short-term mortality risk among patients with cancer starting chemotherapy. *JAMA network open*, *1*(3), e180926-e180926.
[15] Brooks, G. A., Kansagra, A. J., Rao, S. R., Weitzman, J. I., Linden, E. A., & Jacobson, J. O. (2015). A clinical prediction model to assess risk for chemotherapy-related hospitalization in patients initiating palliative chemotherapy. *JAMA oncology*, *1*(4), 441-447.
[16] Yeo, H., Mao, J., Abelson, J. S., Lachs, M., Finlayson, E., Milsom, J., & Sedrakyan, A. (2016). Development of a nonparametric predictive model for readmission risk in elderly adults after colon and rectal cancer surgery. *Journal of the American Geriatrics Society*, *64*(11), e125-e130
[17] Vogel, J., Evans, T. L., Braun, J., Hanish, A., Draugelis, M., Regli, S., ... & Berman, A. T. (2017). Development of a trigger tool for identifying emergency department visits in patients with lung cancer. *International Journal of Radiation Oncology, Biology, Physics*, *99*(2), S117
[18] Rajkomar, A., Oren, E., Chen, K., DAI, A. M., Hajaj, N., Hardt, M., ... & Dean, J. (2018). Scalable and accurate deep learning with electronic health records. *NPJ Digital Medicine*, *1*(1), 1-10
[19] Furlow, B. (2019). Predictive analytics reduces chemotherapy-associated hospitalizations. *Managed Healthcare Executive.* https://www.managedhealthcareexecutive.com/mhe-articles/predictive-analytics-reduces-chemotherapy-associated-hospitalizations.Accessed March,13.
[20] Yu, K. H., Zhang, C., Berry, G. J., Altman, R. B., Ré, C., Rubin, D. L., & Snyder, M. (2016). Predicting non-small cell lung cancer prognosis by fully automated microscopic pathology image features. *Nature communications*, *7*(1), 1-10.
[21] Sooriakumaran, P., Lovell, D. P., Henderson, A., Denham, P., Langley, S. E. M., & Laing, R. W. (2005). Gleason scoring varies among pathologists and this affects clinical risk in patients with prostate cancer. *Clinical oncology*, *17*(8), 655-658
[22] Bejnordi, B. E., Veta, M., Van Diest, P. J., Van Ginneken, B., Karssemeijer, N., Litjens, G., ... & CAMELYON16 Consortium. (2017). Diagnostic assessment of deep learning algorithms for detection of

lymph node metastases in women with breast cancer. *Jama*, *318*(22), 2199-2210.

[23] Bi, W. L., Hosny, A., Schabath, M. B., Giger, M. L., Birkbak, N. J., Mehrtash, A., ... & Aerts, H. J. (2019). Artificial intelligence in cancer imaging: clinical challenges and applications. *CA: a cancer journal for clinicians*, *69*(2), 127-157.

[24] Chan, H. P., Hadjiiski, L., Zhou, C., & Sahiner, B. (2008). Computer-Aided diagnosis of lung cancer and pulmonary embolism in computed tomography—A review. *Academic radiology*, *15*(5), 535-555.

[25] Song, S. E., Seo, B. K., Cho, K. R., Woo, O. H., Son, G. S., Kim, C., ... & Kwon, S. S. (2015). Computer-Aided detection (CAD) system for breast MRI in assessment of local tumour extent, nodal status, and multifocality of invasive breast cancers: preliminary study. *Cancer Imaging*, *15*(1), 1-9.

[26] Sorace, A. G., Wu, C., Barnes, S. L., Jarrett, A. M., Avery, S., Patt, D., ... & Virostko, J. (2018). Repeatability, reproducibility, and accuracy of quantitative mri of the breast in the community radiology setting. *Journal of Magnetic Resonance Imaging*, *48*(3), 695-707.

[27] Hravnak, M., DeVita, M. A., Clontz, A., Edwards, L., Valenta, C., & Pinsky, M. R. (2011). Cardiorespiratory instability before and after implementing an integrated monitoring system. *Critical care medicine*, *39*(1), 65.

[28] Parikh, R. B., Kakad, M., & Bates, D. W. (2016). Integrating predictive analytics into high-value care: the dawn of precision delivery. *Jama*, *315*(7), 651-652.

[29] Burki, T. K. (2016). Predicting lung cancer prognosis using machine learning. *The Lancet Oncology*, *17*(10), e421.

[30] van den Akker, J., Mishne, G., Zimmer, A. D., & Zhou, A. Y. (2018). A machine learning model to determine the accuracy of variant calls in capture-based next generation sequencing. *BMCgenomics*, *19*(1), 1-9.

[31] Raza, S. A., Barreira, C. M., Rodrigues, G. M., Frankel, M. R., Haussen, D. C., Nogueira, R. G., & Rangaraju, S. (2019). Prognostic importance of CT ASPECTS and CT perfusion measures of infarction in anterior emergent large vessel occlusions. *Journal of neurointerventional surgery*, *11*(7), 670-674.

[32] Pashayan, N., Morris, S., Gilbert, F. J., & Pharoah, P. D. (2018). Cost-effectiveness and benefit-to-harm ratio of risk-stratified screening for breast cancer: a life-table model. *JAMA oncology*, *4*(11), 1504-1510.

[33] Lee, C. H., & Yoon, H. J. (2017). Medical Big Data: promise and challenges. *Kidney research and clinical practice*, *36*(1), 3.

[34] Rumsfeld, J. S., Joynt, K. E., & Maddox, T. M. (2016). Big Data analytics to improve cardiovascular care: promise and challenges. *Nature Reviews Cardiology*, *13*(6), 350-359.

[35] Do Nascimento, I. J. B., Marcolino, M. S., Abdulazeem, H. M., Weerasekara, I., Azzopardi-Muscat, N., Gonçalves, M. A., & Novillo-Ortiz, D. (2021). Impact of Big Data analytics on people's health: Overview of systematic reviews and recommendations for future studies. *Journal of medical Internet research*, *23*(4), e27275.

[36] Lyko, K., Nitzschke, M., & Ngonga Ngomo, A. C. (2016). Big Data acquisition. In *New horizons for a data-driven economy* (pp. 39-61). Springer, Cham.

[37] Luan, H., Geczy, P., LAI, H., Gobert, J., Yang, S. J., Ogata, H., & Tsai, C. C. (2020). Challenges and future directions of Big Data and artificial intelligence in education. Frontiers in psychology, 2748.

[38] Computing Community Consortium. (2011). Advancing Discovery in Science and Engineering. Computing Community Consortium, Computing Research Association, Springer.

[39] Makler, A., & Narayanan, R. (2016). Big Data analytics and cancer. *MOJ Proteomics Bioinformatics*, *4*(2), 196-199

[40] Weinstein, J. N., Collisson, E. A., Mills, G. B., Shaw, K. R., Ozenberger, B. A., Ellrott, K., ... & Stuart, J. M. (2013). The cancer genome atlas pan-cancer analysis project. *Nature genetics*, *45*(10), 1113-1120

[41] International Cancer Genome Consortium. Hudson TJ, Anderson W, Artez A, Barker AD, Bell C, Bernabe RR, Bhan MK, Calvo F, Eerola I, et al. 2010. *International network of cancer genome projects. Nature*, *464*, 993-998.

[42] Cline, M. S., Craft, B., Swatloski, T., Goldman, M., Ma, S., Haussler, D., & Zhu, J. (2013). Exploring TCGA pan-cancer data at the UCSC cancer genomics browser. *Scientific reports*, *3*(1), 1-6.

[43] Forbes, S. A. (2015). beare D. *Gunasekaran P, Leung K, bindal N, boutselakis H, Ding M, bamford S, Cole C, Ward S, et al: COSMIC: exploring the world's knowledge of somatic mutations in human cancer. Nucleic Acids Res*, *43*, D805-D811.

[44] Rhodes, D. R., Kalyana-Sundaram, S., Mahavisno, V., Varambally, R., Yu, J., Briggs, B. B., ... & ChinnAIyan, A. M. (2007). Oncomine 3.0: genes, pathways, and networks in a collection of 18,000 cancer gene expression profiles. *Neoplasia*, *9*(2), 166-180.

[45] Fleurence, R. L., Curtis, L. H., Califf, R. M., Platt, R., Selby, J. V., & Brown, J. S. (2014). Launching PCORnet, a national patient-centered clinical research network. *Journal of the American Medical Informatics Association*, *21*(4), 578-582.,

[46] Clarke, L., Zheng-Bradley, X., Smith, R., Kulesha, E., Xiao, C., Toneva, I., ... & Flicek, P. (2012). The 1000 Genomes Project: data management and community access. *Nature methods*, *9*(5), 459-462.

[47] Goecks, J., Nekrutenko, A., & Taylor, J. (2010). Galaxy: a comprehensive approach for supporting accessible, reproducible, and transparent computational research in the life sciences. *Genome biology*, *11*(8), 1-13.

Biography of Authors

Ayush Chandra Mishra

Ayush Chandra Mishra is a Pharmacy Graduate student from Dr. M. C. Saxena College of Pharmacy, Lucknow. During my graduation, he developed an interest in writing papers. He has been participating in various academic activities like attending international and national seminars, conferences, and gave poster and oral presentations. He has also done some conference papers. Learning new technologies and devising unique techniques to address chronic diseases have always been top priorities for him. One of the most significant impacts on my interest in technology has been this book chapter along with different faces of the research area and future venture.

Ratnesh Chaubey

Ratnesh Chaubey is a versatile Pharmacy graduate Student from Dr. M. C. Saxena College of Pharmacy, Lucknow. He have been participating in various academic activities like attending international and national seminars, conferences, and gave oral and poster presentations. Learning new technology and devising unique techniques to address chronic diseases have always been top priorities for him. One of the significant influences on my interest in technology has been this book chapter.

9

Dose Prediction in Oncology using Big Data

Akash Chauhan[1], Ayush Dubey[1], Md. Aftab Alam[1], Rishabha Malviya[1*], and Mohammad Javed Naim[2]

[1]Department of Pharmacy, School of Medical and Allied Sciences, Galgotias University, Greater Noida, India
[2]Department of Pharmaceutical Chemistry, Faculty of Pharmacy, Tishk International University, Iraq
*Corresponding Author: Department of Pharmacy, School of Medical and Allied Sciences, Galgotias University, Greater Noida, India, EMail: rishabhamalviya19@gmail.com

Abstract

In this era of information technology, big data analysis is entering biomedical sciences. Big data is moving from hype to reality in medicine. Patient-specific dose prediction improves the efficiency and quality of radiation treatment planning and reduces the time required to find the optimal plan. Big data and predictive analytics have immense potential to improve risk stratification, particularly in data-rich fields like oncology. We argue that current predictive analytic interventions could address sizable gaps in risk stratification strategies in oncology. Burgeoning applications of predictive analytics in pathology interpretation, drug development, and population health management provide a way forward for future tools to move into clinical practice. However, to achieve this potential, clinicians, developers, and policymakers must address the research, technical, and regulatory barriers that hamper the applications of analytics in oncology.

Keywords: Big Data; Dose Prediction; Artificial Intelligence; Patient Monitoring.

9.1 Introduction

Large-scale randomized controlled trials are the foundation of Level I evidence-based medicine. To achieve precision medicine, additional clinical and biological aspects must be explored, necessitating the development of specialized research [1]. New therapeutic alternatives are needed for patients of various ages and demographics. Medical imaging, blood testing, and other diagnostic instruments of genetics can all be used to identify the optimum combination of treatment choices (radiotherapy, chemotherapy, targeted therapy, and immunotherapy) for a specific patient. A unique collection of molecular abnormalities is implicated in the etiology of the disease or is connected with how effectively the therapy works in each patient. Finding and utilizing a patient's specific aberrations is critical to tailoring their therapy. Because of this shift to molecular oncology, oncology research has achieved EGFR inhibitors which have made tremendous headway in illnesses with poor prognoses, such as lung cancer that is not tiny cells [2] and melanoma during the last 25 years (through the use of immunotherapy) [3]. Variant mutations (i.e. a recombinant combining with genetic material from two separate variations to form a new version), have the potential to alter hundreds of in a tumor, there are genes. Next-Generation Sequencing (NGS) may be used to target specific areas, as well as the whole exome (which contains all coding genes), or the complete genome (all DNAs are sequenced). In the same way, the transcriptome may be analyzed. It will be critical to do as many genetic studies as possible as we begin to comprehend the complicated molecular circuits that mediate both primary and secondary treatment resistance, as well as the reaction to radiation [4]. Because of the complexities, it is very hard to construct trials that are unique to each case. It was formerly considered that human brains could integrate up to five elements into a decision. A single patient's decision will be dependent on up to 10,000 criteria by 2020 [5]. All that stands in the way of discovering traits that impact sickness outcome is a dearth of big phenotyped cohorts, although costs of sequencing have decreased significantly [6–8], while computing power has continually grown (Figure 9.1). We now have a wonderful opportunity to establish relevant phenotypes as a result of the broad usage of EHRs. Data science has a role to play when it comes to generating predictive models from big databases. Furthermore, it has been questioned if comorbidities, severity, a time before therapy, and tumor features are comparable across clinical research participants and patients in regular care [9]. Data-driven strategies that use routine healthcare data to improve decision-making are becoming increasingly common. In the future, clinical decision support systems, according to I. S. Kohane, will be solely data-driven. As more data are gathered, it will be feasible to make conclusions from observations free of "unknown confounders" [10]. To build accurate models, we must first overcome the challenge of incorporating such a huge and diverse set of information. The purpose of this research is to identify the most major informatics challenges that radiation oncologists confront when attempting to develop a precision medicine program. We will go over the various methods for creating models that predict how effectively radiotherapy or chemoradiation will perform.

9.1.1 Data should be reviewed and organized

A comprehensive list of the qualities that should be considered while developing a prediction model has been provided by Lambin et al. [11].

They consist the following:

- Clinical characteristics (Patient status, cancer grade and stage, Hematological analysis results, and patient information surveys).
- Medication characteristics include dose distribution (both geographically and temporally) and chemotherapeutic support.
- Data analysis might be as easy as removing it from the record-and-verify software.
- Carcinoma size and volume, as well as metabolic uptake, are all imaging features (more broadly covered in the study topic of "radiomics").
- Radiosensitivity [12], hypoxia [13], proliferation, and normal cells response are all factors considered [14]. Genomic investigations are crucial in defining these characteristics.

9.1.2 Information database management

Patients may be able to view exactly how their treatment was carried out in a digitalized format, thanks to breakthroughs in radiation oncology. We maintain track of the actual radiation treatments that each patient receives. This information is stored digitally for each patient and therapy session, so that we may access it at any time for any patient. On-board imaging compensates for daily changes, allowing us to pinpoint exactly where the dose is administered. These technologies can offer information on the time and location of the therapy. Every patient's data are collected prospectively by each department's record-and-verify software. As a consequence, the process of providing care might be quantified and analyzed. Most other medical specialties have a far lower requirement for data quality. Different levels of extraction can be employed to include these data into a hospital's clinical data warehouse (CDW). The raw data include dose–volume histograms, treatment volumes, interactional intervals, and total treatment duration, as well as images generated by onboard equipment. [15] Another approach would be to extract only relevant data before incorporating it into the CDW, which would severely restrict the amount of data available [15]. Radiation oncology and medicine, in general, rely significantly on the previously described data, together with continuous testing to determine the level of toxicity. An increasing number of wearable and mobile gadgets should be used in this sector. Patients will no longer have to postpone their next radiation oncologist visit to report any adverse effects they experience during or following treatment. It has been proven that patient-reported results improve follow-up in many studies [16, 17]. Every day, more and more data must be collected and managed. A single patient's genetic data, which accounts for around 70% of the total, is expected to be over 7 GB (Table 9.1). Maintaining safety and accessibility of patient health data is a significant concern for any organization. They should be available with ease and quickness wherever they are without jeopardizing their safety. Remote data access needs an architecture that complies with strong security requirements, such as tight user authentication and technologies that allow you to follow the path each piece of data takes as it is processed. Login processes for important healthcare professionals necessitate a scalable process that comes at a high cost, however, they cannot be ignored [18]. In most cases, medical record linking and data anonymization are required tasks when obtaining data for research. They typically employ the assistance of a trusted third party to perform these tasks. Data from the care zone, where the patient and provider have a mutually trusted relationship, must be made public, to the non-care zone, where data governance organizations have power, to be anonymized and made available for research purposes. One possibility for conserving and keeping care available is to use translational research platforms. On these platforms, large clinical datasets may be combined with omics data. Other authors anticipate that, despite technological developments, hospitals will be unable to keep up with the growing volume of data [20]. One alternative is to shift the oldest and biggest data to external storage, as most hospitals do with old medical records. To enable rapid and simple access to the most data, we have to relocate it to a new storage platform for digital data. As demonstrated in Figure 9.1, data from the hospital and patients may be combined into a single system.

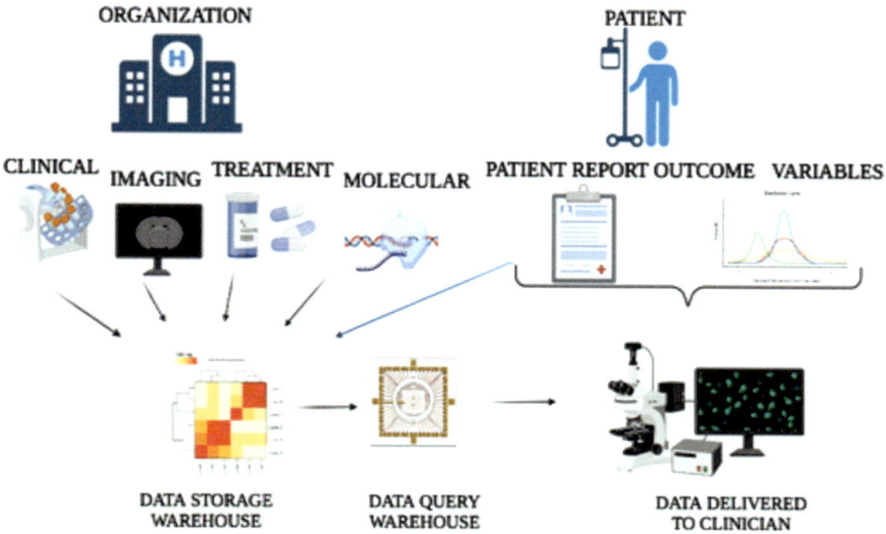

Figure 9.1 Data collection and management system (PRO = patient-reported outcomes).

9.2 Significance of "Big Data"

Several individuals and corporations take advantage of the phrase "big data" interchangeably, although it does not always imply the same thing. Even though most of us have a hazy understanding of what big data is, it is more than just "a lot of data." "Big data's five Vs" provides a definition of big data in healthcare (Anil Jain, September 17, 2016). The phrase "big data" encompasses the following concepts (Figure 9.2):

Volume: The term "volume" refers to the sheer number of data points and records that comprise "big data." Therapy information (surgical procedure, response data, systemic treatment, radiotherapy, and their combinations) and repercussions are given in addition to clinical, radiological, and pathological diagnostic data.

- Velocity: The velocity of big data has two components: (1) it is being created at an increasing pace, and (2) it must be computed/digested at a rate similar to the rate of generation. Patients with cancer are enjoying longer lives as the disease's incidence rises across the globe. As technology and monitoring equipment progress, more data must be processed concurrently.

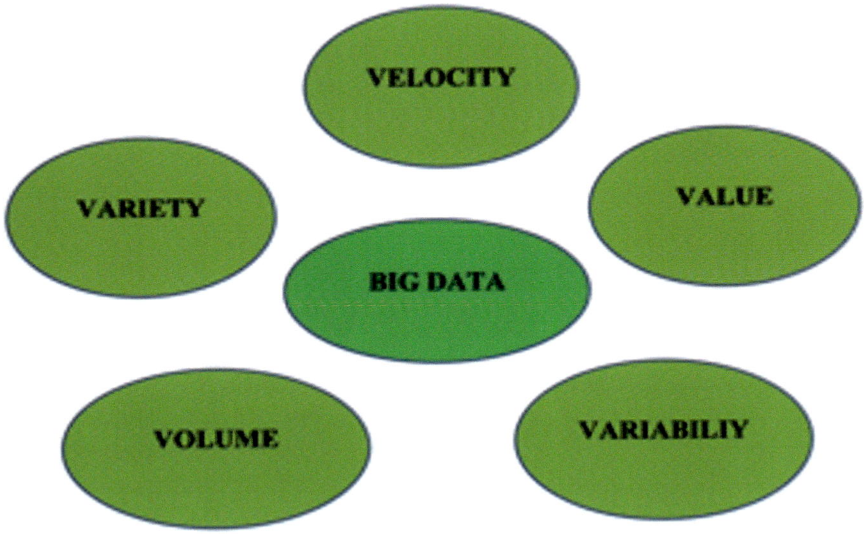

Figure 9.2 Significance of "big data."

- Variety: Big data encompasses a broad variety of information forms. This diversity offers both prospects and challenges (several data formats improve data quality and usability) (synoptical reporting).
- Variability: It is important to remember that data collecting occurs all over the place and at various times of the day. To get the bulk of the (synoptic) provided data, a (predefined) minimum dataset must be obtained. To make this work, there must be agreement on the key facts as well as uniform terminology (e.g., recurrence vs residual disease).
- Value: The construction of exact data-derived findings or metrics that can be utilized to make real-world adjustments is the only way to accomplish healthcare improvement and impact.

More important than just being concerned about the sheer volume of data is the reality that data resources are often dispersed throughout the globe and stored in a way that makes data integration difficult. Even if scaling concerns arise as a result of the data being sent over the internet, efforts at standardization and harmonization are required if the data are to be pooled and used in a consistent way to cover a wide variety of subjects.

9.2.1 Requirement of big data (BD)

BD can be useful in many scenarios. Location monitoring may help logistics companies reduce transit risks, expedite deliveries, and ensure delivery reliability. The SEC's use of big data network analytics uncovers predicted financial crimes. For example, Netflix and YouTube depend on prior watching patterns and other internet data to increase user engagement and profitability. Advertising firms are unquestionably the most important players in the big data space. Advertisers may target consumers based on their interests and purchases by using data from social media sites such as Facebook, Twitter, and Google. Drones are used by breeding companies to fly over agricultural areas and relay visual data back to their headquarters throughout the breeding process. Using big data, hospitals can better monitor patients in critical care. Various treatment techniques can be analyzed, and pandemic breakouts can be forecast in advance. Disease prevention strategies might benefit from a better knowledge and use of big data in the medical industry. Through the integration of large genetic and environmental datasets, it will be feasible to determine whether individuals or groups are prone to certain (chronic) disorders such as cancer. These discoveries might pave the way for customized medicines focused on lowering biological consequences related to healthiness in certain groups. Big data might be used to assess recent preventative measures and find fresh insights that can be utilized to improve them. It is also conceivable to use big data in a therapeutic setting to monitor the effect of specialized drugs, such as high-priced oncology procedures, in context to sufferer and tumor (genetical) traits. This will aid to enhance précised medication by giving critical information for calculating the cost-effectiveness of various treatment regimens.

9.2.2 Medical big data analysis

Massive volumes of data are accessible in a variety of forms and sizes. In the subject of cancer, patient-generated data are the most prominent. For therapeutic purposes, a broad range of data points/subjects are often recorded in computerized patient files. Many different types of data are stored here, including demographics like age and gender as well as symptoms, family history of the disease (including a history of the disease running in the family), imaging data (such as CT, PET), histopathology (such as entire genome BAM data (up to 100 GB), DNA/RNA sequencing, blood analysis, and immunohistochemistry. The findings of in vitro studies have importance since they may give useful information. The second source of vast volumes of big data is the computerized study of this information. Among the processed data are radiological and digital picture analytical determination as well as genomic expression and gene alteration studies. Machine learning is a rising source of processed data, which often comprises substantial processor database documents including defined information. A tertiary resource of big data is patient outcomes and patient experience measures (PREMs), which are collected by patients using software on computers and mobile devices to record various metrics offered by their providers (eHealth, telemedicine) or on their proposal. A quaternary possible resource is the published literature (IBM project). Only one doctor on the planet has the time to study even a fraction of the biomedical articles produced each year, much alone all related textbooks and online resources. The amount (depth) of data provided by each patient is a significant factor in cancer research. Nonetheless, although the average patient cohort size is small, oncologists are continually generating and preserving observables (thousands to millions). In uncommon diseases like head and neck melanoma, the discrepancy between the amount of data collected on an individual patient and the size of the cohort is particularly pronounced. Machine learning and neural networks have recently made significant methodological improvements, which may be quite advantageous if there are examples to learn from. There are several ways to improve object identification in photos, such as using hundreds, if not millions, of images to train a system. This indicates that to utilize the data for designing personalized medicines, the sample size will need to be increased [4]. Strong data management, standardization, data sharing, provenance, and data transfer standards are required in the Oncology of the head and neck.

9.2.3 The application of big data in the therapy of head and neck cancer/melanoma (Hd.Nk.C)

Because of the vast amount of data and the diversity of data sources, consistency in data collection is crucial. Because datasets will be more consistent and full as a result of this standardization, they will be much simpler to relate to more information sources. Data standardization is necessary for information assimilation, which is essential for information analysis and value development. To measure the quality of treatment following particular surgical procedures, for example, the patient mix in conditions of tumor stage, (neo-) adjuvant therapy, co-morbidity, and so on must be known. Clinical, pathological, genetic/genomic, and other data have been meticulously documented in the Netherlands' several national databases PROM/PREMs as

Table 9.1 Data sources for most common cancer types in The Netherlands, including HNC.

Cancer type	Clinical	Pathological	Chromosomal	PROM/PREM
HNSCC	DHNA	PALGA	PALGA/HMF	NET-QUIBIC
Pulmonary melanoma	DLCA/NVALT	PALGA	PALGA/HMF	
Prostate tumor		PALGA	PALGA/HMF	
Breast melanoma	NBCA	PALGA	PALGA/HMF	
Melanoma	DMTR	PALGA	PALGA/HMF	
CRC	DSCA	PALGA	PALGA/HMF	

well as data (Table 9.1). The Dutch Head and Neck Audit (DHNA) started collecting clinical data in 2014 and are now part of the Dutch Institute for Clinical Auditing (DICA), which has formed subgroups for different medical disorders (cancer and non-cancer). More than 20 distinct types of tumor cell-based data, including pathological and genomic/genetic data, have been conscientiously accumulated and are now synoptically available throughout the nation. On the other hand, the absence of a standardized data basis for radiological data persists. Thanks to the efforts of the NET-QUBIC group, PREMs/PROMs for HNC and other malignancies may now be reported on a single national website. HNC in the Netherlands is now using a significant amount of patient-derived data. TCIA [5] and other related programs, such as HNSCC (Hd.Nk.squamous cell melanoma) collection [6], are still in operation across the globe. It is crucial to collect these data on an individual patient basis as soon as possible. In the Netherlands, the Dutch Hd.Nk. (i.e. head and neck) society operates an eight-hospital network of six preferred partners for Oncological therapy for the Hd.Nk.(NWHHT). As a result of this, the Hd.Nk. the oncological area is at an excellent point to pool resources and coordinate efforts to ensure consistent data input and dissemination of the existing data banks. These databases might be connected throughout the nation, enabling researchers in head and neck melanoma to develop algorithms for combining data from all across the country.

9.3 Efficacy of BD

The capability of BD in biomedical investigation/research as a result (BI/R) has yet to be found. Big data may now be used for routine diagnosed data, advanced characteristics of care/life, and BR, among other things (and future). The apps listed below are samples of those that are presently available.

9.3.1 Diagnostics are carried out daily

Using big data in the battle against the disease has already been shown to be beneficial. As an example, Dutch pathologists have near-real-time access to each patient's nationwide histological follow-up. The PALGA foundation has been in charge of all Dutch digital histopathological records since 1971. (www.palga.nl). The PALGA database has about 72 million records from over 12 million Dutch patients, making it one of the world's largest biological databases. All 55 pathology labs in the nation may be located here. Each histology report is signed by a Dutch pathologist and sent to PALGA and the local hospital's information system in two copies. As a consequence, any PALGA member gets access to any patient's pathological follow-up data in this database (pathologist or molecular biologist). As a diagnostic tool, it can reveal previously unknown patient history (oncological) elements, such as resection margins and positive lymph nodes, in cases where pathological analysis was conducted at a separate lab and the patient's history is unknown. Another advantage of utilizing this database is that it may be utilized to look for correlations in low-prevalence situations that seem unconnected at first sight [8]. Electronic patient records provide a wealth of medical data that may be used to predict the prognosis of a patient [9] as one of the early models of prediction for HNC sufferers having therapy in medicinal institutions in developed countries. Models may be automatically updated when new data are available since statistical prognostication techniques are automated [10]. These data might be utilized to construct clinical decision-making tools for better sufferer advising and non-binary results evaluations.

9.3.2 Determining the quality level of care

Data on sufferers' characteristics and treatment may be connected to patient outcome databases, providing unmatched insight into the quality and efficacy of healthcare. NSCLC is the most frequent kind of non-small cell lung cancer (NSCLC), yet there are numerous therapy alternatives tailored to therapy based on molecular testing. This gives you real-time feedback on the best test–treatment correlations. Furthermore, failing labs may be driven to modify their techniques and workflows to improve patient care. By merging information from the national cancer registry with the PALGA database, it was able to examine the vast diversity of approaches to cancer treatment in the Netherlands [12, 13]. Such data must be shared to enhance treatment quality, but it must be done with caution since laboratories and healthcare may be afraid of reputational harm or name and embarrassment [14] if they remark on such data. In reality, many healthcares are willing to engage in parallel feedback as long as it is anonymous to the public and only shared with the patient one-on-one. As a result, in the Netherlands, algorithms by the Dutch Institute for Clinical Auditing, for example, have been created to provide frequent automated feedback on pathology and treatment-related features (www.dica.nl). When mirror data reveal that the facility has a higher recurrence rate than similar facilities, there might be an inducement to investigate the inherent chain further.

9.3.3 Biological and medical research

In the upcoming time, big data is projected to have the greatest influence on scientific discoveries. This move from "genome-wide association studies" (GWAS) to "big data wide association studies" (DWAS) is occurring, and it is all due to big data. Data scientists and bioinformaticians have unrivaled access to the increasing fields of imaging and molecular research, as well as the integration of these data with other information. Big data satisfies a commercial demand in biological research. One of the most difficult problems in medicine, for example, is our incapacity to understand disease biology. To collect all critical multisource variable quantities like Deoxyribonucleic acid, Ribonucleic acid, polypeptides, and metabonomic data, enormous volumes of big data must be pooled and combined into more accurate models to estimate how tumors would act and which patients will help the most from particular therapies. With this integrated multi-omics data, researchers will be able to get a better knowledge of the molecular behavior and processes that underpin HNSCC growth patterns, metastatic potential, and response to (targeted) treatment.

9.3.4 Personalized medication

Personalized therapy is based on actionable insights gained from enormous volumes of data, which is where big data comes into play [15]. When new technologies like sequencing and imaging create terabytes of data, the quantity of information accessible to the biomedical community expands quickly. Rather than patient-specific records, the bulk of data is produced,

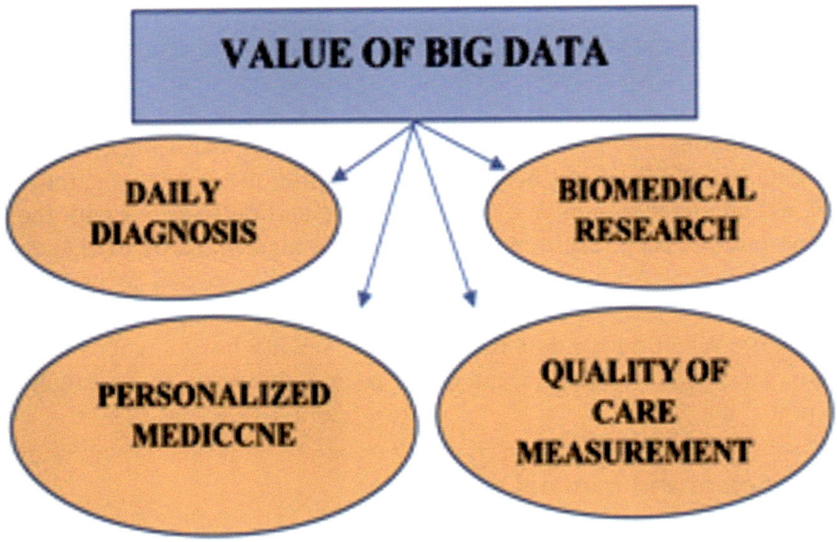

Figure 9.3 Personalized medication.

radiology and digital image analysis are examples of calculated automated data analytics. Head and neck tumors provide a distinctive diagnostic and therapeutic challenge due to their intricate architecture and unpredictability. With the aid of Radiomics [16], we may be able to resolve these challenges. According to S. M. Willems and colleagues, et al., radiomics are one of the non-invasive and inexpensive method that could be used to gather and mine medical imaging features. The use of predictive and trustworthy machine-learning approaches allowed by radiation omics in precision oncology and cancer care may make it simpler to treatment results for people with head and neck cancer can be predicted cancer [17]. Among these strategies is stratification (or customization), which entails detecting variations among (groups of) patients based on (expected/predicted) survival. As a result, medical and radiation oncologists may be able in some patient groups, to (de-)escalate systemic therapy and radiation dosages. Value of big data in terms of personalized medicine is illustrated in Figure 9.3.

9.4 Fair Data

To make sure that information can be used in secondary research, the FAIR (Findable, Accessible, Interoperable, and Reusable) principles must be followed. The notion of FAIR data was initially described in 2014 [18]. Since then, the G20 (2016) and G7 (2017) have recognized and accepted the FAIR principles, and the EU has made FAIR data a key component of the European Open Science Cloud (EOSC). The use of a community-recognized ontology is essential for findability (F), accessibility (A), and interoperability (I). The specification also mentions data privacy requirements. Many variables influence data reusability, including the data's origins and the accuracy of the meta-data, and its completeness (R).

9.5 Ontologies are used to extract high-quality data

When variables and the data used to generate models are standardized, as are the terminologies applied in the EHR, therapy protocols and genetic explanations improve in quality and comparability. Due to the vast number of characteristics, obtaining and aggregating high-quality data is particularly challenging. The ontology or set of common ideas, of a data-gathering system, is a critical component of any predictive modeling attempt. There are now about 400 biological ontologies. Other prominent thesaurus resources, in addition to SNOMED and the NCI Thesaurus, include CTC AE, UMLS, and SNOMED. These ontologies, for example, do not include concepts like "area of interest," "target volume," and "dose–volume histogram." The Radiation Oncology Ontology (ROO) [38] was created as a result of this, and it utilized existing ontologies, while adding RO terminology including "area of interest," "target volume," and "dose–volume histogram" (DVH). The usage of standard ontologies will make available information from multiple sources to be extracted and integrated. The quality of the data and the characteristics chosen are essential. When possible, there should be a second curator or data checker hired to confirm the accuracy of the data. To ensure the validity of their findings, the physician and data scientist working on the project must collaborate.

9.5.1 Procedure for developing a predictive model

It is a two-step procedure that begins with qualifying and ends with validation in predictive modeling. To qualify, the data must be proven to be indicative of a result. Any newly identified predictive or prognostic characteristics should be confirmed using a distinct dataset. More research must be conducted once a model has been validated and qualified to determine whether therapy recommendations based on the model genuinely improve patient outcomes.

- Kang et al. offered seven modeling principles in radiation oncology.
- Dosimetric and non-dosimetric predictors should be considered.
- Before doing automated analysis, predictors should be manually selected.
- Make the process of selecting predictors more automated.
- Assume multicollinearity in the predictor model.

Cross-validation should be used correctly to give model genericity when using external datasets, to increase prediction performance and generalization to external data.

Investigate several models and compare them to existing models.

These factors should be considered while developing and validating predictive models for use in the medical business. Our models are created using either statistical or machine learning methods. The major emphasis of our machine learning research will be oncological radiation oncology.

9.6 Standard Statistical Techniques

While Cox regression is commonly used to analyze survival information, logistic regression is less commonly employed is less generally employed should be investigated for models that predict qualitative outcomes (such as toxicity). Logistic regression (LR) is a statistical approach that plots the likelihood of a given event against a collection of factors using a sigmoidal logistic function. LR is the best approach when only a few unrelated predictors are being evaluated (age, sex, tumor size, etc.). For lung SBRT, for example, the optimum radiation dosage may be established using one-dimensional data as a predictor or even the GTV size (two-dimensional data). The model's features are introduced sequentially and progressively. One-dimensional data less than the number of parameters assessed can be used to form a decision boundary (one-dimensional line for two predictors, two for three, etc.). In several studies [39–41], after lungs or head and neck surgery, LR has been focused to foresee acid reflux disease and radiation-induced xerostomia.

9.6.1 Machine learning techniques (ML)

Numerous ML sets of rules have been developed used in oncology, which include the following:

- An algorithm can construct mutually exclusive categories by responding to questions consecutively [42].
- Using Naive Bayes (NB) classifiers, it is feasible to derive probabilistic connections between variables [43, 44].
- K-nearest neighbors (k-NN) are classification and regression; it is used to categorize a feature based on its dataset's nearest neighbor.
- Data may be categorized into categories using a previously trained SVM (support vector machine) [45].
- There are two kinds of artificial neural networks (ANNs): ANNs based on biological neural network models and ANNs with several neuronal layers [46].

Each of these methods has its own set of benefits and drawbacks downsides in terms of processing power (Table 9.1). When deciding on a technique for a data analysis project, these should be considered. Two techniques used in radiation oncology research will be thoroughly discussed: SVM, ANN, and a variant known as DL [47].

9.6.2 Support vector machines

The linear threshold is defined for LR's restriction on the number of characteristics that can be defined. SVM may be used to detect complicated patterns if the model must incorporate a large number of variables that are not linearly separable. To change the data, similarity functions (or kernels) are utilized, and "support vectors" are chosen. Patient comparisons and prognostication based on a range of vectors are performed on patients with varying histories. Radiation pneumonitis after conformal radiotherapy [48], local control after lung SBRT [49], and chemo- and radiosensitivity in esophageal melanoma have all been predicted using SVMs in multiple investigations. In these research works, DHVs, EUDs, and BEDs were classified as dosage input parameters, whereas the authors classified the non-dosage characteristics (clinical or biological features). The precise number and kind of characteristics used may limit the influence and applicability of the results, which should be recognized.

9.6.3 Artificial neuron network

An artificial neural network is made up of numerous layers of neurons. The weight of each "neuron" determines its importance. This system is organized into layers, with each taking input from the previous layer and computing a score before passing it on to the next layer. The neurons and connections of an ANN must be appropriately weighted for it to function. To do this, random weights may be assigned to neurons, and these weights can be computed and changed over time to enhance the correlation. Sufferers with advanced Hd.Nk.C who have not been able to find a cure have had irradiation and/or chemotherapy may benefit from utilizing ANNs to predict their prognosis [50]. After a series of experiments, a three-layer feed-forward neural network comprising 14 clinical parameters was built using 1,000-iteration training method. Bryce and his colleagues used more predicting parameters than the LR model [51], and used artificial neural networks (ANNs) six years later to improve the understanding of radiation's long-term effects on prostate cancer. To screen for nocturia, rectal bleeding, and PSA, they employed dosimetric parameters (DVH) and three distinct ANNs (analytical neural networks). ANNs were proven to be efficient in predicting with higher than 55% accuracy; biochemical control and particular bladder and rectum abnormalities may be diagnosed. Other studies with bigger datasets [52, 53] increased their sensitivity and specificity. In patients undergoing lung radiation, ANNs have also been applied to envisage pneumonitis [54, 55]. V16, gEUD for exponent $a = 1$, gEUD for exponent $a = 3.5$, FEV1 and DLCO percent, and whether or not the patient took chemotherapy before radiation was the six input characteristics used by Chen et al.. The model was then stripped of all features to assess their utility. Aside from those two characteristics, all of them were required for an accurate prediction of the patient's prognosis. ANNs were also used to predict survival following irradiation therapy for uterine cervical cancer in another investigation [56]. The prediction model used in this investigation was based exclusively on seven factors (age, performance status, hemoglobin, total protein, FIGO stage, histology, and radiation effect grade as determined by periodic biopsy exams), the capacity to learn at a high level.

9.7 Deep Learning

An ANN variant is a subset of that technique. DL differs from ANN in that it can do supervised or unsupervised learning and has more hidden layers than ANN typically does. If there are one or two hidden layers, it falls under the category of supervised machine learning [57]. DL is rapidly being utilized in medical imaging as a categorization or segmentation technique; however, it has yet to be applied to forecast the consequence of radiation. Approaches to education that are learner-centered or learner-agnostic (UL) sustained learning is intended to forecast a predetermined conclusion [58]. This technology is commonly used to recognize images and documents. Supervised algorithms assess a training dataset to generate a utility that is best suited to the training scenarios (where each example is a pair comprising a response feature and the desired output value). The computer will extend this technique to predict couples with unknown output values [59]. The data in UL are unlabeled, and the program will seek naturally occurring patterns or groups in the data. This will entail giving numerical values to each patient's clinical features in the form of vectors in medicine. Higher-level features that might be predictive or prognostic would not have been discovered with human intervention. UL might highlight new groups of patients to find new physiopathology. Because of the expanding relevance of graphics processing units, researchers may pay more attention to UL (GPUs) [60]. Unsupervised machine learning may have a basic flaw in that it does not provide any insight into the meaning of the correlations and connections it discovers between data pieces, even if it is capable of doing so without human supervision. If this is the case, UL may come into a link that no human can grasp. Almost of the subject of predictive oncology, machine learning algorithm research is focused on supervised learning [61].

9.7.1 Big data in the field of radioactivity oncology

The four Vs of BD are volume, variety, velocity, and veracity [62]. The total size of a cancer patient's EHR will be around 8 GB, with genetic data taking up the majority of that space (volume). In radiation oncology, constructing a predictive model entails taking into account a wide range of data sources (variety). This is a significant challenge in and of itself. As a result, if we apply these models, we will need a decision support system that can swiftly analyze massive volumes of data (velocity). Ultimately, information excellence in radiation oncology is very high since all departments employ record-and-verify systems to capture all information about the therapy provided and how the treatments are administered were done, and any potential deviations (veracity). Everything points to the conclusion that big data is a wonderful match for radiation oncology. Artificial intelligence and ML are being employed in a variety of ways in cancer research that can be beneficial.

9.7.2 ML and AI in oncology research

It is not always practical to separate the data testing the strategy on a different dataset, divide the dataset into a training and test set. To do internal validation on a model, 10%–20% of the data must be eliminated during the initial testing [63]. External validation using a unique (large enough) dataset will be used to eliminate or at least mitigate any bias caused by the data used. A patient-per-feature ratio of at least 5–10 is required to include a substantial quantity of data (especially genomics) into a model [64]. If the ratio is too low, an overfitting model will be formed, resulting in the formation of a model for explaining random errors or noise in a dataset but is unreliable when applied to a new dataset or population of data. Multiple ranges of ML algorithms were used to assess the model's performance, and algorithms should be used to generate different prediction models. A new category should, on average, outperform the earlier ones. Only around 17% of cancer ML research reported thus far [65] evaluated more than one ML technique. Although ANN is the most often used technique in cancer prediction modeling, deep learning (DL) is gaining traction in many fields. We may see more deep learning initiatives in the future, thanks to open-source software frameworks like Google's TensorFlow [67].

9.8 Correction of the Oncology Risk Stratification Gap

The challenge of stratifying oncology risk is exacerbated by a scarcity of important prognostic data, the requirement for time-consuming manual data entry, a data gap, and a lack of full data, an overreliance on clinical intuition. Consider the following scenario: a cancer patient with a fatal prognosis and metastatic illness. Prospective data show that doctors, especially for patients with advanced solid tumors, are bad at predicting outcomes [68]. It is critical to recognize cancer sufferers who are at high risk of dying to prevent needless end-of-life treatment or acute care [69]. Oncologists employ prognostic aids for specific cancers, but since they do not apply to the vast majority of tumors, they are seldom used [70, 71], and may lack genetic data, fail to identify the majority patients who are expected to die within a year and require time-consuming human data entry [72]. Interclinician variability and bias may have an impact on performance status and other predictive measures [73]. There is even little published evidence on estimating the probability of other crucial outcomes for cancer patients, including hospitalization or side effects.

The pursuit of more patient-centered treatment is driving advancements in cancer risk categorization systems. A shift in payment mechanisms, such as bundled payments, will encourage appropriate treatment for the correct patient rather than just pushing how many services are provided [74]. Oncologists are increasingly making treatment choices based on a patient's formal risk of particular outcomes. Data on demographics, treatment episodes, and particular clinical difficulties must be collected and analyzed. The Centers for Medicare and Medicaid Services (CMS) and EHR providers gathered data on Medicare patients to enhance data collection as part of the oncology care model [75–77]. Despite the increasing availability of detailed data combining clinical and use characteristics, strong prediction algorithms are still required to identify the risk of excessive consumption or other undesirable consequences in areas like readmission risk reduction in general inpatient settings. Predictive analytics-based decision aids have been discovered to improve value-based healthcare decision-making [78, 79]. Similar technologies are critically needed in the area of cancer to enhance clinical judgment and efforts to improve community health.

9.8.1 Current use cases for oncology predictive analysis

Predictive analytics systems, which utilize algorithms based on past patient data, may anticipate an individual's or community's future health outcomes [80]. As the volume of cancer-related EHR, radiographic, genomic, and other data grows, so does the need for data analysis has also grown, several use cases that may be generalized have evolved.

9.8.2 Management of the general population's health

Community health management is highly reliant on tailoring treatments to people who are most likely to have poor outcomes. Predictive algorithms may be used to identify chemotherapy patients who are in danger of dying or need immediate medical attention [81–93]. For instance, this prediction might be used to influence physician behaviour during chemotherapy, colon cancer surgery, or discharge planning [84, 85]. By dealing with these high-risk patients, resources may be saved. Organizations such as Predictive algorithms are used by Penn Medicine and New Century Health to identify cancer patients who are at high risk of being hospitalized or visiting the emergency room, and then use proactive phone calls or visits to target care management solutions [87, 89]. Although EHR data are often difficult to utilize, researchers have reported adopting it. To speed up the time-consuming process of obtaining data from EHRs, the Fast Healthcare Interoperability Resources format was created. Data from the Fast Healthcare Interoperability Resources format were used to create deep learning systems that can anticipate outcomes accurately a variety of medical outcomes, including in-hospital mortality, readmissions, and lengthened stays, as well as patients' diagnoses upon discharge from the hospital or clinic.

9.8.2.1 Radiomics

Oncology is beginning to use predictive analytics models, as seen by the developing area of radiomics. Radiomics, a kind of texture analysis, employs quantitative data from MRI images [90]. They may be beneficial for tumor characterization, monitoring, and early detection [91], CT [92], and MRI [93], for example, may be used to detect malignant lung nodules and prostate lesions utilizing computer-assisted detection [94]. It is quite interesting that AI-based algorithms for lung cancer is a kind of cancer that affects CTs, which may be able to predict important outcomes including mutation status and the probability of distant metastases [95, 96]. When a patient's tumor reacts to therapy, dynamic MRI may identify early responses and alert doctors. Before any known predictors of response can be used, the tumor's reaction must be assessed and utilized [97].

9.8.2.2 Pathology

Another field that will benefit from predictive analytics is pathology, which is an essential element of cancer treatment. Pathologists have differing views on the identification of non-small cell pulmonary malignancy and Gleason Score using bronchoscopic samples [98, 99]. Inaccurate biopsy findings may lead to treatment options that are either unneeded or unsuccessful, depending on the results. Artificial intelligence algorithms can, with a high degree of selectivity, identify pictures of sentinel lymph node biopsies (AUC, 0.99) equivalent to pathologists' findings [100]. Pathologists may be able to devote more time to other activities if they use these models, which improve scanning performance for big tissue sections. Tumor pathologic feature models are often used to forecast outcomes and predict therapeutic responses for specific conditions. The success of breast cancer chemotherapy may be measured using a variety of widely recognized criteria, including the recurrence score of 21 genes and the recurrence score of 70 genes.

9.8.3 Advanced used cases

9.8.3.1 Medical decision support

Once predictive analytics technologies achieve a particular level of effectiveness, oncology practitioners will be more inclined to employ them to improve routine parts of patient care. Prospective experiments have demonstrated that predictive algorithms reduce response times for patients with sepsis and speed up treatment for stroke victims [101, 102]. Analytics might be used to assist clinicians in making judgments regarding chemotherapy side effects, predicted response length, risk of recurrence, and total life expectancy at the time of treatment [103]. Many cancer patients' short-term mortality risk has been estimated using real-time EHR-based algorithms [104]. These algorithms may improve any electronic health record (EHR) since they employ both structured and unstructured EHR data. Even though the future uses of these algorithms are unknown at this time, oncologists may find the capacity to reliably forecast mortality at the site of treatment highly beneficial.

9.8.3.2 Classification of genetic risk

Because cancer patients are increasingly using germline testing and next-generation tumor sequencing, robust algorithms capable of assessing risk based on hundreds of genes discovered are necessary. Because next-generation sequencing is too costly to use as a random screening tool for a wide population, genomic testing may be focused on specific people. Targeted next-generation sequencing panels used to test machine learning algorithms can accurately distinguish between actual variants and artifacts. Because mutations of uncertain importance can generate major misunderstanding in terms of interpretation among clinicians and patients, this is a potentially valuable prognostic tool [105, 106]. Furthermore, genetic risk stratification may be used to determine if a patient may benefit from breast malignancy screening. According to one British research, screening women with a high hereditary risk of breast cancer using mammography rather than the existing age-based screening paradigm reduced overdiagnosis and boosted cost efficiency [107]. Genetics still has a long way to go before it can be effectively used in predicting the chance of having a disease.

9.9 Challenges Faced in Analytics in Cancer

9.9.1 Information gathering

The development of effective risk stratification models is critical to improve costs and outcomes, but there is a key drawback: shortage of high-quality information. Lack of patient data is a major problem for risk-based models, especially in cancer. Trips to the emergency department in claims-based datasets, hospitalizations are frequently not effectively documented and incorporated into easily available large datasets. Predicting mortality may be challenging owing to the difficulties of establishing a precise date of death, for example. Although patients spend the bulk of their time at home, little information on them is gathered there. New methods of collecting real-time data on cancer patients regularly may help detect patterns in patients at the outset of sickness, reducing needless hospitalizations. Real-world data sources are increasingly being used to get EHR data. Real-world data-based algorithms may be more responsive and beneficial than clinical trial-based algorithms, which often ignore key populations [108, 109]. Although real-world datasets like Flatiron Health and ASCO CancerLinQ may be useful for this purpose, their human curation and the range of user interfaces with the medical health record have significant limitations [110, 111].

9.9.2 Algorithm validation in the future

Improved statistical endpoint measurements, area under the receiving operator characteristic curve, or positive predictive value, for example, have competed for a significant part in current FDA certifications of predictive algorithms [112]. There are relatively few algorithms that have been carefully tested for their influence on significant clinical outcomes notably in cancer, such as overall survival or process metrics like time to diagnosis [113]. For future cancer prediction analytic devices to receive regulatory approval, the FDA's Digital Health Innovation Action Plan includes a precertification program that serves as a conduit for accelerated prospective examinations of possible analytic tools utilized in clinical settings [114]. Other criteria for evaluating and approving complex prediction algorithms in the future developed in recent years may be used to standardize predictive algorithms in cancer [115, 118].

9.9.3 Mitigation of bias and representation

If retrospective data are used to train predictive analytic algorithms, existing biases in clinical treatment may be exacerbated. Algorithms that rely on subjective clinical data or who have access to healthcare may unjustly target particular patient groups [119]. Consider the following scenario if you have a cancer-specific forecasting system based on genetic data from tumors: Patients from specific ethnic groupings may be underrepresented in the datasets used to train the algorithm. This might lead to inaccurate tumour genetic variation categorization in minority populations [120]. There may not be enough data from underrepresented groups to uncover predictive genetic variations, putting the generalizability of the prediction model at risk [121]. To safeguard under-represented groups from systematic bias in predicted outputs, all populations of interest in a training set must be represented, and audit processes must be introduced once the prediction tool is developed [122].

9.9.4 Predictive analytics is ready to take the next step in precision oncology

Similar to how breakthroughs in the genomic and cellular categorization of cancers have significantly advanced biological risk stratification, innovations in computational tools for clinical risk stratification of cancer sufferers promise to help oncology. Algorithm advancements that anticipate the risk of usage, costs, and clinical results are expected to play a bigger role in cancer therapy. With the combination of clinical, genomic, and cellular-based data, a new age of high-accuracy cancer risk assessment may emerge, enabling genuine precision oncology.

9.9.5 Machine learning – ML

ML, the division of AI, is the analysis of creating computer systems that can mimic human thinking. Machine learning algorithms (ML algorithms) are computer programs that learn from data rather than being explicitly designed to deliver a certain conclusion. They are "soft-coded," as opposed to hard-coded algorithms, in the sense that when they are used more and more, they learn from their errors and improve their performance. As part of the training process, trainers offer input data as well as expected outputs. The algorithm then learns from its training data and generalizes to new data, enabling it to produce the intended result when presented with new input. To predict the outcomes of new data, an ML model must first be trained applying training data. ML algorithms have been classified based on how labels are applied to data [123]. Supervised (unsupervised clustering and probability density function) methods (classification or regression estimates), and semi-supervised (text/image retrieval systems) are examples of these categories. Radiation oncology research is increasingly dependent on ML methodologies in the age of BD. Therapy response modeling, therapy planning, organ segmentation, imaging guiding, motion tracking, quality assurance, and other applications fall under this category. From diagnosis and evaluation through therapy delivery and follow-up, this chapter delivers an outline of the most recent advances and tries to cut uses of ML procedures in radiology. This analysis emphasized the application of machine learning to enhance effectiveness, i.e., streamlining and automating clinical procedures, as well as quality, i.e., potentials for decision-making support in radiation therapy. The following is how this section is structured: The first section covers the basics of radiation oncology, big data, and machine learning, followed by a process overview of how ML approaches are applied in radioactivity oncology research. Radiation oncology and machine learning procedures have been used in radiation oncology research to cover almost all parts of the radiotherapy process. Artificial intelligence's medicinal and biological applications of streaming data may be a life-or-death scenario, and approaches may be able to compensate for human shortcomings in processing massive amounts of data. Allowing for the real implementation of precision medicine in radiation oncology would improve patient treatment quality as well. This section will walk you through the radiation oncology process and look at some recent research that incorporated models of machine learning. Therapy simulation, therapy planning, quality assurance, and therapy delivery, as well as therapy outcome and follow-up are all part of the radiation oncology process [124].

9.9.6 Diagnosis, assessment, and consultation of patients

The initial radioactivity oncology consultation: discuss the patient's medical status with the radiation oncologist and design a treatment strategy [125]. The first step in the evaluation and consultation procedure is the detection of cancer in sufferers utilizing medical imaging and subsequent pathological confirmation. To find and categorize cancer subtypes, Computer-aided detection/diagnosis toolkits, for example, have been developed using machine learning created (staging). If a patient's x-ray shows a worrisome location, the radiologist may utilize this tool to assist and assess if a biopsy is required. This method may also be used to categorize lesions as benign or malignant, aberrant or benign, and so on. Machine learning is essential, it may provide a "second opinion" to the physician in diagnostic radiography decision-making through computer-aided detection and diagnosis toolkits, and it may provide a "second opinion" to the physician through computer-aided detection and diagnosis toolkits process in the detection of a crime using computer technology. The term "computer-aided detection" (CADe) is used when a physician or radiologist uses the computer output as a "second opinion." CADe is a popular issue in medical imaging research. To solve a problem, the ML classifier discovers the "best." In a multidimensional feature space, boundaries are used to separate classes. It may provide an estimation of the chance of identifying lesions in medical imaging by focusing on detection. Researchers [126] have built ML-based models based on deep learning on MRI data for the identification of lung nodules in CT images, microcalcifications in mammography [127], prostate cancer [126], and brain lesions [127]. These include lung nodules in CT scans and mammographic microcalcifications. Chan et al.'s automated identification of clustered breast microcalcifications on mammograms has a 0.90 AUC (area under the receiver operating characteristic curve). Suzuki et al. have shown that detecting lung nodules in low-dose CT images is more accurate. According to Zhu et al, the detection rate of prostate cancer on MR images was 90.90%, indicating that high-level features acquired by deep learning may beat handmade features in identifying prostate cancer sites. Rezaei et al. discovered that deep learning technology was more successful than earlier methods in identifying brain lesions. Computer-aided detection systems may greatly improve diagnostic performance as a "second opinion" tool for detecting lesion spots in pictures. Detection sensitivity and specificity (accuracy) would improve, and inter- and intraobserver variability, as well as the danger of cancer being overlooked, would be minimized. In many circumstances, computers may help with diagnostics. CADx, or computer-aided diagnosis, gives a "second objective perspective" in the examination and evaluation of medical images by employing a computerized approach. Its goal is the same as CADe's to solve a categorization issue. Diagnostic (characterization) issues, such as automatically recognizing whether a tumor or lesion is malignant or benign, are central to CADx, which tries to offer a probability of diagnosis. Many research works [128–131] have used CADx techniques to diagnose lung and breast illnesses. Deep learning was evaluated in the identification of breast nodules in US pictures and pulmonary nodules in CT scans, according to Cheng et al. Deep learning-based CADx, according to their findings, outperforms comparable approaches across a broad variety of modalities and illnesses. Deep learning-based CADx identifies multiple instances of breast tumors and pulmonary nodules. Feng et al. and Beig et al. utilized logistic regressions to detect lung lesions on endo-bronchoscopic images, while SVM and a neural network were used to distinguish NSCLC adenocarcinomas from granulomas in non-contrast CT. Adenocarcinoma and squamous cell carcinoma were shown to be correctly differentiated by 86 percent of the time. According to the findings, CADx systems beat radiologist readers on non-contrast CT imaging, identifying adenocarcinomas from granulomas in non-small cell lung cancer. Joo et al. created the CADx technique, which uses an ANN to identify breast nodule malignancy on US images. They were able to show that breast lesions might be more accurately diagnosed using ultrasonography. Radiologists can improve their performance by using "second opinion" methods that use computer-aided diagnostic tools to reduce the number of malignant cases that are misinterpreted as cancerous, lowering the number of cases recommended for the surgical biopsy that are incorrectly interpreted as cancerous. Non-invasive (no biopsy), quick (rapid scanning), and low cost (CADx) are just a few of the benefits of CADx (no additional examination cost). Evaluation and recommendations: The radiation oncologist and the patient meet during the patient evaluation phase to discuss the patient's clinical state. When designing a treatment plan, the risks and advantages of the therapy, as well as the patient's treatment objectives, are taken into account. Tumor information, for example, might be collected to assist in estimating the potential value of therapy. Deep learning-based computer-aided diagnostics may be used to detect lung nodules and breast lesions. It demonstrates how difficult it may be to distinguish a non-medical person from a young doctor in practice (reproduced from). Streaming data may be a life-or-death issue in the medical and biological applications of artificial intelligence, and 44 approaches might compensate for human shortcomings in appropriately processing a massive volume of data. Allowing for the real implementation of precision medicine in radiation oncology would improve patient treatment quality as well. This section will walk you through the radiation oncology process (Figure 9.1) and look at some recent research works that incorporated machine learning models. The radiation oncology process includes treatment simulation, treatment planning, quality assurance, and delivery, and treatment outcome and follow-up are all examples of treatment simulation.

9.9.6.1 In the part, diagnose, assess, and consult with patients

The initial consultation in radiation oncology: Discuss the patient's medical status with the radiation oncologist and design a treatment strategy [132]. The first step in the evaluation and consultation process is the detection of cancer in a patient utilizing medical imaging and subsequent pathological confirmation. To find and categorize cancer subtypes, Computer-aided detection/diagnosis toolkits, for example, have been developed using machine learning created (staging). If a patient's x-ray shows a worrisome location, the radiologist may utilize this tool to assist and assess if a biopsy is required. This method may also be used to categorize lesions as benign or malignant, aberrant or benign, and so on. Machine learning is essential it may provide a "second opinion" in computer-aided detection and diagnostic toolkits to the physician in the diagnostic radiography decision-making process.

9.9.6.2 Detection of a crime using computer technology

The term "computer-aided detection" (CADe) is used when a physician or radiologist uses a "second opinion" provided by the computer's output. CADe is a popular issue in medical imaging research. To solve a problem, the ML classifier discovers the "best." In a multidimensional feature space, borders are used to separate classes. It may provide an estimate of the chance of identifying lesions in medical imaging by focusing on detection. Researchers have built ML-based models based on deep learning on MRI data for the identification of lung nodules in CT images, microcalcifications in mammography, prostate cancer, and brain lesions. These include lung nodules in CT scans and mammographic microcalcifications [133]. Chan et al.'s automated identification of the area under the receiver operating characteristic curve (AUC) for clustered breast microcalcifications on mammograms is 0.90. Low-dose CT scans have been demonstrated by Suzuki et al. [134] to be effective in identifying lung nodules. It is more accurate. According to Zhu et al., the detection rate of prostate cancer on MR imaging was 73%. 90.90% indicated that high-level features acquired by deep learning may beat handmade features in identifying prostate cancer sites. Rezaei et al. discovered that deep learning technology was more successful than earlier methods in identifying brain lesions. Computer-aided detection systems may greatly improve diagnostic performance as a "second opinion" tool for detecting lesion spots in pictures. Detection sensitivity and specificity (accuracy) would improve, and inter- and intraobserver variability, as well as the danger of cancer being overlooked, would be minimized.

9.9.6.3 Making use of a computer to help in diagnosing

CADx, or computer-aided diagnosis, gives a "second objective perspective" in the examination and evaluation of medical images by employing a computerized approach. Its goal is the same as CADe's to solve a categorization issue. Diagnostic (characterization) issues, such as automatically recognizing whether a tumor or lesion is malignant or benign, are central to CADx, which tries to offer a probability of diagnosis. Many researchers have used CADx techniques to diagnose lung and breast illnesses. Deep learning was evaluated in breast tumors and pulmonary nodules on CT scans and ultrasound pictures can both be identified. According to Cheng et al., deep learning-based CADx, according to their findings, outperforms comparable approaches across a broad variety of modalities and illnesses. Deep learning-based CADx [135–136] identifies multiple instances of breast tumors and pulmonary nodules. Feng et al. and Beig et al. [137] utilized logistic regressions to detect endobronchoscopic pictures of lung lesions, while SVM and a neural network were used to distinguish non-contrast CT scans of NSCLC adenocarcinomas from granulomas (NN). Adenocarcinoma and squamous cell carcinoma were shown to be correctly differentiated by 86% of the time. According to the findings, CADx systems beat radiologist readers in non-contrast CT scans, distinguishing adenomas from granulation tissue in non-small cell lung cancer. Joo et al. created the CADx technique, which uses an ANN to identify breast nodule malignancy on US images. They were able to show that breast lesions might be more accurately diagnosed using ultrasonography. Radiologists can improve their performance by using "second opinion" methods that use computer-aided diagnostic tools to reduce the number of malignant cases that are misinterpreted as cancerous, lowering the number of cases recommended for the surgical biopsy that are incorrectly interpreted as cancerous. Non-invasive (no biopsy), quick (rapid scanning), and low cost (CADx) are just a few of the benefits of CADx.

9.10 Evaluation and Recommendations

The radiation oncologist and the patient meet during the patient evaluation phase to discuss the patient's clinical state. When designing a treatment plan, the risks and advantages of the therapy, as well as the patient's treatment objectives, are taken into account [138]. Tumor stage, previous and current therapies, post-resection margin status, tolerance to multimodality therapy, and overall performance status are all utilized to assess the potential benefit of treatment [139]. The age of the patient, comorbidities, functional status, tumor closeness to important natural tissues, as well as the ability to collaborate with movement control are all factors that impact treatment risk and tolerance [140]. They are all significant features in treatment outcome and toxicity prediction models are being built. These models might help physicians and patients manage expectations and make risk–benefit trade-off decisions. These models include decision trees, random forests, gradient boosting, and support vector machines are all examples of logistic regressions. Decision trees and logistic regressions are equally successful in assisting doctors and patients in making the optimal decision but they lose the balance between accurate predictions and data interpretability. When accuracy is more essential than interpretability, random forests and gradient boosting, as well as kernel SVMs, are preferred. Radiation oncology continues to be a difficulty in terms of developing standards for data collecting and developing models that may be useful in these settings.

After the doctor and the patient have agreed on radiation treatment, the doctor will deliver therapy simulation Simulator instructions. Simulating a process requires knowledge about the patient's immobilization, scan range, treatment site, and other variables. The simulated preparation for the patient might involve anything from fasting instructions to bladder/rectal filling to kidney function tests for intravenous (IV) contrast. To guarantee the safety of all of our patients, we request elevator assistance or a translator as needed [141]. The patient is set up and immobilized for treatment simulation, image data from 3DCT and/or 4DCT are collected, and image reconstruction/segmentation is carried out. Algorithms for machine learning may play a vital role in improving treatment outcomes and simulation quality.

9.11 Obtaining 3D/4D CT Images

During the simulation, three-dimensional CT anatomical picture data are obtained for treatment planning purposes utilizing a specialized CT scanner (CT simulator). A good CT simulation is essential for a patient's plan to be exact, high quality, resilient, and deliverable. A repeat CT simulation may not be achievable due to the scan's limited range, poor immobilization, artifacts, inadequate bladder/rectal filling, and breath-hold reproducibility [142]. Radiation departments are increasingly using 4DCT scans to follow patients' tumor migration in connection to the patient's breathing cycle. This gadget monitors the patient's breathing cycle, enabling it to acquire CT pictures at specific points in the cycle or during the whole breathing cycle. The ITV includes the mobility of the CTV (clinical target volume) or MIP (maximum intensity projection) images (internal target volume) generated from CT data [143]. 4DCT imaging is required for effective SBRT deployment in early-stage NSCLC stereotactic ablative radiation treatment (SBRT).

Only a few studies [144–147] have used ML-based algorithms to achieve this aim. Fayad et al. [148], for example, used without the need for patient-specific 4DCT data, an ML technique based on PCA to construct a global respiratory motion model that could relate outside patient surface motion to inner structural movement [149] formalized paraphrase. Its findings are promising, but additional study is required to establish the concept's validity. Steiner et al. [150] explored a machine learning-based model that used correlations and linear regressions to see if 4DCT (cone-beam CT) or 4D CBCT (cone-beam CT) matched the genuine motion range after therapy by comparing it to a 4DCT model using Calypso motion data as the "ground truth." As a consequence of these findings, the intra-fraction lung target mobility is overestimated by 4DCT and 4DCBCT. Dick et al. [151] employed an ANN model to track irradiation of the liver without using fiducially tumors to monitor the lung–diaphragm boundary. The findings revealed a link between the diaphragm and tracking volume, and the technology might eventually replace fiducially indicators in clinical practice. Johansson et al. evaluated the use of a machine learning-based PCA model to recreate breathing-compensated pictures depicting gastrointestinal motility phases. According to the findings of the research, 4D GI MRIs may aid in the definition of treatment planning and GI mobility monitoring during radiation, internal target volumes are used. Overall, the ML-based simulation was successful and performed well, methodologies reported here have shown the capacity to increase CT simulation accuracy for patient CTs. Few individuals are interested in using machine learning to replicate the acquisition of 3D/4D CT images. Numerous issues arise when examining the simulation from this viewpoint and ML algorithms are being used to decision-making and workflow efficiency will be aided.

9.11.1 Making an image from the ground up

Our radiation oncology research focuses on the use of machine learning-based image reconstruction algorithms. We demonstrate how CT can be approximated from MRI pictures using machine learning, as well as how a 3 Tesla (3 T) MR image can be reconstructed into a 7 Tesla (7 T)-like MR image. An MR scan can be used to build a CT scan as an example of first-time use. A technique for extracting or reconstructing synthetic CT images from MR pictures is required for the practical application of the MRI-only treatment planning radiation strategy. Dosage calculations in radiation oncology are increasingly supported by computed tomography (CT) scans. However, there are some disadvantages to CT imaging, such as a) CT imaging exposes patients to radiation, which can have negative consequences, whereas MRI is far safer and does not use radiation; and b) CT imaging exposes patients to radiation, which can cause negative effects, whereas MRI is considerably safer and does not use radiation. CT scans were mapped to MR images using deep learning (full CNN) models, boosting-based sampling (Booster), random forest and auto-context models, and the U-net CNN model [152, 153] have all been utilized. Nie et al. tested whether the deep learning approach is helpful in MRI scans, which are used to predict CT pictures. The "ground truth" MRI as well as the deep learning-derived synthetic CT image are created from MRI data. In contrast, the deep learning model we constructed outperformed the competition. According to Bayisa et al. [154], boost algorithm-based approaches outperform current model-based methods in terms of CT estimate quality in the tissues of the brain and bones. A model based on a structured random forest with auto-context has been demonstrated to beat two state-of-the-art algorithms in predicting CT images in a variety of situations [155]. Experimental outcomes: Chen et al. [156] investigated how MRI data may be utilized to create a synthetic CT using a deep CNN. A further examination of their findings utilizing "ground truth" CT pictures produced results that were more than 98% accurate. The dose–volume histogram (DVH) parameters are as follows: planned intensity-modulated radiotherapy (IMRT) for prostate cancer showed a dosimetric accuracy discrepancy of a maximum point dose disparity of less than 0.87% of less than 1.01% within the PTV. Radiation therapy planning and image guiding are now based only on MR scans but emerging technologies for making synthetic CT scans seem to provide an alternative. The second application allows to reconstruct a high-quality picture modality derived from lower-quality image modality photos. A 7 T-like picture, for example, may be generated by combining three T images. The images produced by new ultra-high 7 T MRI scanners are crisper and more contrasted than those produced by standard 3 T scanners. However, owing to the unusually high magnetic field production, 7 T MRI scanners are becoming more costly and less accessible in clinical settings, necessitating stricter safety regulations. These limitations might be overcome by using machine learning-based techniques to generate/reconstruct a 7 T image from a 3 T MR scan, allowing for early disease detection. Researchers [157, 158] employed machine learning to convert a 3 T MR image into a 7 T-like image. Deep learning CNN models, a hierarchical reconstruction based on group sparsity in a unique multi-level canonical correlation analysis (CCA) space, and random forest and sparse representation were used to convert 3 T MR images to deep learning CNNs were utilized to analyze 7 T-like MR images [59–161]. The visual and numerical findings of Bahrami et al. demonstrated that the comparative approaches were exceeded by the deep learning technology. According to the same author's research, a hierarchical reconstruction based on sparseness outperformed earlier approaches for brain structure segmentation as compared to 3 T MRI image processing. When compared to other techniques, Bahrami et al. [162, 163] found that the predicted 7 T-like MR pictures most closely resemble the "ground truth" 7 T MR images using a group sparse representation and a random forest regression model. In addition, the segmentation of 7 T-like MR images predicted was more accurate than the segmentation of 3 T MR images in the experiment on brain tissue segmentation. In general, the anticipated 7 T-like MR pictures outperformed the 3T images in terms of spatial resolution. The segmentation of critical regions such as 7 T-like MR pictures of brain tissue structures was more accurate than that of 3 T MR images. It is also feasible that high-quality 7 T pictures might help with the detection and treatment of a range of diseases.

9.11.2 Image fusion/registration

The technique of matching pictures in radiotherapy image registration makes it simpler to spot specific changes in the images, such as tumor development. With such an alignment, organ deformation, patient weight loss, or tumor shrinking are not taken into account. DIR, a technique for establishing how the points in one picture map to their counterparts in another, may be used to account for such changes. The introduction of DIR at various times during the radiation process may help the procedure. By considering organ deformation, radiation design, delivery, and assessment may be improved. Intra- and inter-fractionated image registration may be identified during image-guided radiotherapy (IGRT). A single patient's images are matched using intra-fractional and inter-fractional registration to optimize positioning the patient and analyzing organ movements in relation to the bones (Organ motion monitoring through the internet). Inter-patient registration, on the other hand, entails comparing photographs from several patients (i.e., an "average" of images taken from many patients, enabling information from the atlas to be transmitted to the newly acquired image). Once two pictures have been registered, the data from both images are combined using a method known as data fusion. The transmission of one use of data transfer between pictures is to transfer contours from a planning image or an atlas to a newly acquired image [164, 165]. Despite the availability of several image registration techniques, the DIR of complex scenes still faces significant challenges, such as large anatomical variances and dynamic appearance changes. These three photos of the same patient (left, center, and right), obtained at separate periods and using different reconstruction techniques, show different zoom levels of the same patient's 3 T MR scan, as well as 7 T MR "ground truth" scans. In terms of anatomical features and tissue contrast, the 7 T MR scan greatly beats the 3 T MR imaging (reproduced from [166]). Artificial intelligence's medical and biological applications of 50 Advances in deep learning and computer vision may be able to assist in overcoming the issues that traditional rigid/deformable image registrations provide. Many image registration algorithms based on machine learning have been created to align anatomical components as well as to enhance the aesthetic difference. A regression forest approach may be used to register two distinct MR images, according to Hu et al. [167]. The learning-based registration strategy proved to be more accurate than the previous registration techniques. Many computer vision applications require comparing picture patches, and Zagoruyko et al. [168] created a universal similarity function to achieve it precisely. The CNN-based model outperformed other cutting-edge approaches by a wide margin. To accomplish rapid and reliable multimodal image registration, a discriminative local derivative pattern was employed by Jiang et al. [169]. The findings suggested that the proposed method might increase multimodal image registration accuracy, while also demonstrating the feasibility of a clinically guided intervention using ultrasound [170]. The relationship between the predicted inaccuracy of a neural network (NN) and "ground reality" was consistently more than 0.90 in their findings for the PTV and the OARs. To increase the robustness of image registrations, Wu et al. [171, 176] developed a deep learning-based image registration system and a NN-based registration quality assessor. The researchers demonstrated that a 2D/3D stiff image registration technique has the potential to improve overall robustness using quality evaluation methodologies and a unique image registration framework [177]. Kearney et al. [178] suggested for CT scanning to CT-deformed image registration, we used a deep unsupervised learning strategy. On all assessment criteria, deep learning outperformed rigid registration, intensity-corrected demons, and landmark-guided deformable image registration. Machine learning-based multimodal picture registration that outperforms conventional methods algorithms described here has exhibited enhanced accuracy. As a consequence, radiation oncologists now have therapeutically feasible alternatives for improving rigid/deformable image registration.

9.11.3 Image segmentation and reshaping software that works automatically

Volume characterization is required for accurate dose reporting and relevant 3D treatment planning [179]. To help in treatment planning and comparing treatment outcomes, the International Commission on Radiation Units and Measurements (ICRU) has defined and specified target volumes (such as intended target volumes) and critical structures/normal tissue/organs at risk. To precisely estimate the dosage received by an organ, its borders must be delineated to identify the organ that is at risk owing to its radiation sensitivity [180]. On CT slices taken during the patient's treatment simulation, image fusion may aid in the detection of tumor and OAR features. Multimodal diagnostic imaging, such as CT, MRI, US, PET/CT, and others, may be used to aid in the identification of tumor and CT slices taken during the patient's treatment simulation showing OAR formations. The process of delineation (auto-contouring) has been substituted in clinical practice by commercially available automated or semi-automatic analytical model-based software (e.g., atlas-based models). These software methods, although effective for recognizing vital organs and OARs, are not yet adequate for mapping tumors and target structures, which is a difficult task. Both procedures might benefit from the deployment of cutting-edge machine learning techniques.

Brain, prostate, rectum, and sclerosis lesion are among the several ML-based techniques for tumor/target segmentation/auto-contouring that have been developed. Deep learning [181, 182] and ensemble learning were used in brain tumor segmentation competitions. ML-based techniques were shown to be the most successful when compared to other ML-based tactics. Osman made glioma brain tumor segmentation simpler by utilizing SVM, discovering that it functioned well on both training and fresh "unseen" testing data, despite the fact that its accuracy on multi-institution datasets was believed to be excellent. The BRATS' 2017 dataset was utilized to segment the whole glioma brain tumor using an SVM model. For segmentation of organs such as these, our technique outperformed other cutting-edge segmentation methodologies and commercially available software, such as the rectum [183], parotid [184], and others. Machine learning-based tumor/target segmentation/auto-contouring algorithms remain difficult to implement owing to a scarcity of big multimodal image datasets including "For training these models, we need "ground truth" annotation data." Deep learning [61], an innovation in computer vision that is especially well suited for segmentation, has been shown to outperform earlier machine learning approaches in tumor and organ segmentation tasks. The MRI segmentation of entire glioma brain tumours is shown in this picture (Dataset from BRATS 2017 [185, 186]. T2-FLAIR MRI, SVM model glioma segmentation [187], and manual "ground truth" glioma segmentation by a board-certified radiation oncologist were used to segment four independent people. Artificial intelligence's medical and biological applications of 50 Advances in computer vision and deep learning may be able to assist in overcoming the issues that traditional rigid/deformable image registrations provide. Many image registration algorithms based on machine learning have been developed to align anatomical components as well as to enhance the aesthetic difference. A regression forest approach may be used to register two distinct MR images, according to Hu et al. [188]. The learning-based registration strategy proved to be more accurate than the previous registration techniques. Many computer vision applications require comparing picture patches, and Zagoruyko et al. [189] created a universal similarity function to achieve it precisely. The CNN-based model outperformed other cutting-edge approaches by a wide margin. To accomplish rapid and reliable multimodal image registration, a discriminative local derivative pattern was employed by Jiang et al. [190]. The findings suggested that the proposed method might increase multimodal image registration accuracy, while also demonstrating the viability of intervention based on clinical data from the United States. The approach that was recommended. For autonomous DIR performance monitoring, Neylon et al. [191] developed a deep neural network. The correlation between the neural network (NN) projected error and "ground reality" was consistently more than 0.90 in their findings for the PTV and the OARs. Wu et al. created a deep learning-based image registration quality assessor and a QA-based registration quality assessor system to improve the reliability of image registrations [192–195]. The researchers have shown potential in a Stiff picture registration technique in 2D/3D for improved overall robustness using quality evaluation methodologies and a unique image registration framework [196]. Kearney et al. suggested a deep unsupervised learning method for CBCT to CT deformable image registration technique. In all assessment measures, deep learning outperformed deformable image registration using landmarks, stiff registration, and intensity-corrected devils. In comparison to the existing approaches, the machine learning-based multimodal image registration algorithms described here have exhibited enhanced accuracy. As a consequence, radiation oncologists now have therapeutically feasible alternatives for improving rigid/deformable image registration [197].

9.12 Treatment Preparation

As previously described in the section on image segmentation, the planning technique begins with a sketch of the target(s) and the OARs. The first phase in planning treatment is to outline/contour the OARs and target volumes; the second step is to define dosimetric objectives for the targets and normal tissues; the third step is to choose an appropriate treatment strategy, and the fourth step is to assess (estimate the treatment dose distribution with the prescribed dose levels in it).

9.12.1 Making use of data to influence treatment planning

Past treatment plans and patient status may be utilized to warn the treating team of an ongoing case as long as a medical dosimeter or a physician is competent. The premise behind knowledge-based treatment planning is that prior treatment plans may be utilized as a reference. Hundreds of earlier therapy regimens employing KBTP techniques like this have benefited several disease locations. The KBTP strategy is motivated by the need to reduce the existing complexity and time required in developing a fresh treatment plan for each new patient, as well as the possibility of using it to help with radiation decision-making. The KBTP technique, which has been the subject of several investigations [198, 199], has been used to develop radiotherapy treatment regimens. The commercial techniques and academic studies presently available for KBTP are limited to predict DVHs outside the authorized limits. In contrast to human preparations, KBTP and AI-assisted plans meet or exceed the prescribed dosage limits in numerous clinical circumstances. Prostate cancer [200], cervical cancer [201], gliomas and meningiomas [202], head and neck cancer [203], and spine SBRT are a few examples [204]. Another more current commercialized tool, Quick Match (Siris Medical, Redwood City, CA, USA), explores forecasts in patient dose trade-offs using gradient enhancing (the strongest effective method on anticipation when standardized information is provided) [205]. Before you begin the treatment planning process, you may quickly and conveniently receive preliminary treatment planning results with this tool. Doctors and patients may use this tool to better decide on the direction of a treatment plan ahead of time, allowing for better coordination of information and goals between the two parties engaged in the treatment planning process before the actual therapy starts. It may be useful, for example, while deciding on the best strategy to take (For example, photons vs. protons). This method has been utilized to verify that DVH data are of high quality after it has been prepared [206, 207]. Because it streamlines the procedure, using an approach like KBTP makes it simpler to construct treatment programs for new patients. According to some experts, a KBTP-based standardization method may aid in increasing the uniformity, efficiency, as well as quality of the strategy. Data-driven planning is still not entirely automated since it requires professional review or involvement to ensure that treatment plans are effective and are properly executed.

9.12.2 Automated planning procedure for self-driving vehicles

If the diametric objectives and method are provided, an automated plan building technique is also available. Attempts have been made to predict optimal beam orientations to address different parts of this challenge [208, 209]. The more difficult job of automated treatment planning, on the other hand, fits itself well with reinforcement learning. Reinforcement is often used in popular culture applications like gaming and self-driving automobiles. Reinforcement learning may be used to teach a computer program. How to traverse a collection of limitations given a set of rules? If the treatment planning system (the simulator) is correct, the program will make a decision (such as raising the weight of a constraint) and learn from the simulator. Google Brain used this strategy to create an algorithm capable of beating a world champion in the game of Go [210]. Our finest dosimetrists' performance can be equaled by a reinforcement strategy, but only if it is skilfully applied. In general, the necessity for strong treatment planning systems (TPSs) is a result of a close integration barrier to applying reinforcement learning to achieve complete automated planning. The long-term goal is to have an automated planning process with human specialists (dosimetrists, physicists, and doctors) analyzing, overseeing, and providing quality assurance for the output.

9.13 Treatment Administration and Quality Control

Quality control is necessary to ensure the delivery of radiation (QA). When it comes to analyzing and interpreting the findings of medical testing, medical physicists play an important role in clinical practice. Machine learning has the potential to solve a broad variety of long-standing issues, while also improving workflow efficiency. This section, for example, discusses the detection and prediction of radiation therapy errors, as well as treatment planning quality assurance and the validation of mechanisms for delivering treatment (e.g., prediction of planning deviations from initial intentions, and prediction of the need for adaptive radiotherapy re-planning).

9.13.1 Quality control and conformance assurance

Machine learning can enhance numerous parts of the radiation quality assurance program, including the detection and avoidance of mistakes, the quality assurance of treatment devices, and the quality assurance of individual patients. Thanks to machine learning, there are various advantages to automating the quality assurance (QA) process and analysis, including increased efficiency and reduced physical labor required to complete the QA process (ML). As a consequence of these investigations [211, 215], computerized methods for quality assurance testing have been established. Machine-based QA and patient-based QA are two separate forms of QA. Medical linear accelerator (Linac) machine-automated QA processes employing ML have been examined by researchers for machine-based QA methodologies. Li et al. studied the use of artificial neural networks (ANNs) to monitor the Linac to enhance patient safety and treatment quality. Dosimetry and quality assurance professionals discovered that it was more accurate and adaptable than other approaches in the early phases of testing, and in certain situations, it even outperformed established clinical tests in terms of detection rates. El Naqa et al. conducted a study on abnormal occurrences in radiation and proposed an automated approaches to increase safety, efficacy, and efficiency, to discover task-driven quality assurance (QA) procedures are essential. To deal with the issue of direct modeling of QA defects and unusual events, an anomaly detection approach was developed [216]. Ford et al. both claim that evaluated how effectively QC processes and preventive maintenance in radiation oncology perform in identifying mistakes. The results demonstrated that the performance of radiation oncology quality control systems is strongly dependent on the checks utilized and in what combinations, as well as a reduction in treatment cancellations are caused by equipment downtime and other technical issues [217]. According to the AAPM Task Group, the ability of these machine learning algorithms to automatically detect outliers allows physicists to focus their attention on the aspects of a process that are most likely to affect patient care [218]. Several research works revealed the use of ML algorithms, multi-leaf collimators (MLCs), and imaging for patient-based QA [219–223]. Valdes et al. studied the use of an SVM-based system to detect flaws in the Linac automatically two-dimensional and three-dimensional imaging systems [2D/3D] to ensure the validity of patient IGRT therapy. There were no significant variations in the amount of labor necessary to execute the bare minimum and best practices of QA programmes after adopting the suggested technique. Poisson regression with LASSO regularization was used by researchers to anticipate customized IMRT QA passing the rates for plan QA and individual patient QA. Virtual IMRT QA was demonstrated to be capable of correctly predicting passing rates as well as failures caused by setup mistakes. Osman et al. used MLC positioning errors which are estimated using NN and a cubist method from Linac-produced data from IMRT and VMAT-supplied plans in a log file. Using information from their investigation, the researchers revealed if the anticipated values were closer to the actual values than expected. TPS dosage estimations are more precise when projected changes in leaf position are taken into account. Despite the tremendous gains in QA procedures brought about by the use of ML, the increased QA needs for the algorithms come at a hidden cost. As a result, a rising number of tests will be necessary to check the efficiency of all deployed machine learning algorithms on a regular basis [224]. Allowing for intelligent resource allocation in favor of less likely-to-fail strategies, virtual QA may have a significant impact on the existing IMRT/VMAT process.

9.14 In This Part, We Will Go Through How to Give the Therapy

Tumor decrease and morphological patient changes (such as weight loss) are possible in the first few weeks following fractionated radiation treatment. Adaptive radiation therapy is defined as radiation therapy that employs regular imaging to account for changes in the anatomical structure after treatment (ART). Photographs are shot on a regular or virtually daily basis. When significant changes become obvious, re-planning becomes a possibility. Images may be employed off-line (i.e., to improve the next fraction) or live (i.e., by using picture data obtained during a fraction of a second to improve the next fraction) to conduct picture-guided adaptation (i.e., modifying a fraction's treatment plan depending on data from another fraction). The reorganization procedure is divided into three stages: To assess the efficacy of a treatment plan, radiation oncologists utilize daily. The tumor's and OARs' dosage metrics to calculate how much radiation was administered, as well as how much was delivered to the tumor and OARs in the prescribed treatment fraction, using the daily CBCT picture collection. It takes a large amount of time and money to integrate adaptive radiotherapy into routine clinical practice. It must be quick to fit into the clinical procedure (time is of the essence). Using machine learning approaches such as deep learning, very complex adaptive therapeutic software solutions may be created. Deep learning has a wide range of applications, video games, computer vision, and pattern recognition are a few examples [225].

Many studies have looked at the use of machine learning, especially deep learning, to adaptive radiation therapy re-planning. Guidi and colleagues used an SVM and PCA model to identify people who might benefit from antiretroviral therapy (ART) and re-planning intervention. There is a link between the best time to re-plan an intervention and the patients who will benefit from antiretroviral medication (ART). Tseng et al. used deep reinforcement learning to minimize radiation pneumonitis, while increasing tumor local control. Deep reinforcement learning-based automated dosage adjustment was shown to be a feasible and promising technique for reaching outcomes comparable to those obtained by physicians. Patients getting radiation therapy and suffering severe anatomical changes may be recognized using a novel approach developed by Varfalvy et al. who employ hidden Markov models and relative gamma analysis. According to the research, there is evidence that it might be used to improve clinical data gathered throughout therapy and assist in identifying individuals who need plan adjustments. Adaptive radiotherapy requires quick planning and high-quality pictures. Deep learning-based machine learning algorithms have made it feasible to rapidly transform adaptive radiation treatment into an efficient and effective therapeutic practice. When used properly, adaptive radiation therapy has the potential to increase radiotherapy treatment accuracy even more than [226].

9.14.1 Patients are given two and a half follow-ups

It is critical to monitor a patient's development during treatment. If treatment outcomes could be accurately predicted, doctors would be better positioned to make educated decisions regarding expected benefits versus anticipated dangers. Radiation oncologists may be able to employ machine learning to revolutionize the way they monitor patients who have had final radiation treatment. Another possible advantage of applying radiomics "tumor/healthy tissue phenotype" analysis in radiation oncology is the capacity to predict treatment results for specific patients, the therapy's result.

Treatment outcomes are influenced by several variables, including the therapeutic, structural, and patient-related aspects. Tumor control probability (TCP) and normal tissue control probability (NTCP) are very important in radiation oncology research, and their ability to be predicted during treatment planning or fractionated radiation therapy is critical. Recent years have seen an increase in the application of sophisticated bioinformatics and machine learning techniques that incorporate dosage volume parameters with additional patient- or disease-based prognoses to improve outcome prediction [227]. Treatment intensification with extra radiation, the inclusion of systemic therapy sooner rather than later, or the introduction of a novel strategy to treatment are all possibilities that might benefit from improved models built on early assessment. Using machine learning, researchers examined how to predict how people will respond to radiation treatments. Using a Bayesian network ensemble, Lee et al. looked at the likelihood of radiation pneumonitis in NSCLC patients undergoing 3D conformal radiotherapy after using the technique. Using an ensemble technique to predict radiation, pneumonitis may be more accurate than using a single framework.

Poisson-based TCP and the cell death equivalent uniform dosage model were used by Naqa et al. to estimate model parameters for predicting TCP using data mining techniques such as statistical resampling and logistic, SVM, and logistic regression. Results demonstrate that they were able to uncover significant non-linear intricate interactions among model variables using data mining methodologies. They may then forecast unseen data for future clinical applications. The CNN model was created by Zhen et al., who studied and predicted the distribution and toxicity of a rectal dose. An accurate dosimetry prediction model for cervical cancer radiation may be built using transfer learning based on the evaluation results. A study by Deist and colleagues assessed the average discriminative performance of six machine learning ML classifiers for predicting the outcome of radiotherapy treatment. LogitBoost and elastic net logistic regression are two examples of decision trees that may be used in data mining. In terms of chemoradiation outcomes and toxicity, random forest and elastic net logistic regression were the best classifiers, according to the data. A range of statistical-learning algorithms was examined by Yahya et al. to predict urinary symptoms after prostate external beam radiation. Prediction of urinary symptoms was shown to be the most accurate using logistic regression and multivariate adaptive regression splines. Based on dose–volume restrictions, Zhang and his colleagues employed SVMs to predict potential organ-at-risk concerns. Using machine learning, they revealed that it is possible to predict OAR difficulties during treatment planning, enabling the assessment of alternative dose–volume constraint settings inside the IMRT planning framework. Radiation therapy outcome prediction from the doctor's point of view was the subject of Kang et al. A variety of machine learning techniques, including logistic regression, support vector machines, and artificial neural networks, were examined. According to the publication, although the present research is exploratory, the overall methodology has matured, and the region is suitable for larger-scale future investigation. Among the many advantages of using contemporary clinical informatics systems is the ability to learn more about the safety and effectiveness of pharmaceuticals. Radiotherapy treatment regimens and the systems that provide it are now separated by a large period of time as a result of the rapid adoption of new technologies. The use of cutting-edge machine learning algorithms in radiation oncology is necessary in order to better anticipate treatment response and outcomes. The best way to determine if a treatment is effective is to see whether it leads to improved results, such as fewer side effects or better tumour control [229].

9.14.2 Radiomics in radiotherapy for "precise medicine"

When a molecularly targeted pharmaceutical is utilized, treatment decisions are based on genetic variations rather than diseased tissues. A broad variety of phenotypic components that indicate cancer traits or phenotypes may be extracted from medical pictures using radiomics, such as those based on image intensity, texture, voxel connectivity, and fractal patterns (reproduced from [98]). Biomarkers and clinical outcomes might both be predicted using radiomics features. Oncology radiation oncologists may take use of machine learning and radiomics to improve the accuracy of their treatment plans and get more benefits from precision radiotherapy. Radiation outcomes including survival, treatment failure or recurrence, toxicity, and the development of a late consequence may be predicted by machine learning (ML) algorithms employing radiomics data to improve precision medicine decision-making. Arimura et al claim that using radiomic technology and AI may enable for the implementation of precision medicine in radiation therapy. These findings were made by Aerts et al. [98] who analyzed CT scans of individuals with lung or head and neck cancer. The study's results showed that lung and head and neck cancers have a common prognostic feature. To determine if pre-surgical CT intensity and texture information from ground-glass opacities and solid nodule components can be used to predict adenocarcinoma recurrence in the lung, Depeursinge and colleagues used SVMs and LASSOs as well as their survival equivalents, Cox-LASSOs [230]. Using pre-surgical CT scans to identify patients for whom no recurrence is expected with high certainty is recommended by the study's results. It was the goal of this study to create repeatable and automated techniques for extracting additional information from images. There is still a need to test radiomics in multi-centric environments, according to this investigation. There will be an increasing need for preclinical studies in multicenter clinical trials to show that newly discovered imaging markers have clinical validity and value in the future, according to Wu et al. Using the LASSO Cox regression model, Lao et al. studied patients with glioblastoma multiforme to see whether radiomics signals could be used to predict overall survival. The study's findings suggest that the suggested method might provide a biomarker based on deep imaging features for the prognosis and categorization of glioblastoma patients prior to surgical intervention. Imaging data from any imaging source may be used to help predict treatment outcomes via the use of radiomics and radiogenomics. Radiology is concerned about the low repeatability of imaging systems, both inside and between institutions. In other disciplines, such as radiomics research, deep learning for image quantification has shown good results. A more or less intense radiation regimen prescribed to cancer patients may improve their quality of life based on model estimates of local control benefit and toxicity risk, which are taken into account throughout the optimum treatment planning design process]. Using radiomics to improve cancer therapeutic decision support is an once-in-a-lifetime opportunity to keep costs down, while enhancing precision medicine [231].

9.15 General Discussion

In this chapter, the most recent and continuing research in radiation oncology using machine learning approaches in the age of big data for precision treatment is critically examined and addressed in detail.

9.15.1 The issues with big data in radiation oncology

Community-wide initiatives, have made validation frameworks accessible and created, which are applied as a standard for the testing of various methods. Deep learning-based [61] models have been proven to outperform other methods for the majority of radiation oncology prediction problems. However, annotated datasets from many universities are required to increase the system's prediction accuracy (even when transfer learning is applied). It is more challenging in radiation oncology since there are not as many datasets. Oncology taxonomy (i.e., clinical, dosimetric, imaging, and so forth) and data collecting procedure (structure) standards are being developed by the AAPM task group TG-263 in order to train models using datasets from diverse institutions [232].

9.15.2 The use of machine learning algorithms offers both advantages and disadvantages

There is no such thing as a one-size-fits-all answer to every situation ("No Free Lunch"). Every machine learning algorithm has advantages and disadvantages. The advantages and limitations of the very frequently operated machine learning algorithms in radiomic oncology research are highlighted in Table 9.2. By maximizing the use of these models in combination with the available resources, a better solution may be discovered. A difficulty has been identified with the use of machine learning (ML) in medicine, in which the ML algorithm transfers a given input data to output predictions without adding any further information. In order to avoid errors, the algorithms used must be easy to understand (e.g., people must be able to understand why a prediction was made) [233]. When it comes to analyzing huge datasets (as opposed to smaller ones), ensemble techniques and deep neural networks (which are not interpretable and give very little information) often surpass them in terms of accuracy. Various machine learning architectures have been studied recently to see how well they work in creating accurate and understandable models. Like any other algorithm currently used in radiation oncology, ML algorithms must be accepted and commissioned in order to ensure that the correct algorithm or model is used for a given application and model results make sense in a given clinical scenario, as in the case of dose calculation or deformable registrations. There are many challenges ahead for radiation oncology since it is an algorithmic and data-centric field with huge potential [234].

Table 9.2 Strengths and weaknesses of the most machine learning methods discussed here appear in radiation oncology studies.

Method	**Strengths**	**Weaknesses**
Decision tree	Interpretability (due to a format that is compatible with a wide range of clinical paths)	Overgrowing a tree with insufficient leaf node observations
Gradient boosting machines	When compared to random forest, it produces very consistent results.	Overfitting and more tuning parameters (in comparison to random forest)
LASSO regression	When compared to the ridge regularization method, this method has a better interpretability.	Provides a zero-bias (which may or may not be appropriate in particular situations).
Random forest	With little feature engineering, it is often possible to make quite accurate forecasts.	Not easily understandable, and the number of trees is not optimized.
Support vector machines	Very precise, with minimal tuning parameters and kernel possibilities.	Not easily interpretable, and not perfectly optimizing the parameters
K-means	Quick, easy, and adaptable	Specify the number of clusters manually.
Principal component analysis	Flexible, quick, and easy to implement	Not interpretable, and a threshold for a cumulative variance must be manually specified.
Logistic regression	Have a pleasant probabilistic interpretation that can be quickly modified with fresh data	Not adaptable enough to capture increasingly complicated relationships organically
Neural networks or more precisely artificial neural networks	Even if one or a few units fail to reply to the network, it still works.	Models that are referred described as "black boxes" provide relatively little insight and necessitate a vast variety of training datasets.
Ensembles (decision tree)	Work efficient, is resistant to outliers, and are mountable	Overfitting is a risk when you are unconstrained.
Deep learning	It is extremely accurate, adaptable to a wide range of issues, and the hidden layers eliminate the need for feature engineering.	To train, you will need a lot of data and a lot of computing power.
Naive Bayes	It works remarkably well, is simple to set up, and scales with the dataset.	Models that have been properly educated and tweaked have frequently beaten them (algorithms listed)

9.15.3 How accurate are the investigators' findings, according to them?

Several research works have shown that these prediction models are very accurate when applied to real-world data. Because ML models are sensitive to numerous data biases, a lack of generalizability is an issue. Data from beyond the sample set may not be predicted (reproduced) by an AI system trained on a local database alone (revised databases that are not described in the training information). For accurate evaluations of prediction model performance and generalizability, it is necessary to conduct external validation in cohorts that are not part of the discovery cohort (e.g., from another institution). When different algorithms are used in the same dataset, the findings for predictors that are closely related to the outcome of interest may vary. However, in the clinical situation, this might signify a restriction in the self-critical evaluation of published ML models or real confidence levels [235].

9.15.4 What changes would you make to the stated results?

Numerous machine learning (ML)-based radiation oncology prediction models have shown good and improving accuracy results, but actual use of these approaches in everyday clinical practice is still rare. Quick Match (Siris Medical, Redwood City, CA) is one of many commercial solutions that have just entered clinical testing. In its study, the Memorial Sloan Kettering Cancer Center in New York is already using a commercial venture, such as IBM's Watson. Watson Cancer is a cerebral Artificial intelligence computational technique meant to assist cancer doctors in evaluating therapy alternatives for their patients [236]. More multi-institutional training and validation datasets are needed to increase the accuracy of our findings. Comparing these methodologies on standard consent information to give standards for assessing multiple models, will undoubtedly result in improved outcomes and the development of strong toolkits/systems. As machine learning and artificial intelligence (AI) become more extensively applied in clinical practise, sufferer's, population, and the medical career stand to benefit [237].

9.15.5 The influence on healthcare procedure automation

Automated systems that learn on their own have been created and deployed. Driving technologies might be used to pilot-automated clinical practices in radiation oncology. Data-driven planning needs a professional review and/or involvement at this phase to ensure that treatment plans are executed effectively and cannot be automated. Because of the intimate integration and need for strong TPSs, fully autonomous planning via reinforcement learning is problematic. The ultimate objective is to have a completely automated planning process that includes contouring as well as plan generation. Virtual QA based on machines and patients may have a significant impact on the existing IMRT/VMAT procedure in the future. If the procedure was automated, the radiation oncology workflow would undoubtedly be expedited and human engagement would be reduced [238].

9.15.6 The influence of precision medicine on clinical decision-making assistance in radiation oncology

The application of ML tools for computer-aided detection/diagnosis may increase radiologists' performance and diagnostic accuracy. To the maximum degree feasible, new paradigms in radiomics might have a major influence on treatment outcome projections for individual patients, such as patient survival and a reduction in recurrence and late sequelae. Individuals may be classified into subgroups established on radiological biomarkers, which offer data on malignancy characteristics that impact the patient's diagnosis. The use of model estimates of local control assistance and toxic risk, which enable doctors to impose a more or less extreme radiological regimen, may enhance the quality of life for oncology patients undergoing radiotherapy. If ML is utilized to do adaptive radiation therapy, radiation therapy treatments may be more accurate. This may help clinicians and patients get to safer educated choices regarding treatment method and dosage prescription, allowing them to set defined and attainable objectives. Because of the therapeutic implications of personalized cancer therapy, both patients and resources are directed to the right patients [239].

9.15.7 Closing comments

This chapter presents and discusses machine learning approaches from the first consultation with the patient through therapy and follow-up. The utilization of big data in radiation oncology as well as the efforts made and present difficulties are discussed. Radiomic oncology is increasingly relying on machine learning techniques in the age of big data [240]. Machine learning technologies might be used to reimburse for human constraints in successfully managing a high quantity of streaming data, where even little mistakes could mean the distinction between life cycle and mortality. The application of machine learning in the area of radiology, which recognizes tumor image phenotypes and allows for decision-making and precision medicine, enables precision medicine and decision-making in radiation treatment [241].

9.16 Future Opportunities and Difficulties

Not only might big data have a tremendous influence on healthcare, but big data producers are also struggling to make their data meaningful in life sciences. The avalanche of data created by advancing tools such as next-generation sequencing (especially whole exome/genome sequencing) and radiomics is an apparent example. Having a large amount of data may hinder the data interpretation process. Growth in data volume, along with an increase in data heterogeneity (variability), makes deriving solid inferences from data more difficult [242]. This is particularly true when the rise in data volume and velocity is accompanied by an increase in data heterogeneity (variability). Another issue is effectively managing data, especially when it comes from multiple sources. What information is available, and who owns it? How much control does the patient still have over their data? Do the treating physician, researcher, data generator, or data interpreter (e.g., computational biologist or medical bioinformatician) have the last say?

9.17 The learning health system is a future vision

To create and test a fully integrated radiation oncology treatment model, a multicentric interchange of information and researchers will be required. These models and the procedures that generated them can help with tumor localization in general. They will serve as the cornerstone for big data-powered decision support systems in every RO department over the next 10 to fifteen years [243]. To keep these systems up to date in real-time, dynamic programing and strengthening learning approaches will be necessary. They will provide recommendations for the finest therapy alternatives based on the patient's characteristics and level of understanding during the initial consultation. The ideal dose distributions and treatment periods for the patient will be determined not by the physician, but by a computer algorithm. In New York, the Memorial Sloan Kettering Cancer Center is already utilizing private initiatives such as IBM's Watson. Similarly, the same strategy might be used to guide treatment decisions for the management of adverse events as well as the early detection of any recurrence after medication. If this "learning health system" can be implemented, it will have a tremendous influence on the area of cancer. Wearable devices and connected items, which are growing more popular, will need to be considered in the follow-up process. Constant, real-time analysis of aberrant outcomes will go ahead to early diagnosis of recurrence and optimization of the effectiveness and cost of salvage treatment. Such strategies will eventually impact overall survival [244].

9.17.1 Consequences for future clinical research

As a result of improvements in precision medicine, new clinical trial designs have evolved. SHIVA, a clinical trial in people with resistant cancer, compares personalized therapy based on tumor genetic profiling to standard therapy. A comparable comparison might be made between standard radiation therapy and personalized radiation therapy using data-driven algorithms. When applied to this type of data, unsupervised machine learning has the potential to reveal patterns that are unfathomable to humans. In oncology, molecular abnormalities, rather than morphological or histological indicators, are helping to define new patient categories and disorders. As a result of this shift, clinicians will be unable to use the resulting increase and changing amount of knowledge. Another result of this is that clinical trials will turn into increasingly difficult to design, to the fact where statistically sufficient power will be impractical to accomplish [245]. The economical and operational expenses connected with creating new clinical trials may soon render medical research insolvent. In most institutions, it is a straightforward and elegant approach for digitally collecting massive volumes of information on sufferers' traits, therapy aspects, adverse effects, and continuation. A large amount of data access should be used to create revised insights. The quality and type of the information gathered must be considered in order to obtain reliable findings from big data. As the phrase goes, "trash in, garbage out." During clinical research, confounding factors are minimized, and considerable data are obtained. More than a dozen recent SEER investigations have given immediate solutions to critical concerns. Big data, on the other hand, has a significant disadvantage when it comes to analyzing radiation treatments due to a shortage of data on therapy traits. If dosimetric and temporal data are included right away from the database and validate systems, they will be correct. Various groups have previously reported investigations applying prediction to improve tailor radiomic therapies. Not any of these methodologies, however, have yet to be applied in a clinical environment [247]. A user-friendly strategy must be integrated directly into the program to help with treatment planning. Based on the patient's medical history and anatomy, the dosimetrist or physicist would be given the best treatment plan. During therapy, patients would be watched by the same system, and physicians would be contacted if anything went wrong that was not predicted. The model would incorporate the data gathered from each patient and therapy. Before we can achieve this goal, we must overcome a number of methodological challenges. For example, we must enter crucial RO information into the EHR, merge clinical and dosimetric information into a specific model, and test this model on a prospective sample of patients.

9.17.2 Databases and archives of biomolecular patient data should be preserved and disseminated

Despite privacy concerns and security and protection measures, it is typical for researchers in (bio)medical research to gather as much information on their participants as feasible (e.g., the GDPR). Respecting patient-identifying data, like genetic sequences, in a way that enables future study, while protecting people's privacy is a serious concern (www.phgfoundation.org). Open access to massive database in many other professions, such as computational sciences, may be welcomed, but confidentiality troubles must always take precedence over a desire for ultimate openness. EGA, as the name implies, is a database created only to preserve raw sequencing data. A data access committee (DAC) is in charge of assessing whether or not data collected and stored in EGA should be made accessible to other researchers as part of any research project [248]. A second issue occurs for researchers who want to analyze processed biomolecular data but are unable to identify certain markers. This difficulty is addressed by repositories with fine-grained access limitations by summarizing the data without disclosing individual markers. Finally, combining diverse data sources while retaining privacy and being able to track the precise computational processing that has been performed on the data remains a challenge. Big data has enormous promise in biological sciences and the treatment of head and neck malignancies. As a consequence, clinical and research data exchange may be adjusted [249]. Individuals and organizations will find it impossible to physically share information as the amount of data created in the near real-time/streaming age expands. Instead of a single full database, the Dutch Techcentre for Life Sciences (DTL) predicts a more organic, decentralized virtual network in its personal health train. Databases are linked as nodes in these networks and may be accessed under particular circumstances. There will be a need for novel data processing technologies, as well as data conveyance and interpretation back to the particular patient. Finally, before it can be utilized, the information generated by all of these databases must be interpreted into a “modest information” environment suited for the care-dependent patient. However, medical specialists are still required for the final phase, which includes the integration of intuitive and emotional factors. Big data and machine learning will not assist with bedside manners in the near future [250].

9.18 Conclusion

Big data gathering is useful because of its massive, quick, diverse, and genuine datasets. The integration of several sources benefits BR, patient therapy, and quality-of-care monitoring. In the Netherlands, where head and neck melanoma treatment is centralized and national big data technologies are in place, there is a once-in-a-lifetime opportunity to collect, connect, and combine these data. Data input and (bioinformatic) data integration, including FAIRification, should be optimized in such a head and neck cancer infrastructure.

Acknowledgment

The authors are highly thankful to the Department of Pharmacy of Galgotias University, Greater Noida for providing all necessary support for completion of the work.

Funding

There is no source of funding.

Conflict of Interest

There is no conflict of interest in the publication of this chapter content.

References

[1] Chen, M. He, Y. Zhu, L. Shi, X. Wang, Five critical elements to ensure the precision medicine, Cancer Metastasis Rev. 34 (2015) 313–318,

doi:10.1007/ s10555-015-9555-3.

[2] J. G. Paez, P. A. Jänne, J. C. Lee, S. Tracy, H. Greulich, S. Gabriel, et al., EGFR mutations in lung cancer: correlation with clinical response to gefitinib therapy, Science 304 (2004) 1497–1500, doi:10.1126/science.1099314.

[3] F. S. Hodi, S. J. O'Day, D. F. McDermott, R. W. Weber, J. A. Sosman, J. B. Haanen, et al., Improved survival with ipilimumab in patients with metastatic melanoma, N. Engl. J. Med. 363 (2010) 711–723, doi:10.1056/NEJMoa1003466.

[4] A. G. Georgakilas, A. Pavlopoulou, M. Louka, Z. Nikitaki, C. E. Vorgias, P. G. Bagos, et al., Emerging molecular networks common in ionizing radiation, immune and inflammatory responses by employing bioinformatics approaches, Cancer Lett. 368 (2015) 164–172, doi:10.1016/j.canlet.2015.03.021.

[5] A. J. Grossberg, A. S. Mohamed, H. Elhalawani, W. C. Bennett, K. E.Smith, T. S. Nolan, and C. D. Fuller. Imaging and clinical data archive for head and neck squamous cell carcinoma patients treated with radiotherapy. *Scientific Data*, 5(1), (2018), 1–10.

[6] F. Prior, K. Smith, A. Sharma, J. Kirby, L. Tarbox, K. Clark, and J. Freymann. The public cancer radiology imaging collections of The Cancer Imaging Archive. *Scientific data*, 4(1), (2017), 1–7.

[7] M. L. Metzker, Sequencing technologies – the next generation, Nat. Rev. Genet. 11 (2010) 31–46, doi:10.1038/nrg2626.

[8] DNA sequencing costs. n.d. http://www.genome.gov/sequencingcosts/ (accessed 12.03.16).

[9] N. Geifman, A. J. Butte, Do cancer clinical trial populations truly represent cancer patients? A comparison of open clinical trials to the cancer genome atlas, Pac. Symp. Biocomput. 21 (2016) 309–320.

[10] I. S. Kohane, J. M. Drazen, E. W. Campion, A glimpse of the next 100 years in medicine, N. Engl. J. Med. 367 (2012) 2538–2539, doi:10.1056/NEJMe1213371.

[11] P. Lambin, R. G. P. M. van Stiphout, M. H. W. Starmans, E. Rios-Velazquez, G. Nalbantov, H. J. W. L. Aerts, et al., Predicting outcomes in radiation oncology– multifactorial decision support systems, Nat. Rev. Clin. Oncol. 10 (2013) 27–40, doi:10.1038/nrclinonc. 2012.196.

[12] J.-E. Bibault, I. Fumagalli, C. Ferté, C. Chargari, J.-C. Soria, E. Deutsch, Personalized radiation therapy and biomarker-driven treatment strategies: a systematic review, Cancer Metastasis Rev. 32 (2013) 479–492, doi:10.1007/s10555-013-9419-7.

[13] Q.-T. Le, D. Courter, Clinical biomarkers for hypoxia targeting, Cancer Metastasis Rev. 27 (2008) 351–362, doi:10.1007/s10555-008-9144-9.
[14] P. Okunieff, Y. Chen, D. J. Maguire, A. K. Huser, Molecular markers of radiationrelated normal tissue toxicity, Cancer Metastasis Rev. 27 (2008) 363–374, doi:10.1007/s10555-008-9138-7.
[15] J. Kang, R. Schwartz, J. Flickinger, S. Beriwal, Machine learning approaches for predicting radiation therapy outcomes: a clinician's perspective, Int. J. Radiat. Oncol. Biol. Phys. 93 (2015) 1127–1135, doi:10.1016/j.ijrobp.2015.07.2286.
[16] F. Denis, S. Yossi, A.-L. Septans, A. Charron, E. Voog, O. Dupuis, et al., Improving survival in patients treated for a lung cancer using self-evaluated symptoms reported through a web application, Am. J. Clin. Oncol. (2015) doi:10.1097/ COC.00000000000 00189.
[17] A. D. Falchook, G. Tracton, L. Stravers, M. E. Fleming, A. C. Snavely, J. F. Noe, et al., Use of mobile device technology to continuously collect patient-reported symptoms during radiotherapy for head and neck cancer: a prospective feasibility study, Adv, Radiat. Oncol. (2016) doi:10.1016/j.adro.2016.02.001.
[18] M. Li, S. Yu, K. Ren, W. Lou, Securing personal health records in cloud computing: patient-centric and fine-grained data access control in multi-owner settings, in: S. Jajodia, J. Zhou (Eds.), Security and Privacy in Communication Networks, Springer Berlin, Heidelberg, 2010, pp. 89–106. <http://link.springer.com/chapter/ 10.1007/978-3-642-16161-2_6> (accessed 21.05.16).
[19] V. Canuel, B. Rance, P. Avillach, P. Degoulet, A. Burgun, Translational research platforms integrating clinical and omics data: a review of publicly available solutions, Brief. Bioinform. 16 (2015) 280–290, doi:10.1093/bib/bbu006.
[20] V. Huser, J. J. Cimino, Impending challenges for the use of Big Data, Int. J. Radiat. Oncol. Biol. Phys. (2015) doi:10.1016/j.ijrobp.2015.10.060.
[21] Shaikh AR, Butte AJ, Schully SD, et al. Collaborative biomedicine in the age of big data: the case of cancer. J Med Internet Res 2014;16(4):e101. https://doi.org/10.2196/jmir.2496.Apr7.
[22] Roman-Belmonte JM, De la Corte-Rodriguez H, Rodriguez-Merchan EC. How blockchain technology can change medicine. Postgrad Med 2018;130(4):420–7.
[23] Bourne PE. What Big Data means to me. J Am Med Inform Assoc 2014;21(2):194.\ https://doi.org/10.1136/amiajnl-2014-002651.

[24] Zhang C, Bijlard J, Staiger C, Scollen S, et al. Systematically linking tranSMART, S. M. Willems, et al. *Oral Oncology 98 (2019) 8–1211* galaxy and EGA for reusing human translational research data. F1000Res. 2017;6. https://doi.org/10.12688/f1000research.12168.1.Aug16ELIXIR-1488.

[25] Grossberg AJ, Mohamed ASR, El Halawani H, et al. Sci. Data 2018;5:180173. https://doi.org/10.1038/sdata.2018.173.Sep4.

[26] Prior F, Smith K, Sharma A, et al. The public cancer radiology imaging collections of The Cancer Imaging Archive. Sci Data. 2017;19(4):170124https://doi.org/10.1038/sdata.2017.124.Sep.

[27] Bousfield D, McEntyre J, Velankar S, et al. Patterns of database citation in articles and patents indicate long-term scientific and industry value of biological data resources. F1000Research 2016;5(160). https://doi.org/10.12688/f1000research.7911.1.

[28] Ooft ML, van Ipenburg J, Braunius WW, et al. A nation-wide epidemiological study on the risk of developing second malignancies in patients with different histological subtypes of nasopharyngeal carcinoma. Oral Oncol 2016;56:40–6.

[29] Datema FR, Ferrier MB, Vergouwe Y, et al. Update and external validation of a head and neck cancer prognostic model. Head Neck 2013;3(9):1232–7.

[30] Datema FR, Moya A, Krause P, et al. Novel head and neck cancer survival analysis approach: random survival forests versus Cox proportional hazards regression. Head Neck 2012;34(1):50–8.

[31] Barlesi F, Mazieres J, Merlio JP, et al. Routine molecular profiling of patients with advanced non-small-cell lung cancer: results of a 1-year nationwide programme of the French Cooperative Thoracic Intergroup (IFCT). Lancet 2016;387(10026):1415–26.

[32] Petersen JF, Timmermans AJ, van Dijk BAC. Trends in treatment, incidence and survival of hypopharynx cancer: a 20-year population-based study in the Netherlands. Eur Arch Otorhinolaryngol 2018;275(1):181–9.

[33] Timmermans AJ, van Dijk BA, Overbeek LI, et al. Trends in treatment and survival for advanced laryngeal cancer: A 20-year population-based study in The Netherlands. Head Neck 2016;38(Suppl 1):E1247–55.

[34] De Ridder M, Balm AJ, Smeele LE, et al. An epidemiological evaluation of salivary gland cancer in the Netherlands (1989–2010). Cancer Epidemiol 2015;39(1):14–20. Feb.

[35] Govers TM, Rovers MM, Brands MT, et al. Integrated prediction and decision models are valuable in informing personalized decision making. J Clin Epidemiol 2018. Aug 28 pii: S0895-4356(18)30447-5.

[36] Wong AJ, Kanwar A, Mohamed AS. Radiomics in head and neck cancer: from exploration to application. Transl Cancer Res 2016;5(4): 371–82.

[37] Parmar C, Grossmann P, Rietveld D, et al. Radiomic machine-learning classifiers for prognostic biomarkers of head and neck cancer. Front Oncol 2015;3(5):272.

[38] Wilkinson MD, Dumontier M, Aalbersberg IJ, et al. The FAIR guiding principles for scientific data management and stewardship. Sci Data 2016;15(3):160018https://doi.org/10.1038/sdata.2016.18.

[39] National cancer institute thesaurus – summary | NCBO BioPortal. n.d. <https:// bioportal.bioontology.org/ontologies/NCIT> (accessed 07.03.16).

[40] Common terminology criteria for adverse events – summary | NCBO BioPortal. n.d. < https://bioportal.bioontology.org/ontologies/CTC AE>(accessed07.03.16).

[41] Fact SheetUMLS® Metathesaurus®. n.d. <https://www.nlm.nih.gov/ pubs/factsheets/umlsmeta.html>(accessed07.03.16).

[42] Radiation oncology ontology – summary | NCBO BioPortal. n.d. <http://bioportal.bioontology.org/ontologies/ROO> (accessed 07.03.16).

[43] I. El Naqa, J. Bradley, A. I. Blanco, P. E. Lindsay, M. Vicic, A. Hope, et al., Multivariable modeling of radiotherapy outcomes, including dose-volume and clinical factors, Int. J. Radiat. Oncol. Biol. Phys. 64 (2006) 1275–1286, doi:10.1016/j.ijrobp.2005.11.022.

[44] T.-F. Lee, P.-J. Chao, H.-M. Ting, L. Chang, Y.-J. Huang, J.-M. Wu, et al., Using multivariate regression model with least absolute shrinkage and selection operator (LASSO) to predict the incidence of Xerostomia after intensity modulated radiotherapy for head and neck cancer, PLoS ONE 9 (2014) e89700, doi:10.1371/journal.pone.0089700.

[45] T.-F. Lee, M.-H. Liou, Y.-J. Huang, P.-J. Chao, H.-M. Ting, H.-Y. Lee, et al., LASSO NTCP predictors for the incidence of xerostomia in patients with head and neck squamous cell carcinoma and nasopharyngeal carcinoma, Sci. Rep. 4 (2014) 6217, doi:10.1038/srep06217.

[46] J. R. Quinlan, Induction of decision trees, Mach. Learn. 1 (1986) 81–106.

[47] P. Langley, W. Iba, K. Thompson, An analysis of Bayesian classifiers, in: AAAI, 1992, pp. 223–228.
[48] P. Langley, S. Sage, Induction of selective Bayesian classifiers, in: Proceedings of the Tenth International Conference on Uncertainty in Artificial Intelligence, Morgan Kaufmann Publishers Inc., 1994, pp. 399–406. <http://dl.acm.org/citation.cfm?id=2074445>(accessed12.03.16).
[49] E. A. Patrick, F. P. Fischer III, A generalized k-nearest neighbor rule, Inf. Control 16 (1970) 128–152, doi:10.1016/S0019-9958(70)90081-1.
[50] V. Vapnik, Estimation of Dependences Based on Empirical Data, Springer Verlag, New York, 1982.
[51] D. E. Rumelhart, J. McClelland, Parallel Distributed Processing: Explorations in the Microstructure of Cognition, MIT Press, Cambridge, 1986.
[52] Y. LeCun, Y. Bengio, G. Hinton, Deep learning, Nature 521 (2015) 436–444, doi:10.1038/nature14539.
[53] S. Chen, S. Zhou, F.-F. Yin, L. B. Marks, S. K. Das, Investigation of the support vector machine algorithm to predict lung radiation-induced pneumonitis, Med. Phys. 34 (2007) 3808–3814.
[54] R. J. Klement, M. Allgäuer, S. Appold, K. Dieckmann, I. Ernst, U. Ganswindt, et al., Support vector machine-based prediction of local tumor control after stereotactic body radiation therapy for early-stage non-small cell lung cancer, Int. J. Radiat. Oncol. Biol. Phys. 88 (2014) 732–738, doi:10.1016/j.ijrobp.2013.11.216.
[55] Y. Hayashida, K. Honda, Y. Osaka, T. Hara, T. Umaki, A. Tsuchida, et al., Possible prediction of chemoradiosensitivity of esophageal cancer by serum protein profiling, Clin. Cancer Res. 11 (2005) 8042–8047, doi:10.1158/1078-0432.CCR-05-0656.
[56] T. J. Bryce, M. W. Dewhirst, C. E. Floyd, V. Hars, D. M. Brizel, Artificial neural network model of survival in patients treated with irradiation with and without concurrent chemotherapy for advanced carcinoma of the head and neck, Int. J. Radiat. Oncol. Biol. Phys. 41 (1998) 339–345.
[57] S. L. Gulliford, S. Webb, C. G. Rowbottom, D. W. Corne, D. P. Dearnaley, Use of Martificial neural networks to predict biological outcomes for patients receiving radical radiotherapy of the prostate, Radiother. Oncol. 71 (2004) 3–12, doi:10.1016/j.radonc.2003.03.001.
[58] A. Pella, R. Cambria, M. Riboldi, B. A. Jereczek-Fossa, C. Fodor, D. Zerini, et al., Use of machine learning methods for prediction of acute

toxicity in organs at risk following prostate radiotherapy, Med. Phys. 38 (2011) 2859–2867.

[59] S. Tomatis, T. Rancati, C. Fiorino, V. Vavassori, G. Fellin, E. Cagna, et al., Late rectal bleeding after 3D-CRT for prostate cancer: development of a neural-networkbased predictive model, Phys. Med. Biol. 57 (2012) 1399–1412, doi:10.1088/0031-9155/57/5/1399.

[60] S. Chen, S. Zhou, J. Zhang, F.-F. Yin, L. B. Marks, S. K. Das, A neural network model to predict lung radiation-induced pneumonitis, Med. Phys. 34 (2007) 3420–3427.

[61] M. Su, M. Miften, C. Whiddon, X. Sun, K. Light, L. Marks, An artificial neural network for predicting the incidence of radiation pneumonitis, Med. Phys. 32 (2005) 318–325.

[62] T. Ochi, K. Murase, T. Fujii, M. Kawamura, J. Ikezoe, Survival prediction using artificial neural networks in patients with uterine cervical cancer treated by radiation therapy alone, Int. J. Clin. Oncol. 7 (2002) 294–300, doi:10.1007/s101470200043.

[63] K.-L. Hua, C.-H. Hsu, S. C. Hidayati, W.-H. Cheng, Y.-J. Chen, Computer-aided classification of lung nodules on computed tomography images via deep

[64] Y. Guo, Y. Gao, D. Shen, Deformable MR prostate segmentation via deep feature learning and sparse patch matching, IEEE Trans. Med. Imaging (2015) doi:10.1109/TMI.2015.2508280.

[65] R. Raina, A. Madhavan, A. Y. Ng, Large-scale deep unsupervised learning using graphics processors, in: Proceedings of the 26th Annual International Conference on Machine Learning, ACM, New York, NY, USA, 2009, pp. 873–880, doi:10.1145/1553374.1553486.

[66] J. A. Cruz, D. S. Wishart, Applications of machine learning in cancer prediction and prognosis, Cancer Inform. 2 (2006) 59–77.

[67] K. Kourou, T. P. Exarchos, K. P. Exarchos, M. V. Karamouzis, D. I. Fotiadis, Machine learning applications in cancer prognosis and prediction, Comput. Struct. Biotechnol. J. 13 (2015) 8–17, doi:10.1016/j.csbj.2014.11.005.

[68] C. U. Lehmann, B. Séroussi, M.-C. Jaulent, Big3. Editorial, Yearb. Med. Inform. 9 (2014) 6–7, doi:10.15265/IY-2014-0030.

[69] R. L. Somorjai, B. Dolenko, R. Baumgartner, Class prediction and discovery using gene microarray and proteomics mass spectroscopy data: curses, caveats, cautions, Bioinformatics 19 (2003) 1484–1491.

[70] Christakis NA, Lamont EB, Parkes CM, et al. Extent and determinants of error in doctors' prognoses in terminally ill patients: prospective cohort study. BMJ. 2000; 320:469-472.

[71] Sborov K, Giaretta S, Koong A, et al. Impact of accuracy of survival predictions on quality of end-of-life care among patients with metastatic cancer who receive radiation therapy. J Oncol Pract. 2019;15:e262-e270.
[72] Fong Y, Evans J, Brook D, et al. The Nottingham prognostic index: five- and ten-year data for all-cause survival within a screened population. Ann R Coll Surg Engl. 2015;97:137-139.
[73] Alexander M, Wolfe R, Ball D, et al. Lung cancer prognostic index: a risk score to predict overall survival after the diagnosis of non-small-cell lung cancer. Br J Cancer. 2017;117: 744-751.
[74] Lakin JR, Robinson MG, Bernacki RE, et al. Estimating 1-year mortality for high-risk primary care patients using the "surprise" question. JAMA Intern Med. 2016;176:1863-1865.
[75] Morita T, Tsunoda J, Inoue S, et al. The Palliative Prognostic Index: a scoring system for survival prediction of terminally ill cancer patients. Support Care Cancer. 1999;7:128-133.
[76] Chow R, Chiu N, Bruera E, et al. Inter-rater reliability in performance status assessment among health care professionals: a systematic review. Ann Palliat Med.2016;5:83-92.
[77] Burwell SM. Setting value-based payment goals: HHS efforts to improve U.S. health care. N Engl J Med. 2015;372:897-899.
[78] Center for Medicare & Medicaid Innovation. Oncology care model. https://innovation.cms.gov/initiatives/oncology-care/.AccessedOctober17,2018.
[79] Kline R, Adelson K, Kirshner JJ, et al. The Oncology Care Model: perspectives from the Centers for Medicare & Medicaid Services and participating oncology practices in academia and the community. Am Soc Clin Oncol Educ Book. 2017;37:460-466.
[80] Kline RM, Bazell C, Smith E, et al. Centers for Medicare and Medicaid Services: using an episode-based paymentmodel to improve oncology care. J Oncol Pract. 2015;11:114-116.
[81] Ostrovsky A, O'Connor L, Marshall O, et al. Predicting 30- to 120-day readmission risk among Medicare fee-for-service patients using non-medical workers and mobile technology. Perspect Health Inf Manag. 2016;13:1e.
[82] Conn J. Predictive analytics tools help hospitals reduce preventable readmissions. Mod Healthc. 2014;44:16-17.
[83] Elfiky AA, Pany MJ, Parikh RB, et al. Development and application of a machine learning approach to assess short-term mortality risk among patients with cancer starting chemotherapy. JAMA Netw Open. 2018;1:e180926.

[84] Bertsimas D, Dunn J, Pawlowski C, et al. Applied informatics decision support tool for mortality predictions in patients with cancer. JCO Clin Cancer Informatics. 2018;2:1-11.

[85] Brooks GA, Kansagra AJ, Rao SR, et al. A clinical prediction model to assess risk for chemotherapy-related hospitalization in patients initiating palliative chemotherapy. JAMA Oncol. 2015;1: 441-447.

[86] Yeo H, Mao J, Abelson JS, et al. Development of a nonparametric predictive model for readmission risk in elderly adults after colon and rectal cancer surgery. J Am Geriatr Soc. 2016;64:e125-e130.

[87] Fieber JH, Sharoky CE, Collier KT, et al. A preoperative prediction model for risk of multiple admissions after colon cancer surgery. J Surg Res. 2018; 231:380-386.

[88] Manning AM, Casper KA, Peter KS, et al. Can predictive modeling identify head and neck oncology patients at risk for readmission? Otolaryngol Head Neck Surg. 2018;159:669-674.

[89] Vogel J, Evans TL, Braun J, et al. Development of a trigger tool for identifying emergency department visits in patients with lung cancer. Int J Radiat Oncol Biol Phys. 2017;99:S117.

[90] Furlow B. Predictive analytics reduces chemotherapy-associated hospitalizations. Managed Healthcare Executive. https://www.managedhealthcareexecutive.com/mhe-articles/predictive-analytics-reduces-chemotherapy-associated-hospitalizations.AccessedMarch13, 2019

[91] Rajkomar A, Oren E, Chen K, et al. Scalable and accurate deep learning with electronic health records. npj Digital Med. 2018;1.

[92] Wong AJ, Kanwar A, Mohamed AS, et al. Radiomics in head and neck cancer: from exploration to application. Transl Cancer Res. 2016;5: 371-382.

[93] Bi WL, Hosny A, Schabath MB, et al. Artificial intelligence in cancer imaging: clinical challenges and applications. CA Cancer J Clin. 2019;69:127-157.

[94] Chan H-P, Hadjiiski L, Zhou C, et al. Computer-aided diagnosis of lung cancer and pulmonary embolism in computed tomography-a review. Acad Radiol. 2008; 15:535-555.

[95] Wang S, Burtt K, Turkbey B, et al. Computer aided-diagnosis of prostate cancer on multiparametric MRI: a technical review of current research. BioMed Res Int. 2014;2014:789561.

[96] Song SE, Seo BK, Cho KR, et al. Computer-aided detection (CAD) system for breast MRI in assessment of local tumor extent, nodal status, and multifocality of invasive breast cancers: preliminary study. Cancer Imaging. 2015;15:1.

[97] Aerts HJWL, Velazquez ER, Leijenaar RTH, et al. Decoding tumour phenotype by noninvasive imaging using a quantitative radiomics approach [published correction appears in Nat Commun. 2014;5:4644]. Nat Commun. 2014;5:4006.

[98] Coroller TP, Grossmann P, Hou Y, et al. CT-based radiomic signature predicts distant metastasis in lung adenocarcinoma. Radiother Oncol. 2015;114:345-350.

[99] Sorace AG, Wu C, Barnes SL, et al. Repeatability, reproducibility, and accuracy of quantitative MRI of the breast in the community radiology setting. J Magn Reson Imaging. 2018;48:695-707.

[100] Yu K-H, Zhang C, Berry GJ, et al. Predicting non-small cell lung cancer prognosis by fully automated microscopic pathology image features. Nat Commun. 2016;7:12474.

[101] Sooriakumaran P, Lovell DP, Henderson A, et al. Gleason scoring varies among pathologists and this affects clinical risk in patients with prostate cancer. Clin Oncol (R Coll Radiol). 2005;17:655-658.

[102] Ehteshami Bejnordi B, Veta M, Johannes van Diest P, et al; CAMELYON16 Consortium. Diagnostic assessment of deep learning algorithms for detection of lymph node metastases in women with breast cancer. JAMA. 2017;318:2199-2210.

[103] Hravnak M, Devita MA, Clontz A, et al. Cardiorespiratory instability before and after implementing an integrated monitoring system. Crit Care Med. 2011;39:65-72.

[104] Raza SA, Barreira CM, Rodrigues GM, et al. Prognostic importance of CT ASPECTS and CT perfusion measures of infarction in anterior emergent large vessel occlusions. J Neurointerv Surg. Epub 2018 Dec 7.

[105] Parikh RB, Kakad M, Bates DW. Integrating predictive analytics into high-value care: the dawn of precision delivery. JAMA. 2016;315: 651-652.

[106] Burki TK. Predicting lung cancer prognosis using machine learning. Lancet Oncol. 2016;17:e421.

[107] Van den Akker J, Mishne G, Zimmer AD, et al. A machine learning model to determine the accuracy of variant calls in capture-based next generation sequencing. BMC Genomics. 2018;19:263.

[108] Welsh JL, Hoskin TL, Day CN, et al. Clinical decision-making in patients with variant of uncertain significance in BRCA1 or BRCA2 genes. Ann Surg Oncol. 2017; 24:3067-3072.
[109] Pashayan N, Morris S, Gilbert FJ, et al. Cost-effectiveness and benefit-to-harm ratio of risk-stratified screening for breast cancer: a life-table model. JAMA Oncol. 2018;4:1504-1510.
[110] Karanis TB, Bermudez Canta FA, Mitrofan L, et al. Research' vs 'real world' patients: the representativeness of clinical trial participants. Annals Oncol. 2016;27:1570P.
[111] O'Connor JM, Fessele KL, Steiner J, et al. Speed of adoption of immune checkpoint inhibitors of programmed cell death 1 protein and comparison of patient ages in clinical practice vs pivotal clinical trials. JAMA Oncol. 2018;4:e180798.
[112] Flatiron Health Database. https://flatiron.com/real-world-evidence/.AccessedAugust26,2018.
[113] ASCO. CancerLinQ Discovery data access toolkit. https://cancerlinq.org/sites/cancerlinq.org/files/ASC-1725-CancerLinQ-Discovery-ToolkitUpdate-v4.pdf.AccessedAugust26,2018.
[114] Parikh RB, Obermeyer Z, Navathe AS. Regulation of predictive analytics in medicine. Science. 2019;363:810-812.
[115] Topol EJ. High-performance medicine: the convergence of human and artificial intelligence. Nat Med. 2019;25:44-56.
[116] U.S. Department of Health and Human Services. Digital Health Software Precertification (Pre-Cert) Program.https://www.fda.gov/MedicalDevices/DigitalHealth/UCM567265.AccessedFebruary28,2019.
[117] Wolff RF, Moons KGM, Riley RD, et al; PROBAST Group. PROBAST: a tool to assess the risk of bias and applicability of prediction model studies. Ann Intern Med. 2019;170:51-58.
[118] Mullainathan S, Obermeyer Z. Does machine learning automate moral hazard and error? Am Econ Rev. 2017;107:476-480.
[119] Manrai AK, Funke BH, Rehm HL, et al. Genetic misdiagnoses and the potential for health disparities. N Engl J Med. 2016;375:655-665.
[120] Struewing JP, Hartge P, Wacholder S, et al. The risk of cancer associated with specific mutations of BRCA1 and BRCA2 among Ashkenazi Jews. N Engl J Med. 1997;336:1401-1408.
[121] El Naqa I, Li R, Murphy M. Machine Learning in Radiation Oncology: Theory and Applications. Cham: Springer; 2015. 336 p. DOI: 10.1007/978-3-319- 18305-3

[122] Kansagra AP, Yu JP, Chatterjee AR, Lenchik L, Chow DS, Prater AB, et al. Big data and the future of radiology informatics. Academic Radiology. 2016;**23**(1):30-42. DOI: 10.1016/j.acra.2015.10.004

[123] Lustberg T, van Soest J, Jochems A, Deist T, van Wijk Y, Walsh S, et al. Big data in radiation therapy: Challenges and opportunities. The British Journal of Radiology. 2017;**90**(1069):20160689. DOI: 10.1259/bjr.20160689

[124] Oberije C, Nalbantov G, Dekker A, Boersma L, Borger J, Reymen B, et al. A prospective study comparing the predictions of doctors versus models for treatment outcome of lung cancer patients: A step towards individualized care and shared decision making. Radiotherapy and Oncology. 2014;**112**:37-43. DOI: 10.1016/j.radonc.2014.04.012

[125] El Naqa I, Ruan D, Valdes G, Dekker A, McNutt T, Ge Y, et al. Machine learning and modeling: Data, validation, communication challenges. Medical Physics. 2018;**45**(10):e834-e840. DOI: 10.1002/mp.12811

[126] Mayo CS, Kessler ML, Eisbruch A, Weyburne G, Feng M, Hayman JA, et al. The big data effort in radiation oncology: Data mining or data farming? Advances in Radiation Oncology. 2016;**1**(4):260-271. DOI: 10.1016/j.adro.2016.10.001

[127] Chen RC, Gabriel PE, Kavanagh BD, McNutt TR. How will big data impact clinical decision making and precision medicine in radiation therapy? International Journal of Radiation Oncology, Biology, Physics. 2016;**95**(3):880-884. DOI: 10.1016/j. ijrobp.2015.10.052

[128] Benedict SH, Hoffman K, Martel MK, Abernethy AP, Asher AL, Capala J, et al. Overview of the American Society for Radiation Oncology- National Institutes of Health-American Association of Physicists in Medicine Workshop 2015: Exploring opportunities for radiation oncology in the era of big data. International Journal of Radiation Oncology, Biology, Physics. 2016;**95**:873-879. DOI: 10.1016/j.ijrobp.2016.03.006

[129] ACR Data Science InstituteTM to Guide Artificial Intelligence Use in Medical Imaging. 2017. Available at: https://www.acrdsi.org/-/media/DSI/Files/Strategic-Plan.pdf?la=en

[130] Alpaydin E. Introduction to Machine Learning. 3rd ed. Cambridge, MA: The MIT Press; 2014

[131] Ao S-I, Rieger BB, Amouzegar MA. Machine Learning and Systems Engineering. Dordrecht, NY: Springer; 2010

[132] Friedman J, Hastie T, Tibshirani R. The Elements of Statistical Learning. Vol. 1. Berlin: Springer; 2001
[133] Feng M, Valdes G, Dixit N, Solberg TD. Machine learning in radiation oncology: Opportunities, requirements, and needs. Frontiers in Oncology. 2018;**8**:110. DOI: 10.3389/fonc.2018.00110
[134] Suzuki K, Armato SG 3rd, Li F,\Sone S, Doi K. Massive training artificial neural network (MTANN) for reduction of false positives in computerized detection of lung nodules in lowdose computed tomography. Medical Physics. 2003;**30**(7):1602-1617. DOI: 10.1118/1.1580485
[135] Chan HP, Lo SC, Sahiner B, Lam\KL, Helvie MA. Computer-aided detection of mammographic microcalcifications: Pattern recognition with an artificial neural network. Medical Physics. 1995;**22**(10):1555-1567. DOI: 10.1118/1.597428
[136] Zhu Y, Wang L, Liu M, Qian C, Yousuf A, Oto A, et al. MRI-based prostate cancer detection with high-level representation and hierarchical classification. Medical Physics. 2017;**44**(3):1028-1039. DOI: 10.1002/mp.12116
[137] Rezaei M, Yang H, Meinel C. Deep neural network with l2-norm unit for brain lesions detection. In: Liu D, Xie S, Li Y, Zhao D, El-Alfy ES, editors. Neural Information Processing. ICONIP 2017. Cham: Springer; 2017. pp. 798-807. DOI: 10.1007/978-3-319-70093-9_85
[138] Cheng JZ, Ni D, Chou YH, Qin J, Tiu CM, Chang YC, et al. Computeraided diagnosis with deep learning architecture: Applications to breast lesions in US images and pulmonary nodules in CT scans. Scientific Reports. 2016;**6**:24454. DOI: 10.1038/srep24454
[139] Feng PH, Lin Y, Lo GM. A machine learning texture model for classifying lung cancer subtypes using preliminary bronchoscopic findings. Medical Physics. 2018;**45**(12):5509-5514. DOI: 10.1002/mp.13241
[140] Beig N, Khorrami M, Alilou M, et al. Perinodular and intranodular radiomic features on lung CT images distinguish adenocarcinomas from granulomas. Radiology. 2018;**290**(3):1-10. https://doi.org/10.1148/radiol.2018180910
[141] Joo S, Yang YS, Moon WK, Kim HC. Computer-aided diagnosis of solid breast nodules: Use of an artificial neural network based on multiple sonographic features. IEEE Transactions on Medical Imaging. 2004;(10): 1292-1300. DOI: 10.1109/ TMI.2004.834617
[142] Valdes G, Luna JM, Eaton E, Simone CB II, Ungar LH, Solberg TD. MediBoost: A patient stratification tool for interpretable decision making in the era of precision medicine. Scientific Reports. 2016;**6**:37854. DOI: 10.1038/ srep37854

[143] Caruana R, Lou Y, Gehrke J, Koch P, Sturm M, Elhadad N. Intelligible models for healthcare: Predicting pneumonia risk and hospital 30-day readmission. In: Proceedings of the 21th ACM SIGKDD International Conference on Knowledge Discovery and Data Mining. New York, USA: ACM; 2015. pp. 1721- 1730. DOI: 10.1145/2783258.2788613

[144] Caruana R, Niculescu-Mizil A. An empirical comparison of supervised learning algorithms. In: Proceedings of the 23rd International Conference on Machine Learning. Pittsburgh: ACM; 2006. pp. 161-168. DOI: 10.1145/1143844.1143865

[145] Fernández-Delgado M, Cernadas E, Barro S, Amorin D. Do we need hundreds of classifiers to solve real world classification problems. Journal of Machine Learning Research. 2014;**15**:3133-3181

[146] Fayad H, Gilles M, Pan T, Visvikis D. A 4D global respiratory motion model of the thorax based on CT images: A proof of concept. Medical Physics. 2018;**45**(7):3043-3051. DOI: 10.1002/mp.12982

[147] Steiner E, Shieh C, Caillet V, O'Brien R, et al. WE-HI-KDBRB1-10: 4DCT and 4DCBCT under-predict intrafraction lung target motion during radiotherapy. Medical Physics. 2018;**45**(6):e646-e647. DOI: 10.1002/mp.12938 *DOI:* http://dx.doi.org/10.5772/intechopen.84629

[148] Dick D, Wu X, Zhao W. MO-E115- GePD-F5-3: Fiducial-less tracking for the radiation therapy of liver tumors using artificial neural networks. Medical Physics. 2018;**45**(6):e415-e415. DOI: 10.1002/mp.12938

[149] Johansson A, Balter J, Cao Y. WE-AB-202-5: Gastrointestinal 4D MRI with respiratory motion correction. Medical Physics. 2018;(6):e583-e583. DOI: 10.1002/mp.12938

[150] Nie D, Cao X, Gao Y, Wang L, Shen D. Estimating CT image from MRI data using 3D fully convolutional networks. In: Carneiro G et al., editors. Deep Learning and Data Labeling for Medical Applications. DLMIA 2016, LABELS 2016. LNCS. Vol. 10008. Cham: Springer; 2016. pp. 170-178. DOI: 10.1007/978-3-319-46976-8_18

[151] Bayisa F, Liu X, Garpebring A, Yu J. Statistical learning in computed tomography image estimation. Medical Physics. 2018;**45**(12):5450-5460. DOI: 10.1002/mp.13204

[152] Huynh T, Gao Y, Kang J, et al. Estimating CT image from MRI data using structured random forest and auto-context model. IEEE Transactions on Medical Imaging. 2016;**35**(1):174-183. DOI: 10.1109/TMI.2015.2461533

[153] Chen S, Qin A, Zhou D, Yan D. Technical note: U-net-generated synthetic CT images for magnetic resonance imaging-only prostate intensity-modulated radiation therapy treatment planning. Medical Physics. 2018;**45**(12):5659-5665. DOI: 10.1002/ mp.13247

[154] Bahrami K, Shi F, Rekik I, Shen D. Convolutional neural network for reconstruction of 7T-like images from 3T MRI using appearance and anatomical features. In: Carneiro G et al., editors. Deep Learning and Data Labeling for Medical Applications. Cham, Switzerland: Springer, Verlag; 2016. pp. 39-47. DOI: 10.1007/978-3-319-46976-8_5

[155] Bahrami K, Shi F, Zong X, Shin HW, An H, Shen D. Reconstruction of 7T-like images from 3T MRI. IEEE Transactions on Medical Imaging. 2016;**35**(9):2085-2097. DOI: 10.1109/TMI.2016.2549918

[156] Bahrami K, Rekik I, Shi F, Gao Y, Shen D. 7T-guided learning framework for improving the segmentation of 3T MR images. Medical ImageComputing and Computer-Assisted Intervention. 2016;**9901**:572-580. DOI: 10.1007/978-3-319-46723-8_66

[157] Bahrami K, Shi F, Rekik I, Gao Y, Shen D. 7T-guided super-resolution of 3T MRI. Medical Physics.2017;**44**(5):1661-1677. DOI: 10.1002/ mp.12132

[158] Guerrero T, Zhang G, Huang TC, Lin KP. Intrathoracic tumour motion estimation from CT imaging using the 3D optical flow method. Physics in Medicine and Biology. 2004;**49**(17):4147-4161

[159] Zhang T, Chi Y, Meldolesi E, Yan D. Automatic delineation of on-line head-and-neck computed tomography images: Toward on-line adaptive radiotherapy. International Journal of Radiation Oncology, Biology, Physics. 2007;**68**(2):522-530. DOI: 10.1016/j.ijrobp. 2007.01.038

[160] Hu S, Wei L, Gao Y, Guo Y, Wu G, Shen D. Learning-based deformable image registration for infant MR images in the first year of life. Medical Physics. 2017;**44**(1):158-170. DOI: 10.1002/ mp.12007

[161] Zagoruyko S, Komodakis N. Learning to compare image patches via convolutional neural networks. IEEE Conference on Computer Vision and Pattern Recognition. 2015:4353-4361. DOI: 10.1109/ CVPR.2015.7299064

[162] Jiang D, Shi Y, Chen X, Wang M, Song Z. Fast and robust multimodal image registration using a local derivative pattern. Medical Physics. 2017;**44**(2):497- 509. DOI: 10.1002/mp.12049

[163] Neylon J, Min Y, Low DA, Santhanam A. A neural network approach for fast, automated quantification of DIR performance. Medical Physics. 2017;**44**(8):4126-4138. DOI: 10.1002/mp.12321

[164] Wu J, Su Z, Li Z. A neural networkbased 2D/3D image registration quality evaluator for pediatric patient setup in external beam radiotherapy. Journal of Applied Clinical Medical Physics. 2016;**17**(1):22-33. DOI: 10.1120/jacmp.v17i1.5235

[165] Wu G, Kim M, Wang Q, Munsell BC, Shen D. Scalable highperformance image registration framework by unsupervised deep feature representations learning. IEEE Transactions on Biomedical Engineering. 2016;**63**(7):1505-1516. DOI: 10.1109/TBME.2015.2496253

[166] Kearney V, Haaf S, Sudhyadhom A, Valdes G, Solberg TD. An unsupervised convolutional neural network-based algorithm for deformable image registration. Physics in Medicine and Biology. 2018;**63**(18):185017. DOI: 10.1088/1361-6560/aada66

[167] International Commisssion of Radiation Units and Measuremets. The ICRU Report 83: Prescribing, Recordingand Reporting Photon-beam Intensity Modulated Radiation Therapy (IMRT). Oxford University Press; 2010. 107 p. DOI: 10.1093/jicru/ndq002

[168] Podgorsak EB. Radiation Oncology Physics: A Handbook for Teachers and Students. International Atomic Energy Agency (IAEA): IAEA; 2005. 657 p

[169] Menze BH, Jakab A, Bauer S, Kalpathy-Cramer J, Farahani K, et al. The multimdal brain tumor image segmentation benchmark (BRATS). IEEE Transactions on Medical Imaging. 2015;**34**(10):1993-2024. DOI: 10.1109/ TMI.2014.2377694

[170] Bakas S, Akbari H, Sotiras A, Bilello M, Rozycki M, Kirby J, et al. Segmentation Labels and radiomic features for the pre-operative scans of the TCGA-GBM collection. The Cancer Imaging Archive. 2017. DOI: 10.7937/ K9/TCIA.2017.KLXWJJ1Q

[171] Osman AFI. Automated brain tumor segmentation on magnetic resonance images and patient's overall survival prediction using support vector machines. In: Crimi A, Bakas S, Kuijf H, Menze B, Reyes M, editors. Brainlesion: Glioma, Multiple Sclerosis, Stroke and Traumatic Brain Injuries. Cham: Springer; 2018. pp. 435-449. DOI: 10.1007/978-3-319-75238-9_37

[172] Kamnitsas K, Bai W, Ferrante E, et al. Ensembles of multiple models and architectures for robust brain tumour segmentation. In: Crimi A, Bakas S, Kuijf H, Menze B, Reyes M, editors.Brain Les 2017. . LNCS. Vol. 10670. Cham: Springer; 2018. pp. 450-462. DOI: 10.1007/978-3-319-75238-9_38

[173] Kamnitsas K, Ledig C, Newcombe VF, et al. Efficient multi-scale 3D CNN with fully connected CRF for accurate brain

lesion segmentation. Medical Image Analysis. 2017;**36**:61-78. DOI:10.1016/j.media.2016.10.004

[174] Pereira S, Pinto A, Alves V, Silva CA. Brain tumor segmentation using convolutional neural networks in MRI images. IEEE Transactions on Medical Imaging. 2016;**35**(5):1240-1251. DOI: 10.1109/TMI.2016.2538465

[175] Guo Y, Gao Y, Shen D. Deformable MR prostate segmentation via deep feature learning and sparse patch matching. IEEE Transactions on Medical *DOI:* http://dx.doi.org/10.5772/intechopen.84629 Imaging. 2016;**35**(4):1077-1089. DOI: 10.1109/TMI.2015.2508280

[176] Men K, Dai J, Li Y. Automatic segmentation of the clinical target volume and organs at risk in the planning CT for rectal cancer using deep dilated convolutional neural networks. Medical Physics. 2017;**44**(12): 6377-6389. DOI: 10.1002/mp.12602

[177] Carass A, Roy S, Jog A, Cuzzocreo JL, Magrath E, Gherman A, et al. Longitudinal multiple sclerosis lesion segmentation: Resource and challenge. NeuroImage. 2017;**148**:77-102. DOI: 10.1016/j.neuroimage.2016.12.064

[178] Yang X, Wu N, Cheng G, Zhou Z, Yu DS, Beitler JJ, et al. Automated segmentation of the parotid gland based on atlas registration and machine learning: A longitudinal MRI study in head-and-neck radiation therapy. International Journal of Radiation Oncology, Biology, Physics.2014;**90**(5):1225-1233. DOI: 10.1016/j. ijrobp. 2014.08.350

[179] Ibragimov B, Xing L. Segmentation of organs-at-risks in head and neck CT images using convolutional neural networks. Medical Physics. 2017;**44**(2):547-557. DOI: 10.1002/mp.12045

[180] LeCun Y, Bengio Y, Hinton G. Deep learning. Nature. 2015;**521**(7553): 436-444. DOI: 10.1038/nature14539

[181] Thompson RF, Valdes G, Fuller CD, Carpenter CM, Morin O, Aneja S, et al. Artificial intelligence in radiation oncology: A specialty-wide disruptive transformation? Radiotherapy and Oncology. 2018;**129**(3):421-426. DOI: 10.1016/j.radonc.2018.05.030

[182] Nwankwo O, Mekdash H, Sihono DS, Wenz F, Glatting G. Knowledge-based radiation therapy (KBRT) treatment planning versus planning by experts: Validation of a KBRT algorithm for prostate cancer treatment planning.Radiation Oncology. 2015;**10**:111. DOI: 10.1186/s13014-015-0416-6

[183] Li N, Carmona R, Sirak I, Kasaova L, Followill D, Michalski J, et al. Highly efficient training, refinement, and validation of a knowledge-based planning quality-control system for radiation therapy clinical trials. International Journal of Radiation Oncology, Biology, Physics. 2017;**97**(1):164-172. DOI: 10.1016/j.ijrobp.2016.10.005

[184] Chatterjee A, Serban M, Abdulkarim B, Panet-Raymond V, Souhami L, Shenouda G, et al. Performance of knowledge-based radiation therapy planning for the glioblastoma disease site. International Journal of Radiation Oncology, Biology, Physics. 2017;**99**(4):1021-1028. DOI: 10.1016/j.ijrobp.2017.07.012

[185] Tol JP, Delaney AR, Dahele M, Slotman BJ, Verbakel WF. Evaluation of a knowledge-based planning solution for head and neck cancer. International Journal of Radiation Oncology, Biology, Physics. 2015;**91**(3):612-620. DOI: 10.1016/j.ijrobp.2014.11.014

[186] Foy JJ, Marsh R, Ten Haken RK, Younge KC, Schipper M, Sun Y, et al. An analysis of knowledge-based planning for stereotactic body radiation therapy of the spine. Practical Radiation Oncology. 2017;**7**(5):e355-e360. DOI: 10.1016/j.prro.2017.02.007

[187] Valdes G, Simone CB, Chen J, Lin A, Yom SS, et al. Clinical decision support of radiotherapy treatment planning: A data-driven machine learning strategy for patient-specific dosimetric decision making. Radiotherapy and Oncology. 2017;**125**(3):392-397. DOI: 10.1016/j.radonc.2017.10.014

[188] Zhu X, Ge Y, Li T, Thongphiew D, Yin FF, Wu Q J. A planning quality evaluation tool for prostate adaptive IMRT based on machine learning. Medical Physics. 2011;**38**(2):719-726. DOI: 10.1118/1.3539749

[189] Moore KL, Schmidt R, Moiseenko V, Olsen LA, Tan J, Xiao Y, et al. Quantifying unnecessary normal tissue complication risks due to suboptimal planning: A secondary study of RTOG 0126. International Journal of Radiation Oncology, Biology, Physics. 2015;**92**(2):228-235. DOI: 10.1016/j.ijrobp.2015.01.046

[190] Rowbottom CG, Webb S, Oldham M. Beam-orientation customization using an artificial neural network. Physics in Medicine and Biology. 1999;**44**:2251. DOI: 10.1088/0031-9155/44/9/312

[191] Llacer J, Li S, Agazaryan N, Promberger C, Solberg TD. Noncoplanar automatic beam orientation selection in cranial IMRT: A practical methodology. Physics in Medicine and Biology. 2009;**54**(5):1337-1368. DOI: 10.1088/0031-9155/54/5/016

[192] Silver D, Huang A, Maddison CJ, Guez A, Sifre L, van den Driessche G, et al. Mastering the game of go with deep neural networks and tree search. Nature. 2016;**529**:484-489. DOI: 10.1038/nature 16961

[193] Li Q, Chan MF. Predictive timeseries modeling using artificial neuralnetworks for Linac beam symmetry: An empirical study. Annals of the New York Academy of Sciences. 2017;**1387**(1): 84-94. DOI: 10.1111/nyas.13215

[194] El Naqa I. SU-E-J-69: An anomaly detector for radiotherapy quality assurance using machine learning. Medical Physics. 2011;**38**:3458. DOI:10.1118/1.3611837

[195] Ford EC, Terezakis S, Souranis A, Harris K, Gay H, Mutic S. Quality control quantification (QCQ): A tool to measure the value of quality control checks in radiation oncology. International Journal of Radiation Oncology, Biology, Physics. 2012;**84**(3):e263-e269. DOI: 10.1016/j.ijrobp.2012.04.036

[196] Hoisak JD, Pawlicki T, Kim GY, Fletcher R, Moore KL. Improving linear accelerator service response with a realtime electronic event reporting system. Journal of Applied Clinical Medical Physics. 2014;**15**(5):4807. DOI: 10.1120/ jacmp.v15i5.4807

[197] Huq MS, Fraass BA, Dunscombe PB, Gibbons JP Jr, Ibbott GS, et al. The report of task group 100 of the AAPM: Application of risk analysis methods to radiation therapy quality management. Medical Physics. 2016;**43**(7):4209. DOI: 10.1118/1.4947547

[198] Osman A, Maalej N, Jayesh K. SU-KKDBRA1- 01: A novel learning approach for predicting MLC positioning during dynamic IMRT delivery. Medical Physics. 2018;**45**(6):e357-e358. DOI:10.1002/mp.12938

[199] Valdes G, Morin O, Valenciaga Y, Kirby N, Pouliot J, Chuang C. Use of TrueBeam developer mode for imaging QA. Journal of Applied Clinical Medical Physics. 2015;**16**(4):322-333. DOI: 10.1120/jacmp.v16i4.5363

[200] Valdes G, Scheuermann R, Hung CY, Olszanski A, Bellerive M, Solberg TD. A mathematical framework for virtual IMRT QA using machine learning. Medical Physics. 2016;**43**(7):4323. DOI: 10.1118/1.4953835

[201] Valdes G, Chan MF, Lim SB, Scheuermann R, Deasy JO, Solberg TD. IMRT QA using machine learning: A multi-institutional validation. Journal of Applied Clinical Medical Physics. 2017;**18**(5):279-284. DOI: 10.1002/ acm2.12161

[202] Carlson JN, Park JM, Park SY, Park JI, Choi Y, Ye SJ. A machine learning *DOI:* http://dx.doi.org/10.5772/intechopen.84629 approach to the accurate prediction of multi-leaf collimator positional errors. Physics in Medicine and Biology. 2016;**61**(6):2514-2531. DOI: 10.1088/0031-9155/61/6/2514

[203] Liu C, Kim J, Kumarasiri A, Mayyasa E, Browna S, Wen N, et al. An automated dose tracking system for adaptive radiation therapy. Computer Methods and Programs in Biomedicine. 2018;**154**:1-8. DOI: 10.1016/j.cmpb.2017.11.001

[204] Guidi G, Maffei N, Meduri B, D'Angelo E, Mistretta GM, et al. A machine learning tool for re-planning and adaptive RT: A multicentre cohort investigation. Physica Medica. 2016;**32**(12):1659-1666. DOI: 10.1016/j.ejmp.2016.10.005

[205] M. Feng, G. Valdes, N. Dixit, and T. D. Solberg. Machine learning in radiation oncology: opportunities, requirements, and needs. *Frontiers in Oncology*, 8, (2018), 110.

[206] Tseng HH, Luo Y, Cui S, Chien JT, Ten Haken RK, Naqa IE. Deep reinforcement learning for automated radiation adaptation in lung cancer. Medical Physics. 2017;**44**:6690-6705. DOI: 10.1002/mp.12625

[207] Varfalvy N, Piron O, Cyr MF, Dagnault A, Archambault L. Classification of changes occurring in lung patient during radiotherapy using relative γ analysis and hidden Markov models. Medical Physics. 2017;**44**: 5043-5050. DOI: 10.1002/mp.12488

[208] Lee S, Ybarra N, Jeyaseelan K, Faria S, Kopek N, Brisebois P, et al. Bayesian network ensemble as a multivariate strategy to predict radiation pneumonitis risk. Medical Physics. 2015;**42**(5):2421-2430. DOI: 10.1118/1.4915284

[209] Naqa IE, Deasy JO, Mu Y, Huang E, Hope AJ, Lindsay PE, et al. Datamining approaches for modelling tumor control probability. Acta Oncologica. 2010;**49**(8):13631373.DOI:10.3109/02841861003649224

[210] Zhen X, Chen J, Zhong Z, Hrycushko B, Zhou L, Jiang S, et al. Deep convolutional neural network with transfer learning for rectum toxicity prediction in cervical cancer radiotherapy: A feasibility study. Physics in Medicine and Biology. 2017;**62**(21):8246-8263. DOI: 10.1088/1361-6560/aa8d09

[211] Deist TM, Dankers FJWM, Valdes G, Wijsman R, Hsu IC, Oberije C, et al. Machine learning algorithms for outcome prediction in (chemo)radiotherapy: An empirical comparison of classifiers. Medical Physics. 2018;**45**(7):3449-3459. DOI: 10.1002/mp.12967

[212] Yahya N, Ebert MA, Bulsara M, House MJ, Kennedy A, Joseph DJ, et al. Statistical-learning strategies generate only modestly performing predictive models for urinary symptoms following external beam radiotherapy of the prostate: A comparison of conventional and machine-learning methods. Medical Physics. 2016;**43**(5):2040. DOI: 10.1118/1.4944738

[213] Zhang HH, D'Souza WD, Shi L, Meyer RR. Modeling plan-related clinical complications using machine learning tools in a multiplan IMRT framework. International Journal of Radiation Oncology, Biology, Physics. 2009;**74**(5):1617-1626. DOI: 10.1016/j.ijrobp.2009.02.065

[214] Kang J, Schwartz R, Flickinger J, Beriwal S. Machine learning approaches for predicting radiation therapy outcomes: A Clinician's perspective. International Journal of Radiation Oncology, Biology, Physics. 2015;**93**(5):1127-1135. DOI: 10.1016/j.ijrobp.2015.07.2286

[215] Baumann M, Krause M, Overgaard J, Debus J, Bentzen SM, Daartz J, et al. Radiation oncology in the era of precision medicine. Nature Reviews. Cancer. 2016;**16**(5):234-249. DOI: 10.1038/nrc.2016.18

[216] El Naqa, J. Irrer, T. A. Ritter, J. DeMarco, H. AlâĂŘHallaq, J. Booth, and J. M. Moran. Machine learning for automated quality assurance in radiotherapy: a proof of principle using EPID data description. *Medical Physics*, 46(4), (2019), 1914–1921.

[217] E. C. Ford, S. Terezakis, A. Souranis, K. Harris, H. Gay, and S. Mutic. Quality control quantification (QCQ): a tool to measure the value of quality control checks in radiation oncology. *International Journal of Radiation Oncology* Biology* Physics*, 84(3), (2012), e263–e269.

[218] Depeursinge A, Yanagawa M, Leung AN, Rubin DL. Predicting adenocarcinoma recurrence using computational texture models of nodule components in lung CT. Medical Physics. 2015;**42**(4):2054-2063. DOI: 10.1118/1.4916088

[219] Lambin P, Rios-Velazquez E, Leijenaar R, Carvalho S, van Stiphout RG, Granton P, et al. Radiomics: Extracting more information from medical images using advanced feature analysis. European Journal of Cancer. 2012;**48**(4):441-446. DOI: 10.1016/j.ejca.2011.11.036

[220] Wu J, Tha KK, Xing L, Li R. Radiomics and radiogenomics for precision radiotherapy. Journal of Radiation Research. 2018;**59**(suppl_1): i25-i31. DOI: 10.1093/jrr/rrx102

[221] Lao J, Chen Y, Li ZC, Li Q , Zhang J, Liu J, et al. A deep learning-based radiomics model for prediction of survival in glioblastoma multiforme. Scientific Reports. 2017;**7**:10353. DOI: 10.1038/s41598-017-10649-8

[222] Gulshan V, Peng L, Coram M, Stumpe MC, Wu D, Narayanaswamy A, et al. Development and validation of a deep learning algorithm for detection of diabetic retinopathy in retinal fundus photographs. The Journal of the American Medical Association. 2016;**316**:2402-2410. DOI: 10.1001/jama.2016.17216

[223] Mayo CS, Moran JM, Bosch W, Xiao Y, McNutt T, Popple R, et al. American Association of Physicists in Medicine Task Group 263: Standardizing nomenclatures in radiation oncology. International Journal of Radiation Oncology, Biology, Physics. 2018;**100**(4):1057-1066. DOI: 10.1016/j.ijrobp.2017.12.013

[224] Luo W, Phung D, Tran T, et al. Guidelines for developing and reporting machine learning predictive models in biomedical research: A multidisciplinary view. Journal of Medical Internet Research. 2016;**18**(4):e323. DOI: 10.2196/jmir.5870

[225] Parodi S, Riccardi G, Castagnino N, Tortolina L, Maffei M, Zoppoli G, et al. Systems medicine in oncology: Signaling network modeling and new generation decision-support systems. Methods in Molecular Biology. 2016;**1386**:181-219. DOI: 10.1007/978-1-4939-3283-2_10

[226] Memorial Sloan Kettering Cancer Center. Watson oncology. n.d. https://www.mskcc.org/about/innovativecollaborations/watson-oncology [Accessed:30January2019]

[227] Kohn MS, Sun J, Knoop S, Shabo A, Carmeli B, Sow D, et al. IBM's health analytics and clinical decision support. Yearbook of Medical Informatics. 2014;**9**:154-162. DOI: 10.15265/IY-2014-0002

[228] Bibault JE, Giraud P, Burgun A. Big data and machine learning in radiation oncology: State of the art and future prospects. Cancer Letters. 2016;**382**(1):110-117. DOI: 10.1016/j.canlet.2016.05.033

[229] S. Parodi, G. Riccardi, N. Castagnino, L. Tortolina, M. Maffei, G. Zoppoli, et al., Systems medicine in oncology: signaling network modeling and new-generation decision-support systems, Methods Mol. Biol. 1386 (2016) 181–219, doi:10.1007/978-1-4939-3283-2_10.

[230] Memorial Sloan Kettering Cancer Center, Watson oncology. n.d. <https://www.mskcc.org/about/innovative-collaborations/watson-oncology> (accessed 10.03.16).

[231] The coming era of human phenotyping, Nat. Biotechnol. 33 (2015) 567, doi:10.1038/nbt.3266.

[232] N. Savage, Mobile data: made to measure, Nature 527 (2015) S12–S13, doi:10.1038/527S12a.

[233] N. Servant, J. Roméjon, P. Gestraud, P. La Rosa, G. Lucotte, S. Lair, et al., Bioinformatics for precision medicine in oncology: principles and application to the SHIVA clinical trial, Front. Genet. 5 (2014) 152, doi:10.3389/ fgene.2014.00152.

[234] R. C. Chen, P. E. Gabriel, B. D. Kavanagh, T. R. McNutt, How will big data impact clinical decision making and precision medicine in radiation therapy, Int. J. Radiat. Oncol. Biol. Phys. (2015) doi:10.1016/j.ijrobp.2015.10.052.

[235] A. Berrington de Gonzalez, R. E. Curtis, S. F. Kry, E. Gilbert, S. Lamart, C. D. Berg, et al., Proportion of second cancers attributable to radiotherapy treatment in adults: a cohort study in the US SEER cancer registries, Lancet Oncol. 12 (2011) 353–360, doi:10.1016/S1470-2045(11)70061-4.

[236] B. A. Virnig, J. L. Warren, G. S. Cooper, C. N. Klabunde, N. Schussler, J. Freeman, Studying radiation therapy using SEER-Medicare-linked data, Med. Care 40 (2002) IV–49–54,doi:10.1097/01.MLR.0000020940.90270.4D.

[237] S. C. Darby, P. McGale, C. W. Taylor, R. Peto, Long-term mortality from heart disease and lung cancer after radiotherapy for early breast cancer: prospective cohort study of about 300,000 women in US SEER cancer registries, Lancet Oncol. 6 (2005) 557–565, doi:10.1016/S1470-2045(5)70251-5.

[238] X. Du, J. L. Freeman, J. S. Goodwin, Information on radiation treatment in patients with breast cancer: the advantages of the linked medicare and SEER data. Surveillance, epidemiology and end results, J. Clin. Epidemiol. 52 (1999) 463–470.

[239] Y. Song, W. Wang, G. Tao, W. Zhu, X. Zhou, P. Pan, Survival benefit of radiotherapy to patients with small cell esophagus carcinoma – an analysis of Surveillance Epidemiology and End Results (SEER) data, Oncotarget (2015) doi:10.18632/ oncotarget.6764.

[240] B. Wu, T. McNutt, M. Zahurak, P. Simari, D. Pang, R. Taylor, et al., Fully automated simultaneous integrated boosted-intensity modulated radiation therapy treatment planning is feasible for head-and-neck cancer: a prospective clinical study, Int. J. Radiat. Oncol. Biol. Phys. 84 (2012) e647–e653, doi:10.1016/ j.ijrobp.2012.06.047.

[241] B. Wu, F. Ricchetti, G. Sanguineti, M. Kazhdan, P. Simari, R. Jacques, et al., Data-driven approach to generating achievable dose-volume histogram objectives in intensity-modulated radiotherapy planning, Int. J. Radiat. Oncol. Biol. Phys. 79 (2011) 1241–1247, doi:10.1016/j.ijrobp.2010.05.026.

[242] S. F. Petit, B. Wu, M. Kazhdan, A. Dekker, P. Simari, R. Kumar, et al., Increased organ sparing using shape-based treatment plan optimization for intensity modulated radiation therapy of pancreatic adenocarcinoma, Radiother. Oncol. 102 (2012) 38–44, doi:10.1016/j.radonc.2011.05.025.

[243] L. M. Appenzoller, J. M. Michalski, W. L. Thorstad, S. Mutic, K. L. Moore, Predicting dose-volume histograms for organs-at-risk in IMRT planning, Med. Phys. 39 (2012) 7446–7461, doi:10.1118/1.4761864.

[244] X. Zhu, Y. Ge, T. Li, D. Thongphiew, F.-F. Yin, Q. J. Wu, A planning quality evaluation tool for prostate adaptive IMRT based on machine learning, Med. Phys. 38 (2011) 719–726.

[245] S. P. Robertson, H. Quon, A. P. Kiess, J. A. Moore, W. Yang, Z. Cheng, et al., A data-mining framework for large scale analysis of dose-outcome relationships in a database of irradiated head and neck cancer patients, Med. Phys. 42 (2015) 4329–4337, doi:10.1118/1.4922686.

[246] Klonowska K, Czubak K, Wojciechowska M, et al. Oncogenomic portals for the visualization and analysis of genome-wide cancer data. Oncotarget 2016;7(1):176–92. Jan 5.

[247] Christoph J, Knell C, Bosserhoff A, et al. Usability and suitability of the omicsintegrating analysis platform tranSMART for translational research and education. Appl Clin Inform. 2017;8(4):1173–83.

[248] He S, Yong M, Matthews PM, et al. TranSMART-XNAT connector tranSMART-XNAT connector-image selection based on clinical phenotypes and genetic profiles. Bioinformatics 2017;33(5):787–8. Mar 1.

[249] Hoogstrate Y, Zhang C, Senf A, et al. Integration of EGA secure data access into galaxy. F1000Res 2016;5. https://doi.org/10.12688/f1000 research.10221.1.Dec12pii:ELIXIR-2841.eCollection2016.

[250] Eijssen L, Evelo C, Kok R, et al. The Dutch techcentre for life sciences: enabling dataintensive life science research in the Netherlands. F1000Research 2015.https://doi.org/10.12688/f1000research.6009.2 . S. M. Willems, et al. *Oral Oncology 98 (2019) 8–1*212

10

Big Data in Drug Discovery, Development, and Pharmaceutical Care

Dhanalekshmi Unnikrishnan Meenakshi[1*], Shah Alam Khan[1], and Arul Prakash Francis[2]

[1]College of Pharmacy, National University of Science and Technology, Sultanate of Oman
[2]Centre for Molecular Medicine and Diagnostics (COMManD), Department of Biochemistry, Saveetha Dental College and Hospitals, Saveetha University, India
*Corresponding Author: College of Pharmacy, National University of Science and Technology, P.O. Box 620, P.C. 130 Muscat, Oman, EMail: dhanalekshmi@nu.edu.om, Phone: [+968] 24235000, Fax: [+968] 24504820, ORCID ID: 0000-0002-2689-4079.

Abstract

The role of big data (BD) in the medical arena extends from the discovery and development of drugs to diagnosis and pharmaceutical care. The scope of BD's use in drug discovery and development, predictive modeling, identification and discovery of new drug targets, and the development process are all covered in this book chapter. The impact of BD information in the field of oncology, the reason for the integration of various sources of clinical, pathological, and quality-of-life data for the precise clinical decision have also been touched upon. The critical role of BD in clinical trial management and data analysis is briefly covered. It also spotlights the significance of drug development and considers how BD might enhance several aspects such as drug development, clinical research methodology, and identification of toxicities. Artificial intelligence (AI), machine learning (ML), and BD analytics will all play a bigger role in drug development in the coming years, transforming drug composition and molecular building. Organizations that really can collect such large datasets and use them for drug research and development, and also scale-up and produce pharmacological drugs, would have a significant strategic advantage. Data mining will increase our knowledge in this domain and result in better therapeutics for a larger spectrum of patients; therefore the search for the proper computation technology is the competition in the bio-pharmaceutical industry.

Keywords: AI, Drug Discovery, Drug Development, Big Data, Prediction, Data Analysis, Patients, Oncology, Target, Proteins, Clinical Trials.

10.1 Introduction

The term BD in healthcare commonly refers to large patient-level data sources like computerized patient records, scanning data, patient clinical data, therapeutic data, tracking devices' data, and certain other challenging data. BD is working toward another comprehensive integration of the chemistry subspace to the target area in drug discovery. Analysis of BD could be able to generate evidence-based data that may address many of the questions of drug discovery and development related especially to cancer care and therapeutic improvement. The data elements are pooled or processed to glean new insights. BD derived from electronic health records (EHR) offers a wealth of real-world evidence for computer-aided diagnosis and prescriptive analysis. Traditionally, finding new targeted medications has been a complicated journey that has cost thousands of dollars and has taken over a decade. In the early days, scientists used the traditional experimental methods to identify a therapeutic drug target. Later, the drug target identification was made easy as structural biologists interpreted three-dimensional (3D) architectures and associated ligand-binding features. With the help of high-throughput screening techniques, medicinal chemists and pharmacologists successfully screened promising lead chemical constituents for future risk evaluation and clinical studies. The strong track record of protein kinase inhibitor design has aided in the advancement of structure-based pharmaceutical design and docking methods. Oncology researchers, with the help of tumor cell line sequencing and sketching, were requested to shed some insights on chemical processes as well as physiological processes which produce susceptibility in malignant progenitors. This database boom, fueled by technology, is indeed speeding up biological research, although it is quickly evolving the cancer arena into data science. Such development seems difficult due to the enormous complexity and species diversity of individuals and malignancies. Although clinical trials are vital to the drug development process, they have certain limitations, viz. they are expensive, involve a long duration, and face hurdles in the approval process.

The *in vivo* drug target (enzymes, ion channels, nuclear receptors, and G protein-coupled receptors (GPCRs)) interactions demonstrate the drug's efficacy. For that reason, identification of drug–target interactions (DTI) is vital in clinically associated areas like drug discovery, drug repositioning, side-effect prediction, and drug resistance [1]. Computer-aided drug design (CADD), is an interdisciplinary technique that combines advanced bioinformatics techniques and sophisticated computational algorithms. CADD is the most innovative and cost-effective method for target identification and screening lead compounds compared to traditional drug discovery techniques. CADD selectively chooses the best candidates through molecular docking by introducing the ligand into the ligand promoter regions of a possible molecular target. Molecular docking studies calculate each docked compound's binding energy. The Protein Data Bank (PDB) has developed exponentially and delivered access to more than 1,44,000 entries of macromolecules, including proteins and nucleic acids. Over 84,000 of them seem to have tiny active ingredients crosslinked with themselves, like cofactors, inhibitors, and medications [1, 2]. CADD became the foundation for commercial drug development, and scientific work resulted in high drug-screening success ratios (over 100-fold, ranging between 0.01% and 2%) [4].

Machine learning (ML), a computer technology intended for building predictive models for data analysis using datasets, has become a significant resource in modern biological research. ML has become a mainstream process involved in DTI prediction studies for analyzing and solving problems [5–7]. The key factors promoting ML as a mainstream method of DTI prediction are the actual data backdrop, sophisticated production technology, present state, and demands. A large set of genes were mapped and several molecules have been generated with the development of sequencing technology, high-throughput technology, and CADD. The databases have been constructed and linked data has been structured based on the prevailing associated mechanisms and accrued practice. The statistics available in these databases delivers an excellent data substance for ML to solve problems associated with DTI prediction.

BD analytics uses the investigation reports of numerous clinical trials as a basis to predict the treatment effects. Publicly available clinical trial data at no cost is a vital resource for health services researchers. Clinical studies frequently focus on the entire therapeutic outcomes of the total research sample, rather than the homogeneity of the subgroups. These flaws can be addressed by utilizing systematic review and/or meta-regression analysis to combine data from many clinical studies. This informatics aids in the interpretation of enormous amounts of data and the resolution of numerous concealed issues. Predicting the fate of illnesses using BD in medical information systems improves treatment and life quality of people by averting unnecessary deaths and disease onset [8]. It is useful in medical practice for prediction and diagnosis, as well as in epidemiology investigation, because BD provides a large quantity of information [9]. Tailored medicines often have a lower toxic effect than standard chemotherapy; nevertheless, continuous administration leads to less harm that persists. Researchers were able to accurately forecast treatment-related efficacy or death by merging data from various clinical studies that underpinned the approval of novel cancer medicines [10]. BD identifies the right population or target group through data management, EHR, and data analysis. The presence of BD enables the formulation or adaption of such a strategy or treatment to target a specific medical issue by providing a more detailed portrait of the community and overall health history. Clients, caretakers, merchants, and research scientists have all contributed to the pharmaceutical industry's expansion. Pharmaceutical businesses could benefit from BD's help in identifying novel prospective and successful medications and getting them to patients faster. A considerable amount of information engendered through numerous foundations, such as electronic records and patient health-related datasets, should be analyzed effectively to face new challenges in the healthcare industry. Moreover, updatable technology is necessary for taking attention to this extensive dataset in the future. With the assistance of multiple ML algorithms, BD yields valuable findings based on data configurations and evaluates the link between distinct datasets. Analytical methods help to extract meaningful information from the data. A new revolution can be created in the healthcare industry using BD [11]. The application of numerous analytical techniques, data gathering, and machine learning approaches in BD discovery give many chances for developing disease cures employing different analytic equipment. This chapter examines the use of BD in drug discovery and development, predictive modeling, and identification as well as innovation of new therapeutic targets. It also emphasizes the aspects of the drug formulation and establishment process. The impacts of BD information in cancer as well as the rationale of combining diverse sources of clinical, pathological, and quality-of-life data for precise treatment decisions, have been discussed. It also examines the current status of drug development and considers how BD might help with several aspects of the process, including drug discovery, clinical trial design, and adverse drug reaction (ADR) identification. AI, ML, and BD analytics will all play a bigger role in medication research shortly, transforming drug design, discovery, and formulation.

10.2 Role of BD in Drug Candidate Selection, Drug Discovery, and Development

Despite tremendous advancement in technology and understanding of the pathogenesis of diseases at the molecular level, the drug discovery and development process remains a multifaceted, challenging, laborious, and expensive affair. A new drug usually requires around 13 years and an expenditure of more than 1 billion dollars to reach the clinical setting, i.e., from bench to bedside [12]. The majority of drug candidates showing promise in preclinical studies fail in phase I clinical trials mainly due to pharmacokinetic incompatibilities, inflicting a significant financial burden on manufacturers [13]. *In vitro* and *in silico* studies have the potential to increase the success rate of identifying the therapeutic agents by eliminating the toxic and inactive compounds with a poor pharmacokinetic profile. However, results of *in vitro* and *in silico* studies show considerable variation upon *in vivo* testing, and thus therapeutic actions and efficacy are poorly correlated. QSAR (quantitative structure–activity relationship) is a computational modeling method that predicts biological activities and pharmacokinetic profiles of new chemical substances based on the physicochemical properties of chemical fragments database, but this approach alone is not enough to overcome the challenges of conventional drug discovery [14]. In the late 20th century, the high-throughput screening (HTS) technique of combinatorial chemistry that evaluates hundreds to millions of compounds from the large chemical libraries using a rapid and standardized protocol became quite a popular drug discovery tool in pharmaceutical industries [15]. The enormous amount of biological (chemical response on-target) data obtained through HTS is growing at a very fast pace and constitutes a part of the BD era for drug development. BD is a term that describes an extremely large dataset or a vast collection of data that is bigger than one terabyte (TB) and requires special computer-based algorithmic programs for analysis to divulge patterns, trends, and associations. AI is used to construct models to extract the information from BD to expedite the drug discovery process, although handling this big data poses some serious challenges. Four Vs, namely, volume (memory to store the data), variety (structured/unstructured data format), velocity (rate of new data availability), as well as veracity (accuracy of data) are the attributes of BD. An enormous volume of data on chemical compounds and their bioactivity is available to the scientific community for data mining, exploration, analysis, and to discover drugs to prevent and cure diseases. However, as stated earlier, the vastness of BD poses a major challenge in its analysis for innovation. Some of the BD-sharing projects available for drug screening are listed in Table 10.1.

10.2.1 CADD, QSAR, and chemical data-driven techniques

Computer-aided drug design (CADD) is an example of rational drug design. Virtual screening (VS) is considered the most important and materialized tool of CADD for drug discovery. It involves the identification of a set of targeted lead candidates through the search of small-molecule databases based on their interaction or QSAR studies with the target protein [16]. Target selection among the four large macromolecules, viz., proteins (GPCR, ion channels, and enzymes), polysaccharides, lipids, and nucleic acids is the first important step in the VS. In the next step, small molecules from the chemical databases such as ZINC and PubChem are retrieved and then docked with the selected target using a suitable molecular docking software such as Autodock, MolDock, Chimera, Maestro, DOCK, Gold, AADS, FlexX, etc. Ligands with high docking scores calculated based on binding energy are selected, and *in vitro* and *in vivo* biological experiments are performed to verify their potential. The most potent compound(s) could be further evaluated in clinical trials.

Meticulous and unerring analysis of BD bases can generate evidence-based data that may address many of the questions of drug discovery and development, especially related to cancer care and therapeutic development. Companies that can curate this information and utilize the information extracted from these data to quicken the progression of drug discovery as well as development will have an extremely competitive advantage in many aspects. It will decrease the attrition and improve the success rate of developing a new drug which will result in lowering the development cost as well as efficient scale-up and manufacturing of pharmacotherapeutic agents for better therapies. Zhavoronkov *et al.* (2019) used DL methods and progressed from computational methods to preclinical animal studies within seven weeks to identify potent DDR1 kinase inhibitors [17]. BD performs a pivotal part in the refining of clinical trials by data analysis, quick and correct clinical diagnosis through omics data management, identification/selection of drug candidates, and drug repurposing using complex AI algorithms. Nowadays, companies

Table 10.1 Some selected BD-sharing projects for drug screening.

S. No	Name of the database	Description	Webpage
1	PubChem	A publicly accessible big data repository containing more than 97 million chemical substances and over a million bioassays providing target response information.	https://pubchem.ncbi.nlm.nih.gov/
2	ChEMBL	It provides manually curated interactions and pharmacokinetic (ADMET) parametric information from literature for a large number of compounds. It enlists bioactivity data for approximately 18 million chemical substance-target pairs arising out of the testing of 2.1 million compounds against 14,500 targets. The European Bioinformatics Institute (EBI) is the data custodian.	https://www.ebi.ac.uk/chembl/
3	DrugBank	Publicly available dataset listing all approved drugs with their mechanisms of action, interactions, and therapeutic targets. The most recent released version contains 12,110 medication records comprising officially marketed drugs (2,553), sanctioned biotechnology-based medications (1,280), biologics and nutraceuticals (130), and investigational medications (>5,842).	https://go.drugbank.com/
4	DrugMatrix	It provides toxicogenomic data of drugs from tissues, especially the liver, obtained by administering over 600 drugs to experimental rodents. It helps reduce the time to formulate a xenobiotic's potential for toxicity.	https://ntp.niehs.nih.gov/data/drugmatrix/

Table 10.1 (Continued.)

5	Binding Database	It provides drug–target binding (protein/enzymes) information. It contains 1.59 million binding data points obtained through the binding of 0.71 million small molecules to the 7,235 protein targets.	https://www.bindingdb.org/bind/index.jsp
6	PubMed	Detailed information of billions of biomedical literatures.	https://pubmed.ncbi.nlm.nih.gov/
7	SureChEMBL	It is a very good tool for patent literature. It automates the excavation of biochemical data assemblies from the documents of intellectual property rights.	https://www.surechembl.org/search/
8	ZINC	It is among the most comprehensive collections of commonly produced active compounds and is freely accessible. The information of compounds in this database is used for virtual screening.	https://zinc.docking.org/
9	CSD, COD	Cambridge structural database (CSD) and Crystallography Open Database (COD) contain crystal structures of experimentally derived small molecules.	https://www.ccdc.cam.ac.uk/solutions/csd-core/components/csd/ http://www.crystallography.net/cod/
10	PDB	Protein data bank (PDB) provides free access to more than a hundred thousand protein structures. It is very useful for structure-based drug design.	https://www.rcsb.org

are leveraging the power of BD to screen and shortlist the drug candidates using available data. Information mined from BD is used to build models to predict drug-likeness, toxicity, ligand-protein interaction, metabolism, drug interaction, bioavailability, side effects, etc. NuMedii Company established in 2010 in Menlo Park, California, is one of the early advocates of the use of BD in the drug discovery process [18]. The proprietary technology of this bioinformatics company uses distinctive features of diseases to predict drug efficacy by correlating drug data with disease information, thereby helping pharmaceutical industries identify drugs in the pipeline for further development and improving the chances of these drugs entering the market. It employs network-based algorithms to mine billions of points of disease, pharmacological, and clinical data from its database.

Another company, GenoKey, uses GPU processing to analyze healthcare BD to extract information to provide patterns in the data and solve very large combinatorial problems in case-control data. MedChemica, a UK-based company, has collaborated with leading pharmaceutical industries to speed up drug development using data mining of preclinical, precompetitive-shared data (more than 1.2 million data points). It uses matched molecular pair analysis software to analyze the BD obtained during the iterative development process to reduce the steps between hit and drug candidate. The identified closely matched pairs are further analyzed and mapped for differences based on *in vitro* data to tag the biological activity with the structural changes. The result would then be utilized to forecast the simulated substances' biological processes and toxicities. Some early cheminformatics predictive models developed based on ML were exploited to acquire information on ADMET characteristics of molecules. These models employed homologous structural compounds to study QSAR via simple linear regression analysis of training datasets [19]. Merck recently used a DL-based multitask deep feed-forward neural network for QSAR studies that yielded much better results than those produced by conventional QSAR predictive models including random forests [20]. DL methods such as convolutional neural networks (CNN) have also produced much better results in the case of structure-based virtual screening in the prediction of drug–protein interaction [21, 22].

The result is credited to the use of ML to correctly forecast secondary and tertiary protein structures using ultra-DL models [23]. The successful automation and streamlining of many analytical processes have resulted from the process of addressing universal data management concerns. Insights from varied datasets/data-sharing projects, as mentioned earlier, speed up research and early development of drugs by allowing for better decision-making, and also smoothing the repurposing of current medications for new therapeutic areas. Integration of multilayer omics databases, mining of different kinds of literature, drug targets, and drug combination analysis reports, and statistical and computational risk assessments and reports are some of the data-driven technological aspects. Despite depending on preconceived biological explanations, genomic sequence investigations that link germline DNA polymorphisms to clinical characteristics have enabled the detection of several ailment-susceptibility genes and the faster development of therapeutic methods.

ML techniques are equally useful in genomics to predict genome segmentation, the role of coding regions in pathogenesis of rare and common diseases [24, 25], splice sites prediction [26], transcriptional promoters and enhancers [27, 28], microarray data from analysis of RNA-Seq gene expression data [29], transcription factor-binding predictions [30], and most importantly, protein druggability [31], etc. Overall, the processes used for drug target identification are expensive and tiresome. In November 2018, research was been carried out to estimate the overall cost of experiments required to implement innovative Food and Drug Administration (FDA)-authorized medications. Remarkably, the study results showed that the average price utilized for the efficacy trials of 59 new FDA-approved drugs during 2015–2016 was $19 million [32]. To address the constraints of traditional drug target identification strategies, effective, minimal-cost, and analytical methodologies are required. Among more than 20,000 genes that can express one or more peptides, only a few were identified as pharmacologically active candidates for already licensed medications. The PubChem database currently contains 111 million chemicals; however, proteins that potentially interact with the majority of such molecules are nevertheless unidentified.

10.2.2 Biological BD

Accepting the pathophysiology of each ailment at the molecular level, identification, and validation of new targets are of paramount importance in discovering better therapeutic modalities. Biological BD is complex, huge, and diverse, but systematic analysis of it plays a prominent role in the drug discovery and development process, in the identification of lead candidates, validation, and development of novel therapies for rare diseases including unmet medical conditions. Some data-driven approaches have resulted in the translation of genomics research data from the laboratory to the approval in a clinical setting. Aspirin (ASA), a century-old analgesic drug, has been recommended by the US Preventive Services Task (UPST) Force in the treatment of colorectal cancer. The repurposed use of ASA was based on the integration of information obtained from electronic health records of patients, post-marketing surveillance data, experimental evidence on ASA potential in the treatment of colorectal cancer, and pharmacological analysis [33]. Raloxifene is another example of a repurposed drug. It is an antiosteoporotic drug but authorized by FDA in 2007 for the treatment of invasive breast cancer [34].

Over the past two to three decades, many anticancer drugs have been developed utilizing omics databases of cancer research. The computational data mining of transcriptome data in the cancer genome atlas (TCGA), the human protein atlas (THPA) databases have helped in the documentation of cancer-related proteins as specific drug targets [35, 36]. Liu and colleagues have devised a revolutionary method for discovering chemotherapeutic medicines. By using the same method, their group discovered 73 novel molecules and 12 FDA-approved therapeutic agents acting at more than 30 novel targets at a much lower cost and in a short span in comparison to traditional drug discovery methods. They carried out data mining using TCGA and THPA to obtain information about targets related to cancer prognosis, followed by computational correlation. In the next step, 3D protein structure information was retrieved from PDB and then experimental studies were carried out [37]. Imatinib, a protein kinase inhibitor (PKI) used to treat various cancers, including chronic myeloid leukemia, was also developed using a therapeutic target approach by analysis of BD. The chemical databases were screened by investigators to identify bcr-abl protein inhibitors. High-throughput screening of chemical libraries resulted in the identification of 2-phenylaminopyrimidine as a lead compound. It was further structurally modified by adding methyl and benzamide functionalities, to obtain imatinib with improved binding properties [38].

10.2.3 Applications of BD in drug discovery

BD analytics has helped the pharmaceutical business expedite especially the drug discovery, development, and screening processes (Figure 10.1). BD notably aids scientists in tracing the origins of molecules and, from there, developing novel drugs. This technique is particularly useful in the pharmaceutical sector since it allows researchers to trace a drug's origins step by step and then investigate the attributes of each reaction engaged in the process. The pharmaceutical sector requires predictive analytics because drug discovery, production, and manufacturing are tasks that require extreme precision and excellence. Large volumes of data must be gathered and processed for effective patterns to lead to strategic decisions, and BD enables the creation of precise prediction patterns. Following are some of the common applications of BD in drug discovery:

- Predictions of molecular properties: molecular weight (M.Wt), calculated log*P* (clog*P*), and molecular similarities to identify bioactive compounds

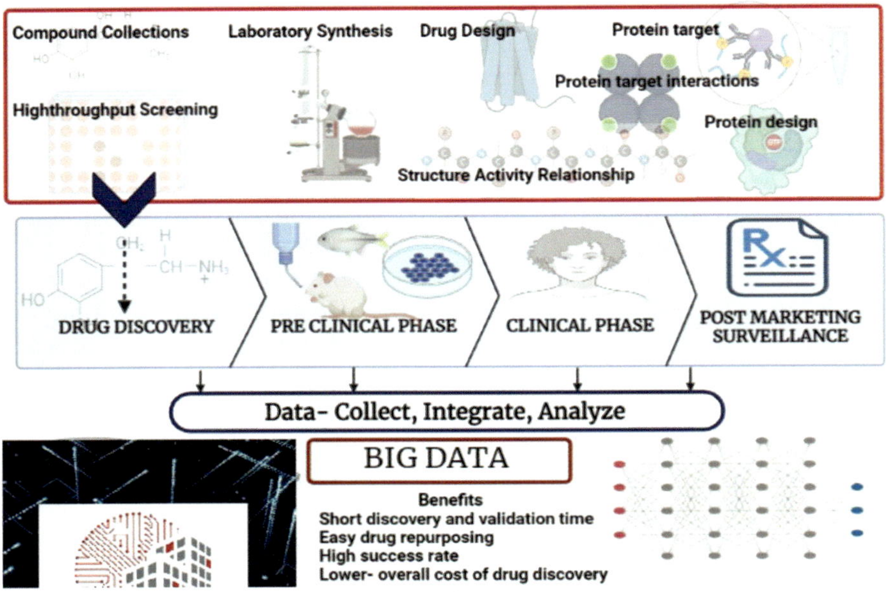

Figure 10.1 Description of Big Data in Drug Discovery, Development, and Screening Process for effective pharmaceutical care.

- De novo molecular design using genetic algorithm
- Optimization of biological activities of drug candidates
- Multiple synthetic routes planning
- Bioisosteric replacement and scaffold-hopping molecular design strategies
- Optimization of lead candidate
- Inference and prediction of clinical toxicity

10.3 Drug–Target Interactions (DTI)

With the rise of protein–protein interaction research, a vast number of interfaces could be evaluated in a high-throughput manner, resulting in massive datasets. In recent trends, the majority of target bio-molecules have been proteins, with four protein categories contributing to 44% of protein targets (kinases, GPCRs, ion channels, and nuclear receptors). Surprisingly, all four protein family members represent their targets of 70% of presently produced medications [39, 40]. The majority of computational approaches use the available datasets to see if a medicine could bind with a specific protein. Drug–target compatibility, which characterizes the interface between both the therapeutic agent and the targeting protein, has indeed been observed in various investigations. For estimating drug–target selectivity, the kinases and KIBA databases are often employed [41, 42]. Researchers are permitted to collect datasets from the database based on their needs. UniProt, PubChem, DrugBank, KEGG, and BindingDB are the few among them [43–45]. Many toolkits and web servers including STITCH, SwissTargetPrediction, RDkit, OpenChem, iFeature, and Pse-in-one have been developed to solve problems in DTI prediction [37, 46, 47]. Furthermore, for several developing, extremely contagious, and lethal novel pathogens, such as H7N9, SARS, Mers, Ebola, and the COVID-19 virus, the standard protocols of wet trials are insufficient [48]. Because of the enormous volumes of existing information on ailments that lead to severe social and mental health risks, computing methods are used to predict DTI and improve the efficiency of the drug. To make forecasts with great results, published works on DTI forecasting utilize various computation or evaluation methodologies for dataset gathering, feature exploration, and extraction, including task algorithms design. Time-consuming long experimental cycles, inaccurate and biased experimental results due to redundant data, and unrepresentative trials can be reduced by using different data acquisition methods on model construction.

PrePPItar is a set of computational algorithms for predicting protein–protein interactions (PPIs) as molecular targets by revealing possible drug–PPI connections. Negative datasets were collected by Wang *et al.* (2010) through arbitrary choice to resolve the data inequity issue [49]. Researchers randomly choose a negative dataset to keep a balance among learning positives and negatives, information seems to be the same size as the training-positive datasets in the support vector machine plyometric training. In another experiment, random selection was used to extract negative examples to reduce the influence of the unverified negative models [50]. On the other hand, Mahmud *et al.* (2020) addressed the data imbalance problem in the Pdti-EssB model using random under-sampling [51].

10.4 BD Predictive Analytics and Analytical Techniques

Finding a few drug candidates for testing against a specific disease from a library of thousands, or perhaps a million, would be like looking for a needle in a haystack. Employing advanced algorithms to sift through massive databases of physiological, biochemical, and clinical information could allow identifying the possible targets for further research. Indeed, modern analytics can assist pharmaceutical businesses in analyzing the clinical urgency and candidate profiles to construct a product development pipeline.

10.4.1 ML- and DL-based analytical techniques

ML technique is a part of AI that produces a model to classify and predict the condition selected from available data. In other words, it enhances decision-making as it comprises numerous parameters than humans [52]. Statistical ML and network-based (NB) methods are among the computational tools used to analyze BD. These technologies have already shown considerable promise in bridging the gap between significant data creation and interpretation, but there is still much opportunity for improvement. ML techniques seem to be a central focus of BD evaluation to its recognized capacity to collaboratively extract (integrate) huge, diversified, and contradictory physiological types of data. This is a significant challenge in medical informatics [53]. The process in ML includes the generation of dependencies for a given dataset and the prediction of new outputs with the help of generated dependencies. ML methods are classified into supervised and unsupervised learning, based on the usage of data. Supervised algorithms use the data mapped to the anticipated production arrangement. Unsupervised learning algorithms, on the other hand, employ unlabeled instances to form a team based on the observed patterns discovered in the database. Sensitivity, specificity, accuracy, and the area-under-curve (AUC) are used to assess a model's performance on a validation sample. However, the utilization of validation methods such as the holdout method, random sampling, cross-validation, and bootstrapping is limited only to the single dataset available for testing [54]. ML methods include artificial neural networks (ANN), Decisional Trees (DTs), Support Vector Machine (SVM) technique, and Bayesian Networks (BNs). ML is limited in information value (missing, duplicated, noise, outliers) that determines the quality of algorithm [55]. Besides, a higher number of features (data type) implies more samples. A prior data processing is essential to enhance the robustness of the developed models as the data samples should be 5–10 times more numerous than the number of features [56]. The most important techniques used in ML are dimensionality reduction, and extraction (generating innovative data types from the unusual dataset, including the pertinent material). Nevertheless, the generation of many signatures with different features for a given question, such as prognosis, leads to a lack of agreement toward the components to be considered [57, 58].

Visual morphological characteristics and radiomic fingerprint detection, therapeutic prognosis, visual dosage measurement, plasma concentration modeling, radiation adaptation, and image production are some of the applications of deep learning (DL) in medicine [59–61]. Convolutional neural networks (CNNs), auto-encoder (AE), deep deconvolution neural network (DDNN), deep belief network (DBN), and transfer learning (TL) are the few techniques included in DL [62–64]. The implementations of DL are primarily confined to feature reduction or feature extraction, after which categorization on features extracted is performed using SVM or logistic regression. The application of virtual networks is that they have a higher model capacity than other machine learning approaches like SVM or logistic regression. Hence, particular caution should be applied while comparing the classification of small datasets of neural networks and other ML techniques [65].

10.4.2 Natural language processing

Natural language ;processing (NLP) is a segment of ML that helps understand, segment, explore, or translate text written in a natural language [66]. NLP plays a significant role in the identification of postoperative complications automatically by repurposing electronic medical records (EMR). EMR is a database created using chest radiographic reports or a clinical summary of the data collected from the patients [67]. NLP is categorized into low-level and high-level based on the tasks. Low-level NLP includes tasks such as sentence boundary recognition, tokenization, complex phrase deconstruction, and language categorization [68]. Greater tasks comprise grammatical error detection, particular phrases disambiguation, negation but also uncertain classification, connection extraction, chronological predictions, and information retrieval. NLP methods that help transform unstructured data into structured databases successfully could be a key factor in BD analysis [69].

10.5 BD and Its Applications in Clinical Trial Design and Pharmacovigilance

Clinical studies in patients for specific disease conditions aim to establish the effectiveness and safety of a pharmaceutical product and take 6–7 years and significant capital expenditure. However, just one out of every ten compounds that undergo such studies is cleared, resulting in a significant expenditure for the company [70]. These failures could be due to poor participant screening, a lack of technological needs, or a lack of resources. However, these failures can be reduced with the available digital medical data through the implementation of AI [71].

10.5.1 Clinical trial design

A clinical trial's effectiveness is based on the enrolment of suitable patients; else, 86% of trials fail [13]. By applying patient-specific genetic information, AI can help choose only a certain patient group for enrolment in Phase 2 and 3 clinical trials that could aid in predicting the occurrence of possible therapeutic targets in the individuals chosen or enrolled in the trials [72]. Clinical development of molecules and also predicting chemical constituents prior to the beginning of clinical trials using several facets of AI, such as predictive ML, as well as other rational methodologies, aid in the earlier detection of lead compounds that will pass medical testing while taking into account the patient's needs and treatment plan [71]. Participants dropping out of clinical studies contribute to 30% of drug trial failures, resulting in substantial enrolment demands for the trial's conclusion and in wasted time and money. It can be avoided by constantly reviewing the participants and assisting them in adhering to the clinical trial's protocol [13]. AiCure designed mobile apps that tracked medication consumption adherence by schizophrenia patients in a phase II trial, resulting in a 25% emphasis on patient compliance and guaranteeing the research trial's accomplishment [72]. While undertaking critical scientific and business judgments regarding drug development, medical scientists, biopharma directors, shareholders, and investment advisors all assess the potential to succeed [73, 74]. A medication candidate may not show statistical significance during a clinical study, and sometimes may be extremely successful in a specific category of the population. BD informatics could help to identify certain subgroups for whom a "failure" drug might have been effective. Individuals with unusual or hereditary illnesses will benefit from this pharmaceutical repositioning strategy, which may not be financially appealing for specialized product innovation but have a high clinical demand. Both the content threat and the contextual risk add to the overall danger of re-identification when clinical trial information is made anonymous. The context risk is calculated by taking into account three different sorts of re-identification attacks on a data source: (a) a deliberate assault by adversaries, (b) an unintentional re-identification by a data scientist who identifies someone whom they know, as well as (c) unauthorized access. The effectiveness of the three attacks is influenced by the preliminary precautions in place. The context is made up of several elements, the first of which are contractual controls that limit context risk. The residual risk is managed by cybersecurity safeguards, both of which are integral components. The scope of these precautions greatly reduces the total risk. Any residual anger is dealt with by modifying the data [75].

Ever since the 1990s, this notion of BD has evolved further into appropriate, broader, and deeper sets of data that are responsible for new medication research, enhanced clinical procedures, and medical finance in the healthcare sector [76]. A vast pool of digitized health records or observational methods, such as medication safety reports, medication prescriptions, and admission and discharge information, can be used for BD analytics [77]. Because there are a limited number of tested persons in a clinical trial, many rare side effects go undiscovered; as a result, it is vital to monitor medications long after they are launched in the public market.

10.5.2 Pharmacovigilance

"Pharmacovigilance" is a term used to describe the process of collecting, analyzing, and disseminating medication-adverse response information received during the post-marketing period [78, 79]. Data mining using drug safety incident archives and scientific journals take time; nevertheless, with the tech transformation, researchers are investigating whether BD might be utilized to investigate and track medication safety. Substantial progress in computing power and speed paved the way for the automation of drug safety surveillance signal detection [80]. BD usage in pharmacovigilance comprises modern electronic approaches for analyzing the enormous and rising quantity of ADEs in spontaneous reporting system (SRS) networks and other internet platforms. SRS databases are repositories for medical professionals, consumers, medical product companies, and other stakeholders to spontaneously report ADEs to regulatory agencies. To find new correlations between medications, ADEs, and health conditions, BD approaches are utilized to examine segments inside datasets. The desirability of employing BD for drug safety surveillance is obvious as this is the goal of patient safety [81].

Implications of BD for pharmacovigilance were explored by researchers from governmental authorities, the pharmaceutical sector, and medication safety organizations. Medication safety monitoring utilizing the BD technique has indeed been proven to be cost-effective, quick, and competent in uncovering previously unknown empirical connections between drug–ADE interactions [82, 83]. The FDA has specified many benefits of data mining for drug safety surveillance. Data mining automation helps to optimize sample selection and processing. Data mining also saves a lot of time by enabling the study of drug–ADE connections throughout a dataset at the same time. By combining reference databases on pharmaceutical biochemistry and physiology, data-mining tools help to identify the biological explanation for molecular signals and pathways [84]. Drug chemical characteristics and impacts on metabolic pathways and major organs can be connected to ADEs, assisting in the identification of ADE processes. In addition, such systematic approaches to drug safety surveillance can be predictive, offering the potential to identify potential ADEs before they are observed [83]. BD has some drawbacks, including a lack of specified requirements and validation methodologies, prejudice, misleading signals, interfering factors, and information discrepancies, despite its important significance in pharmacovigilance. Another significant difficulty in BD analytics is that the data supplied is inaccurate, incomplete, or redundant. Decision-makers may not well understand poor-quality data algorithms to make inferences about particular signal detection [85].

10.6 Assessing the Drug Development Risk Using BD

The processes and protocols of drug development are always a risky endeavor. When data is coupled with ML, an even more unbiased and objective evaluation of drug discovery potential may be made. The difficulty of developing drugs and also the probability that a pharmaceutical product will be authorized by the FDA, are hard to quantify. Because of the intricacies of drug biology and clinical trials, these tasks are extremely difficult. This default risk is frequently misinterpreted and mischaracterized, resulting in poor resource allocation and, as a result, a decline in total research and drug development productivity. Scientists developed a machine learning (ML) technique that delivers a more accurate and unbiased estimate of drug development risk than traditional statistical models [86, 87]. Using a mix of the newest analytical ML technologies as well as a unique IT framework, researchers can find novel correlations with clinical importance that cannot be recognized by individuals alone. The IT infrastructure was created to handle massive amounts of confidential documents. This helps them to find phenotypic and initial markers of clinical symptoms and recurrence in patients [75]. The University of Oxford's Big Data Institute (BDI) has created a project collaboration in order to enhance healthcare and medical innovation by making them more effective and targeted. If biopharma sectors and investors have an improved insight into the role of new medicines approval as well as more precise projections of the probability of clinical study accomplishment, they should be better able to determine the impacts of various drug-development initiatives and thus assign their assets more effectively [73].

10.7 Advantages of BD in Healthcare

In medical informatics, BD can help forecast the severity of diseases and epidemics, improve activity and standard of living, and protect against untimely death and sickness progression. BD provides the data associated with the disease and cautionary clues for the treatment. The usage of BD reduces the expense of treating diseases for quite a significant number of people [88]. BD is useful for clinical treatment and population investigation since it provides a vast amount of data. BD has been employed by a variety of organizations and establishments to develop strategies, programs, and pharmacological procedures such as pharmaceutical development. Patients who provide data connected to treatment decisions are highly popular these days, and they want assistance in the overall health assessment process. In addition, BD's function in maintaining the updated patient history and fulfilling health-related treatment always makes it the best choice for clinical practice [89]. BD has the proficiency to decrease the regency bias, i.e., the present actions are being reviewed more thoroughly than prior efforts in order to regain the situation that could lead to poor decisions.

Moreover, problematic operative issues could be overwhelmed with BD that saves time, value and improves the productivity. The infrastructure might also be updated, making it simpler to deliver accurate patient information and prescribe medications without interruption. It is difficult to discover and discuss treatment issues before they become unmanageable in predictive research. BD assists healthcare professionals in reducing risk and overcoming the challenge of materials implementation [11]. Data classification methods, electronic pharmaceutical records, and record investigation are all services that BD provides for the healthcare industry. It also has a diverse population database as well as specific data that may be used for risk evaluation and dissemination. BD would provide a more accurate picture of the population and medical issues. It assists pharmaceutical companies in identifying new techniques and novel treatments and bringing them to clients as rapidly as feasible [90].

10.8 BD Analytics in the Pharmaceutical Industry

To uncover patterns, test theories, and evaluate the efficacy of treatments, the pharmaceutical industry has traditionally relied on empirical data. The introduction of BD analytics is assisting Borges in the streamlining of numerous complex corporate procedures and increasing overall efficiency. Drug discovery, research and development, clinical trials, precision medicine, and sales and marketing all benefit from BD analytics. BD pharma is growing wiser at analyzing and driving success in its sales and marketing operations as competition in the life sciences industry heats up [91]. Pharma businesses can identify new markets by evaluating data from social media, demography, electronic medical records, and other sources [86]. Furthermore, they may assess the efficiency of their activities and make crucial marketing and sales strategy decisions. It is critical to have a strategic plan for integrating BD analytics with infrastructure and corporate systems. Pharmaceutical makers can acquire far more insight into existing patient behavior thanks to increased volumes of data that corporations can use – including information from distant sensor devices – combined with new analytic models. The organization can then utilize this data to create services tailored to specific demographics or at-risk patient groups to improve treatment efficacy [92, 93]. As mentioned earlier, NuMedii is a renowned bioinformatics and clinical research lab and the first to advocate the use of BD, advanced analytics, and systems biology to speed up drug discovery and development. The company's innovative technology helps pharmaceutical companies establish a pipeline for development by making drug–disease links through the study of unique illness signatures. The Biomedical Intelligence Cloud (BIC) of Data4Cure collects data on genes and genetic variations, diseases, and treatments. Data4Cure can provide information for quick product development to pharmaceutical companies. Bringing a medication to market is a huge undertaking. Even when a drug is approved, work continues: a business must position and promote itself in the market, analyze its drug's performance, foster relationships between patients and scientific and medical communities, and be on the lookout for and manage any adverse effects. To fine-tune their marketing tactics, Voxx Analytics collects market data and generates relevant customer insights regularly [94, 95].

AI and BD continue to upgrade the pharma sector in multiple ways. The growing use of social and digital media platforms by physicians and patients has resulted in an increase in the number and variety of data that businesses may access, gather, and analyze [94, 95]. It has been reported that, by obtaining data from a variety of sources and leveraging the power of data analytics, pharma companies can gain a better understanding of end users' behavior patterns, response to marketing campaigns, product performance, and upcoming industry trends, which, when thoroughly analyzed and interpreted, can result in improved marketing and sales. Aside from research and development and clinical trials, BD has a lot to offer the pharmaceutical business in terms of regulatory compliance, customer support, and sophisticated contract management solutions to generate win-win situations with numerous stakeholders and payer organizations. It is no exaggeration to claim that cloud computing's rapid adoption is transforming the pharmaceutical sector.

10.9 Conclusion

The present healthcare sector is confronted with several complicated challenges, including years of drug development and approval processes, and rising drug and therapy costs. The utility of BD, along with its outstanding tools, is playing a crucial role in continually minimizing obstacles experienced by pharmaceutical businesses, affecting the drug development progression and the total lifespan of the product. Interoperable BD drives the field of life sciences, including pharma and other healthcare enterprises. In recent trends, rather than addressing symptoms, drug discovery and development are focusing on creating personalized therapy for patients. Even while diseases like COVID-19 will never be completely eradicated, scientists' understanding of data science, as well as the usage of AI and other digital platforms, will aid in the development of innovative drugs with high specificity and fewer side effects. Apart from research and drug development, and clinical trials, BD has a lot to offer the pharmaceutical industry in terms of regulatory compliance, customer service, and complex contract management solutions to create win-win scenarios with a variety of stakeholders. The quest for the right algorithm utilizing BD is the new competition in the biotechnology and pharmaceutical industries since it helps to deepen understanding of pathology and treatment outcomes.

Acknowledgment

The authors are thankful to their respective Universities for supporting the successful completion of this chapter.

Conflicts of Interest

The authors declare no conflict of interest.

References

[1] Masoudi-Nejad A, Mousavian Z and Bozorgmehr J H 2013 Drug-target and disease networks: polypharmacology in the post-genomic era *In Silico Pharmacol* **1** 17

[2] Zardecki C, Dutta S, Goodsell D S, Voigt M and Burley S K 2016 RCSB protein data bank: A resource for chemical, biochemical, and structural explorations of large and small biomolecules *J Chem Educ* **93** 569–75

[3] Burley S K, Berman H M, Bhikadiya C, Bi C, Chen L, di Costanzo L, Christie C, Duarte J M, Dutta S, Feng Z, Ghosh S, Goodsell D S, Green R K, Guranovic V, Guzenko D, Hudson B P, Liang Y, Lowe R, Peisach E, Periskova I, Randle C, Rose A, Sekharan M, Shao C, Tao Y P, Valasatava Y, Voigt M, Westbrook J, Young J, Zardecki C, Zhuravleva M, Kurisu G, Nakamura H, Kengaku Y, Cho H, Sato J, Kim J Y, Ikegawa Y, Nakagawa A, Yamashita R, Kudou T, Bekker G J, Suzuki H, Iwata T, Yokochi M, Kobayashi N, Fujiwara T, Velankar S, Kleywegt G J, Anyango S, Armstrong D R, Berrisford J M, Conroy M J, Dana J M, Deshpande M, Gane P, Gáborová R, Gupta D, Gutmanas A, Koča J, Mak L, Mir S, Mukhopadhyay A, Nadzirin N, Nair S, Patwardhan A, Paysan-Lafosse T, Pravda L, Salih O, Sehnal D, Varadi M, Vǎreková R, Markley J L, Hoch J C, Romero P R, Baskaran K, Maziuk D, Ulrich E L, Wedell J R, Yao H, Livny M and Ioannidis Y E 2019 Protein data bank: the single global archive for 3D macromolecular structure data *Nucleic Acids Res* **47** D520–8

[4] da Silva Rocha S F L, Olanda C G, Fokoue H H and Sant'Anna C M R 2019 Virtual screening techniques in drug discovery: Review and recent applications *Curr Top Med Chem* **19** 1751–67

[5] Liu X, Hong Z, Liu J, Lin Y, Rodríguez-Patón A, Zou Q and Zeng X 2020 Computational methods for identifying the critical nodes in biological networks *Brief Bioinform* **21** 486–97

[6] Tang Y J, Pang Y H and Liu B 2021 IDP-Seq2Seq: identification of intrinsically disordered regions based on sequence to sequence learning *Bioinformatics* **36** 5177–86

[7] Wang J, Wang H, Wang X and Chang H 2019 Predicting drug-target interactions via FM-DNN learning *Curr Bioinform* **15** 68–76

[8] Ann Alexander C and Wang L 2017 Big data in healthcare: a new frontier in personalized medicine *Open Access J Transl Med Res* **1** 15–8
[9] Piai S and Claps M 2013 Bigger data for better healthcare *IDC Health Insights* 1–24
[10] Dash S, Shakyawar S K, Sharma M and Kaushik S 2019 Big data in healthcare: management, analysis and future prospects *J Big Data* **6** 54
[11] Fatt Q K and Ramadas A 2018 The usefulness and challenges of big data in healthcare *J Healthc Commun* **3** 21
[12] Pammolli F, Magazzini L and Riccaboni M 2011 The productivity crisis in pharmaceutical R&D *Nat Rev Drug Discov* **10** 428–38
[13] Fogel D B 2018 Factors associated with clinical trials that fail and opportunities for improving the likelihood of success: A review *Contemp Clin Trials Commun* **11** 156–64
[14] Cherkasov A, Muratov E N, Fourches D, Varnek A, Baskin I I, Cronin M, Dearden J, Gramatica P, Martin Y C, Todeschini R, Consonni V, Kuz'Min V E, Cramer R, Benigni R, Yang C, Rathman J, Terfloth L, Gasteiger J, Richard A and Tropsha A 2014 QSAR Modeling: Where have you been? Where are you going to? *J Med Chem* **57** 4977–5010
[15] Szymański P, Markowicz M and Mikiciuk-Olasik E 2012 Adaptation of high-throughput screening in drug discovery-toxicological screening tests *Int J Mol Sci* **13** 427–52
[16] Cerqueira N M F S A, Gesto D, Oliveira E F, Santos-Martins D, Brás N F, Sousa S F, Fernandes P A and Ramos M J 2015 Receptor-based virtual screening protocol for drug discovery *Arch Biochem Biophys* **582** 56–67
[17] Zhavoronkov A, Ivanenkov Y A, Aliper A, Veselov M S, Aladinskiy V A, Aladinskaya A v., Terentiev V A, Polykovskiy D A, Kuznetsov M D, Asadulaev A, Volkov Y, Zholus A, Shayakhmetov R R, Zhebrak A, Minaeva L I, Zagribelnyy B A, Lee L H, Soll R, Madge D, Xing L, Guo T and Aspuru-Guzik A 2019 Deep learning enables rapid identification of potent DDR1 kinase inhibitors *Nat Biotechnol* **37** 1038–40
[18] Anon NuMedii – Disrupting Drug Discovery Using Big Data and Artificial Intelligence
[19] Borman S 1990 New QSAR techniques eyed for environmental assessments *Chem Eng News* **68** 20–3
[20] Ma J, Sheridan R P, Liaw A, Dahl G E and Svetnik V 2015 Deep neural nets as a method for quantitative structure-activity relationships *J Chem Inf Model* **55** 263–74
[21] Ragoza M, Hochuli J, Idrobo E, Sunseri J and Koes D R 2017 Protein-ligand scoring with convolutional neural networks *J Chem Inf Model* **57** 942–57

[22] Wallach I, Dzamba M and Heifets A 2015 AtomNet: A deep convolutional neural network for bioactivity Prediction in structure-based drug discovery *arXiv preprint* 1510.02855
[23] Wang S, Sun S, Li Z, Zhang R and Xu J 2017 Accurate de novo prediction of protein contact map by ultra-deep learning model *PLoS Comput Biol* **13** e1005324
[24] Hoffman M M, Buske O J, Wang J, Weng Z, Bilmes J A and Noble W S 2012 Unsupervised pattern discovery in human chromatin structure through genomic segmentation *Nat Methods* **9** 473–6
[25] Schubach M, Re M, Robinson P N and Valentini G 2017 Imbalance-aware machine learning for predicting rare and common disease-associated non-coding variants *Sci Rep* **7** 2959
[26] Degroeve S, de Baets B, van de Peer Y and Rouzé P 2002 Feature subset selection for splice site prediction *Bioinformatics* **18 Suppl 2** S75–83
[27] Heintzman N D, Stuart R K, Hon G, Fu Y, Ching C W, Hawkins R D, Barrera L O, van Calcar S, Qu C, Ching K A, Wang W, Weng Z, Green R D, Crawford G E and Ren B 2007 Distinct and predictive chromatin signatures of transcriptional promoters and enhancers in the human genome *Nat Genet* **39** 311–8
[28] Liu F, Li H, Ren C, Bo X and Shu W 2016 PEDLA: predicting enhancers with a deep learning-based algorithmic framework *Sci Rep* **6** 28517
[29] Urda D, Montes-Torres J, Moreno F, Franco L and Jerez J M 2017 Deep learning to analyze RNA-Seq gene expression data *Lecture Notes in Computer Science* vol 10306 (Springer, Cham) pp 50–9
[30] Qin Q and Feng J 2017 Imputation for transcription factor binding predictions based on deep learning *PLoS Comput Biol* **13** e1005403
[31] Yildirim M A, Goh K il, Cusick M E, Barabási A L and Vidal M 2007 Drug—target network *Nat Biotechnol* **25** 1119–26
[32] Moore T J, Zhang H, Anderson G and Alexander G C 2018 Estimated Costs of Pivotal Trials for Novel Therapeutic Agents Approved by the US Food and Drug Administration, 2015-2016 *JAMA Intern Medi* **178** 1451–7
[33] Sukumar S R, Natarajan R and Ferrell R K 2015 Quality of big data in health care *Int J Health Care Qual Assur* **28** 621–34
[34] Pushpakom S, Iorio F, Eyers P A, Escott K J, Hopper S, Wells A, Doig A, Guilliams T, Latimer J, McNamee C, Norris A, Sanseau P, Cavalla D and Pirmohamed M 2019 Drug repurposing: progress, challenges and recommendations *Nat Rev Drug Discov* **18** 41–58

[35] Tomczak K, Czerwińska P and Wiznerowicz M 2015 The cancer genome atlas (TCGA): an immeasurable source of knowledge *Contemp Oncol* **19** A68–77

[36] Colwill K, Gräslund S, Persson H, Jarvik N E, Wyrzucki A, Wojcik J, Koide A, Kossiakoff A A, Koide S, Sidhu S, Dyson M R, Pershad K, Pavlovic J D, Karatt-Vellatt A, Schofield D J, Kay B K, McCafferty J, Mersmann M, Meier D, Mersmann J, Helmsing S, Hust M, Dübel S, Berkowicz S, Freemantle A, Spiegel M, Sawyer A, Layton D, Nice E, Dai A, Rocks O, Williton K, Fellouse F A, Hersi K, Pawson T, Nilsson P, Sundberg M, Sjöberg R, Sivertsson Å, Schwenk J M, Takanen J O, Hober S, Uhlén M, Dahlgren L G, Flores A, Johansson I, Weigelt J, Crombet L, Loppnau P, Kozieradzki I, Cossar D, Arrowsmith C H and Edwards A M 2011 A roadmap to generate renewable protein binders to the human proteome *Nat Methods* **8** 551–61

[37] Liu B, Gao X and Zhang H 2019 BioSeq-Analysis2.0: an updated platform for analyzing DNA, RNA and protein sequences at sequence level and residue level based on machine learning approaches *Nucleic Acids Res* **47** e127–e127

[38] Druker B J and Lydon N B 2000 Lessons learned from the development of an Abl tyrosine kinase inhibitor for chronic myelogenous leukemia *J Clin Invest* **105** 3–7

[39] Öztürk H, Özgür A and Ozkirimli E 2018 DeepDTA: deep drug–target binding affinity prediction *Bioinformatics* **34** i821–9

[40] Yamanishi Y, Kotera M, Kanehisa M and Goto S 2010 Drug-target interaction prediction from chemical, genomic and pharmacological data in an integrated framework *Bioinformatics* **26** i246–54

[41] Davis M I, Hunt J P, Herrgard S, Ciceri P, Wodicka L M, Pallares G, Hocker M, Treiber D K and Zarrinkar P P 2011 Comprehensive analysis of kinase inhibitor selectivity *Nat Biotechnol* **29** 1046–51

[42] Tang J, Szwajda A, Shakyawar S, Xu T, Hintsanen P, Wennerberg K and Aittokallio T 2014 Making sense of large-scale kinase inhibitor bioactivity data sets: a comparative and integrative analysis *J Chem Inf Model* **54** 735–43

[43] Zheng L, Liu D, Yang W, Yang L and Zuo Y 2021 RaacLogo: a new sequence logo generator by using reduced amino acid clusters *Brief Bioinform* **22** 1–5

[44] Wang H, Liang P, Zheng L, Long C, Li H and Zuo Y 2021 eHSCPr discriminating the cell identity involved in endothelial to hematopoietic transition *Bioinformatics* **37** 2157–64

[45] Kim S, Chen J, Cheng T, Gindulyte A, He J, He S, Li Q, Shoemaker B A, Thiessen P A, Yu B, Zaslavsky L, Zhang J and Bolton E E 2021 PubChem in 2021: new data content and improved web interfaces *Nucleic Acids Res* **49** D1388–95

[46] Pang Y and Liu B 2020 SelfAT-Fold: protein fold recognition based on residue-based and motif-based self-attention networks *IEEE/ACM Trans Comput Biol Bioinform*

[47] Shao J, Yan K and Liu B 2021 FoldRec-C2C: protein fold recognition by combining cluster-to-cluster model and protein similarity network *Brief Bioinform* **22** 1–11

[48] Cheng L, Han X, Zhu Z, Qi C, Wang P and Zhang X 2021 Functional alterations caused by mutations reflect evolutionary trends of SARS-CoV-2 *Brief Bioinform* **22** 1442–50

[49] Wang Y-C, Yang Z-X, Wang Y and Deng N-Y 2010 Computationally probing drug-protein interactions via support vector machine *Lett Drug Des Discov* **7** 370–8

[50] Wang L, You Z H, Chen X, Xia S X, Liu F, Yan X, Zhou Y and Song K J 2018 A computational-based method for predicting drug-target interactions by using stacked Autoencoder deep neural network *J Comput Biol* **25** 361–73

[51] Mahmud S M H, Chen W, Meng H, Jahan H, Liu Y and Hasan S M M 2020 Prediction of drug-target interaction based on protein features using undersampling and feature selection techniques with boosting *Anal Biochem* **589** 113507

[52] Abernethy A P, Etheredge L M, Ganz P A, Wallace P, German R R, Neti C, Bach P B and Murphy S B 2010 Rapid-learning system for cancer care *J Clin Oncol* **28** 4268–74

[53] Gligorijević V, Malod-Dognin N and Pržulj N 2016 Integrative methods for analyzing big data in precision medicine *Proteomics* **16** 741–58

[54] Kourou K, Exarchos T P, Exarchos K P, Karamouzis M v. and Fotiadis D I 2015 Machine learning applications in cancer prognosis and prediction *Comput Struct Biotechnol J* **13** 8–17

[55] Kang J, Schwartz R, Flickinger J and Beriwal S 2015 Machine Learning Approaches for Predicting Radiation Therapy Outcomes: A Clinician's Perspective *Int J Radiat Oncol Biol Phys* **93** 1127–35

[56] Li X, Xu Y, Cui H, Huang T, Wang D, Lian B, Li W, Qin G, Chen L and Xie L 2017 Prediction of synergistic anti-cancer drug combinations based on drug target network and drug induced gene expression profiles *Artif Intell Med* **83** 35–43

[57] Bibault J-E and Xing L 2020 The Role of Big Data in Personalized Medicine *Precis Med Oncol* ed B Aydogan and J A Radosevich (John Wiley & Sons, Ltd) pp 229–47

[58] Drier Y and Domany E 2011 Do Two Machine-Learning Based Prognostic Signatures for Breast Cancer Capture the Same Biological Processes? *PLOS ONE* **6** e17795

[59] Gernaat S A M, van Velzen S G M, Koh V, Emaus M J, Išgum I, Lessmann N, Moes S, Jacobson A, Tan P W, Grobbee D E, van den Bongard D H J, Tang J I and Verkooijen H M 2018 Automatic quantification of calcifications in the coronary arteries and thoracic aorta on radiotherapy planning CT scans of Western and Asian breast cancer patients *Radiother Oncol* **127** 487–92

[60] Gurbani S S, Schreibmann E, Maudsley A A, Cordova J S, Soher B J, Poptani H, Verma G, Barker P B, Shim H and Cooper L A D 2018 A convolutional neural network to filter artifacts in spectroscopic MRI *Magn Reson Med* **80** 1765–75

[61] Trebeschi S, van Griethuysen J J M, Lambregts D M J, Lahaye M J, Parmer C, Bakers F C H, Peters N H G M, Beets-Tan R G H and Aerts H J W L 2017 Deep Learning for Fully-Automated Localization and Segmentation of Rectal Cancer on Multiparametric MR *Sci Rep* **7** 5301

[62] Tseng H H, Luo Y, Cui S, Chien J T, ten Haken R K and Naqa I el 2017 Deep reinforcement learning for automated radiation adaptation in lung cancer *Med Phys* **44** 6690–705

[63] Lu F, Wu F, Hu P, Peng Z and Kong D 2017 Automatic 3D liver location and segmentation via convolutional neural network and graph cut *Int J Comput Assist Radiol Surg* **12** 171–82

[64] Saltz J, Gupta R, Hou L, Kurc T, Singh P, Nguyen V, Samaras D, Shroyer K R, Zhao T, Batiste R, van Arnam J, Caesar-Johnson S J, Demchok J A, Felau I, Kasapi M, Ferguson M L, Hutter C M, Sofia H J, Tarnuzzer R, Wang Z, Yang L, Zenklusen J C, Zhang J (Julia), Chudamani S, Liu J, Lolla L, Naresh R, Pihl T, Sun Q, Wan Y, Wu Y, Cho J, DeFreitas T, Frazer S, Gehlenborg N, Getz G, Heiman D I, Kim J, Lawrence M S, Lin P, Meier S, Noble M S, Saksena G, Voet D, Zhang H, Bernard B, Chambwe N, Dhankani V, Knijnenburg T, Kramer R, Leinonen K, Liu Y, Miller M, Reynolds S, Shmulevich I, Thorsson V, Zhang W, Akbani R, Broom B M, Hegde A M, Ju Z, Kanchi R S, Korkut A, Li J, Liang H, Ling S, Liu W, Lu Y, Mills G B, Ng K S, Rao A, Ryan M, Wang J, Weinstein J N, Zhang J, Abeshouse A, Armenia J, Chakravarty D, Chatila W K, de Bruijn I, Gao J, Gross B E, Heins Z J, Kundra R, La

K, Ladanyi M, Luna A, Nissan M G, Ochoa A, Phillips S M, Reznik E, Sanchez-Vega F, Sander C, Schultz N, Sheridan R, Sumer S O, Sun Y, Taylor B S, et al 2018 Spatial Organization and Molecular Correlation of Tumor-Infiltrating Lymphocytes Using Deep Learning on Pathology Images *Cell Rep* **23** 181-193.e7

[65] Shin H C, Orton M R, Collins D J, Doran S J and Leach M O 2013 Stacked autoencoders for unsupervised feature learning and multiple organ detection in a pilot study using 4D patient data *IEEE Trans Pattern Anal Mach Intell* **35** 1930–43

[66] Chowdhury G G 2003 Natural language processing *Annu Rev Inf Sci Technol* **37** 51–89

[67] Warner J L, Anick P, Hong P and Xue N 2011 Natural Language Processing and the Oncologic History: Is There a Match? *J Oncol Pract* **7** e15

[68] Spyns P 1996 Natural language processing in medicine: an overview - PubMed *Methods Inf Med* **35** 285–301

[69] Adnan K and Akbar R 2019 An analytical study of information extraction from unstructured and multidimensional big data *J Big Data* **6** 91

[70] Hay M, Thomas D W, Craighead J L, Economides C and Rosenthal J 2014 Clinical development success rates for investigational drugs *Nat Biotechnol* **32** 40–51

[71] Harrer S, Shah P, Antony B and Hu J 2019 Artificial intelligence for clinical trial design *Trends Pharmacol Sci* **40** 577–91

[72] Mak K K and Pichika M R 2019 Artificial intelligence in drug development: present status and future prospects *Drug Discov Today* **24** 773–80

[73] Siah K W, Kelley N W, Ballerstedt S, Holzhauer B, Lyu T, Mettler D, Sun S, Wandel S, Zhong Y, Zhou B, Pan S, Zhou Y and Lo A W 2021 Predicting drug approvals: The Novartis data science and artificial intelligence challenge *Patterns* **2** 100312

[74] Hassanzadeh P, Atyabi F and Dinarvand R 2019 The significance of artificial intelligence in drug delivery system design *Adv Drug Deliv Rev* **151–152** 169–90

[75] Mallon A M, Häring D A, Dahlke F, Aarden P, Afyouni S, Delbarre D, el Emam K, Ganjgahi H, Gardiner S, Kwok C H, West D M, Straiton E, Haemmerle S, Huffman A, Hofmann T, Kelly L J, Krusche P, Laramee M C, Lheritier K, Ligozio G, Readie A, Santos L, Nichols T E, Branson J and Holmes C 2021 Advancing data science in drug development

through an innovative computational framework for data sharing and statistical analysis *BMC Med Res Methodol* **21** 250

[76] Bate A, Reynolds R F and Caubel P 2018 The hope, hype and reality of Big Data for pharmacovigilance *Ther Adv Drug Saf* **9** 5–11

[77] Lee Ventola C 2018 Big data and pharmacovigilance: Data mining for adverse drug events and interactions *Pharm Ther* **43** 340–51

[78] Hussain R and Hassali M A 2019 Current status and future prospects of pharmacovigilance in Pakistan *J Pharm Policy Prac* **12** 14

[79] Hussain R, Hassali M A and Babar Z U D 2020 Medicines safety in the globalized context *Global Pharmaceutical Policy* ed Z-U-D Babar (Palgrave Macmillan, Singapore) pp 1–28

[80] Coloma P M, Trifirò G, Patadia V and Sturkenboom M 2013 Post-marketing safety surveillanceÂă: where does signal detection using electronic healthcare records fit into the big picture? *Drug Saf* **36** 183–97

[81] Price J 2016 What can big data offer the pharmacovigilance of orphan drugs? *Clin Ther* **38** 2533–45

[82] Moore T J and Furberg C D 2015 Electronic health data for postmarket surveillance: A vision not realized *Drug Saf* **38** 601–10

[83] Harpaz R, DuMochel W and Shah N H 2016 Big data and adverse drug reaction detection *Clin Pharmacol Ther* **99** 268–70

[84] Duggirala H J, Tonning J M, Smith E, Bright R A, Baker J D, Ball R, Bell C, Bright-Ponte S J, Botsis T, Bouri K, Boyer M, Burkhart K, Steven Condrey G, Chen J J, Chirtel S, Filice R W, Francis H, Jiang H, Levine J, Martin D, Oladipo T, O'Neill R, Palmer L A M, Paredes A, Rochester G, Sholtes D, Szarfman A, Wong H L, Xu Z and Kass-Hout T 2016 Use of data mining at the Food and Drug Administration *J Am Med Inform Assoc* **23** 428–34

[85] Aitha S R, Marpaka S, Chakradhar T, Bhuvaneshwari E and Kasukurthi S R 2021 Big data analytics in pharmacovigilance - A global trend *Asian J Pharm Clin Res* **14** 19–24

[86] Vergetis V, Skaltsas D, Gorgoulis V G and Tsirigos A 2021 Assessing drug development risk using big data and machine learning *Cancer Res* **81** 816–9

[87] Mason D J, Eastman R T, Lewis R P I, Stott I P, Guha R and Bender A 2018 Using machine learning to predict synergistic antimalarial compound combinations with novel structures *Front Pharmacol* **9** 1096

[88] Raghupathi W and Raghupathi V 2014 Big data analytics in healthcare: promise and potential *Health Inf Sci Syst* **2** 3

[89] Senthilkumar S, Rai B K, Meshram A A, Gunasekaran A and S C 2018 Big data in healthcare management: A review of literature *Am J Theor Appl Bus* **4** 57–69
[90] Kumar Y, Sood K, Kaul S and Vasuja R 2020 *Big Data Analytics in Healthcare* vol 66, ed A J Kulkarni, P Siarry, P K Singh, A Abraham, M Zhang, A Zomaya and F Baki (Cham: Springer International Publishing)
[91] do Nascimento I J B, Marcolino M S, Abdulazeem H M, Weerasekara I, Azzopardi-Muscat N, Goncalves M A and Novillo-Ortiz D 2021 Impact of big data analytics on people's health: Overview of systematic reviews and recommendations for future studies *J Med Internet Res* **23** e27275
[92] Kim G H, Trimi S and Chung J H 2014 Big-data applications in the government sector *Commun ACM* **57** 78–85
[93] Groves P, Kayyali B, Knott D and van Kuiken S 2013 *The big-data revolution in US health care: Accelerating value and innovation*
[94] Alsunaidi S J, Almuhaideb A M, Ibrahim N M, Shaikh F S, Alqudaihi K S, Alhaidari F A, Khan I U, Aslam N and Alshahrani M S 2021 Applications of big data analytics to control COVID-19 pandemic *Sensors* **21** 2282
[95] Galetsi P, Katsaliaki K and Kumar S 2019 Values, challenges and future directions of big data analytics in healthcare: A systematic review *Soc Sci Med* **241** 112533

11

Targeted Drug Delivery in Cancer Tissue by Utilizing Big Data Analytics

Neeraj Kumar[1], Shobhit Prakash Srivastava[1], Ayush Chandra Mishra[1], Amrita Shukla[1], Swati Verma[2], Rajiv Dahiya[4], and Sudhanshu Mishra[3*]

[1]Dr MC Saxena College of Pharmacy, India
[2]Department of Pharmacy, School of Medical and Allied Sciences, Galgotias University, India
[3]Department of Pharmaceutical Science & Technology, Madan Mohan Malaviya University of Technology, Gorakhpur, Uttar Pradesh, India
[4]School of Pharmacy, Faculty of Medical Sciences, The University of the West Indies, Trinidad & Tobago
*Correspondence Author: Department of Pharmacy Birla Institute of Technology & Science, Pilani, EMail: msudhanshu22@gmail.com, Contact: 8377836989.

Abstract

Although conventional chemotherapy has been used to a certain extent to treat cancer, the main drawbacks of chemotherapy are its poor bioavailability, negative side effects, and low doses of treatment. The main goal is to develop a drug delivery vehicle, as well as to reduce drug reactions or side effects, to successfully address drug delivery issues and get the drug to the desired destination. Big data possess an advantage over traditional processing tools by handling broad as well as complex datasets which traditional technologies fail to do. The method of analyzing such broad and complex data is called Big Data Analytics, to open up basic patterns in them. The discovery of drugs is linked to the analysis of big data because the method is based on the extensive experiments and surveys that have been collected via social media, hospitals, pharmaceutical companies, and laboratories. Data storage, management, and analysis may be required. And there are various firms related to it.

Keywords: Chemotherapy, Bioavailability, Low Therapeutic Evidences, Big Data, Big Data Analytics.

11.1 Introduction

Carcinoma is the second leading cause of population decline globally. Overall, the present situation has accelerated cancer spread [1]. Cancer is a category of disorder in which cells produce abnormal growth and spread to other regions of the body. This is related mostly to the lack of a characteristic that restricts cell proliferation. Changes in bowel habits, abnormal bleeding, weight loss, lump development, and a persistent cough are all potential indications and symptoms [2]. According to studies, the big data industry is expected to reach a whopping $6.6 billion by 2021. As technology advances, it has the potential to revolutionize conventional healthcare by recognizing patterns and amplifying illness detection. In fact, in certain areas of the world, it is already being used to diagnose illnesses more reliably, like cancer in its initial stages. Big data and predictive analytics offer huge promise to reduce risk, particularly in data-rich fields like cancer. The goal of predictive analytics is to comprehend the terabytes of data created by electronic health records (EHRs). Analytical algorithms published have been investigated or evaluated to make predictions and, in certain cases, to avert important occurrences such as readmissions from heart failure, cancer, and so on [3]. Large-scale biological investigations, clinical trials, and data-gathering initiatives are creating massive data-driven from many sources including medication development, employing participant medical data. The increased availability of biologically relevant large-scale data has allowed the computational foundation for real-world biomedical research, notably therapy targets and medication discovery for specific illnesses and clinical characteristics [4].

11.2 Application of Big Data in New Drug Discovery

Big data, the data with the greatest diversity, comes in greater volume and at a quicker rate. Big data is a more complicated collection of data, particularly when it comes from new sources. Because these datasets are so enormous, standard data processing software cannot handle or tolerate them, causing them to crash or hang.

Applications for big data in drug discovery research and development span from clinical study design to knowing how to target biological systems to interpreting disease processes. The issue now is, how has big data changed the healthcare system and medicine?

Big data has added a new dimension to disease therapy. Doctors can now better comprehend and cure illnesses, as well as provide precise and individualized therapy. They may also anticipate recurrences and provide preventative measures [5].

11.2.1 Involvement of data science in drug designing

We call for different data science techniques from clinical trial data to make cancer drugs and dosages faster and more accurate for different patients' conditions, to pave the way for clinical trials, and to develop new drugs. BD has sharpened and shortened oncology research to 3–4 years [6].

11.3 Need for This Approach

The specificity of medication action becomes critical in cancer therapy, as chemotherapeutic and radiotherapeutic alternatives are meant to destroy cells. These tactics are founded on the fundamental premise of selectively killing cancer cells while causing no major toxicity to normal cells. To achieve full remission in patients with disseminated illness, all cells prone to becoming cancerous must be eliminated. The elimination of these cells can be either directly as a result of medication action or implicitly as a consequence of the collateral effects of the treatment.

Combination therapy combines high-dose radiation with unceasing ingestion of chemotherapeutic drugs (such as paclitaxel) being considered for the treatment of non-surgically removed locally advanced cancers, as chemotherapy regimens alone are ineffective in advanced carcinomas and may yield only transitory responses. Paclitaxel radio-sensitizes tumor cells, making the combination treatment more effective than either the medication or radiation therapy alone. Obtaining therapeutically effective drug concentrations for the necessary time inside a malignant tumour, especially one that is difficult to treat, so that the medicine may effectively reach the intended therapeutic location. The concentration of drugs at the target site is also a major concern. The inability of these medications to penetrate the physiologically diverse tumor mass results in the presence of cancerous cells even after the long-term use of chemotherapy. This results in the requirement of a high dose of cytotoxic agents to prevent relapse, but such a high concentration produces side effects, prompting a majority of patients to discontinue the treatment. The majority of these side effects have a considerable impact on patients' quality of life. As a consequence of the poor therapeutic indices of various treatment alternatives, there has been a search for efficient delivery methods for currently available medications that may maximize therapeutic effectiveness while minimizing

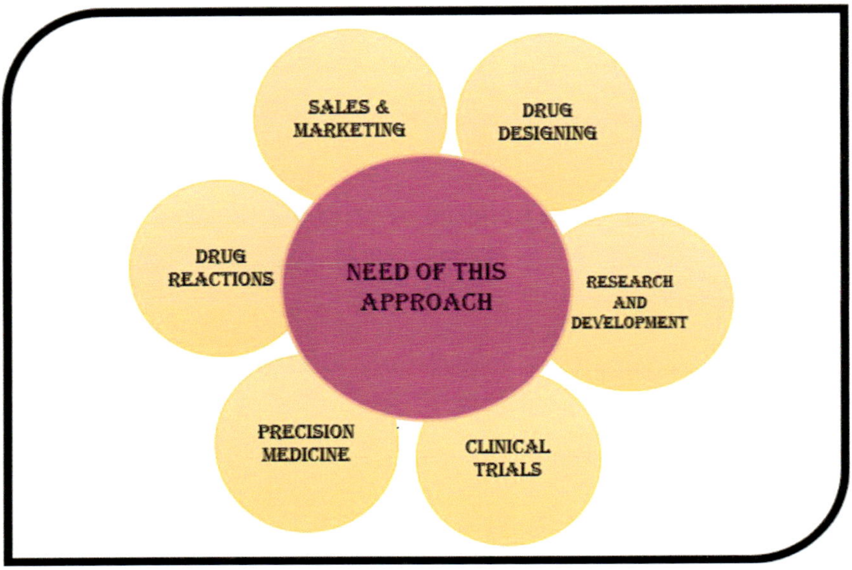

Figure 11.1 Approaches in big data.

unwanted effects. Targeting medications with specifically engineered drug delivery systems is a viable approach for improving therapeutic effectiveness and reducing the occurrence of systemic toxicity in anti-cancer treatments. As a result, the necessity for designing particularly targeted medication delivery systems emerges not only from a therapeutic standpoint but may also aid in the eradication of cancer in the patient before it kills the patient [7].

11.3.1 Drug discovery

A huge amount of time and money is involved in the process of drug discovery and that may be inconvenient to patients suffering from pandemics and other diseases, particularly during epidemics such as Ebola, swine flu, and typhoid.

With the advent of big data, researchers may utilize predictive modeling to identify pharmaceuticals in the pharmaceutical business. Predictive modeling may also be utilized to anticipate drug interactions, toxicity, and inhibition. Sophisticated models of mathematics and simulations are employed to forecast the way a particular substance can react in a specific human body. Historical data from post-market surveillance, clinical research, and medical trials may also be utilized. These statistics, when combined, may aid in anticipating FDA approval and patient outcomes, as well as detecting adverse reactions [7].

11.3.2 Research and development

The pharmaceutical sector may collect a massive amount of data created at different phases of the successful and lucrative chain from medication research to real-world use. As a result, the pharmaceutical sector must discover relevant and trustworthy sources of clinical data for its big data architecture. Using this strategy, businesspeople may connect necessary datasets to improve the quality of research and development [8].

11.3.3 Clinical trial

This procedure is used to determine if a certain medical therapy is beneficial and safe for human consumption. Using data such as genetic information, personality qualities, prior medical conditions, and current disease status, digital data may assist in the identification of suitable candidates for patients. Using this method, scenarios may choose an eligible applicant for a trial. The pharmaceutical sector conducts brief and low-cost clinical studies, which are unusual in the industry [9, 10].

11.3.4 Precision medicine

Big data plays a vital role in the healthcare sector. Pharmaceutical firms may use the information to generate tailored treatments that are suitable and effective for the patient's DNA and present lifestyle, as the disorders are identified and treated by utilizing pertinent data on the patient's genetic composition, environmental circumstances, and behavioral patterns.

Furthermore, precision medicine can predict vulnerability toward a variety of illnesses and adverse side effects, as well as improve problem identification. Precision medicine, using this technique, is more likely to give effective treatment than standard medicine, obviating the need for more money [10].

11.3.5 Drug reactions

In rare cases, medication given to patients for a pharmacological response may have serious or life-threatening consequences on the candidate's health, which is referred to as ADR (adverse drug reaction). The ADRs occur as a consequence of the insufficiency to precisely replicate a real-time event during clinical trials. The reporting systems are mostly based on the statements issued by attorneys, pharmacists, and physicians.

Pharmaceutical businesses might look for ADRs and patient evaluations on social media sites and medical forums. Big data analytics may be used to examine the data obtained by NLP (natural language processing) and sentiment analysis. Pharmaceutical firms may acquire information regarding adverse drug reactions using this method. As a result, to simplify the procedure of assessing medication, big data may be utilized by the pharmaceutical industry [7, 10].

11.3.6 Big data and its pertinence inside the marketing sector

The implication of big data inside various sectors, primarily the pharmaceutical sector, has the potential to boost sales and marketing efforts. Business executives may use big data to evaluate and discover which regions sell the most advertised pharmaceuticals. With this information, firms may decide whether to provide more marketed items in these locations. Similarly, pharmaceutical businesses may receive vital data from a variety of sources, assisting them in making critical choices in their marketing and sales strategies [11].

11.4 Barriers

11.4.1 Cellular defenses

The growth of multidrug resistance (MDR) inside the tumor resulting from the exposure of drug flow proteins over the surface of the cell has raised worries related to the prolonged duration of chemotherapeutic treatment [12, 13]. Cytotoxic pharmaceuticals may be administered to tumor cells by packing them into drug delivery devices, circumventing the challenges related to MDR [14]. A variety of mechanisms related to the resistance of the drug have been proposed. The flow mechanism mediated by membrane-linked p-glycoprotein (Pgp) is known to reduce anticancer drug intracellular accumulation inside the cells that have become resistant [15]. Moreover, a variety of chemicals that are harmful to malignant cells are produced inside cytoplasmic vesicles and secretions, allowing for efficient cytoplasmic transport or localisation in the nucleus, which is the site of action of a number of anticancer drugs (including doxorubicin, cisplatin, and others) [16]. Some additional anticancer medicines, like doxorubicin, are membrane-bound multi-drug resistance protein (MRPs) substrates, which reduce intracellular accumulation [17]. Polymer–drug conjugates, nanoparticles and microparticles, liposomes, and polymeric microelements systems are being explored to overcome drug resistance in cancer therapy [17–19]. However, the aforementioned methods cannot reach the target site directly and can enhance the intracellular delivery of drugs used in cancer treatment compared to drugs in correctionution that target the drug core [20, 21]. The mechanism described above belongs to the category of passive transport and is responsible for the delivery of medications freely available in the blood from the cytoplasm to the nucleus. But such methods may be ineffective in the case of drug targeting. This is because P-gp is expressed on intracellular organelles such as the Golgi apparatus and nuclear membrane envelopes in addition to the plasma membrane. Its Calcabrini *et al.* [21] established the presence of P-gp on the nuclear membrane of multidrug-resistant variants (MCF-7Adr), while another study indicated the flow of doxorubicin from the nucleus, limiting the medications accessible in the nucleus A. As a result, P-gp on the nuclear membrane envelope offers an additional defensive mechanism developed by cells resistant to antineoplastic medicines. As a consequence, just supplying medicine to the cytoplasmic basket will not alleviate the drug resistance problem until maximum drug localization occurs at the center. Furthermore, the effectiveness of several currently used medicine delivery systems may be limited because they get entangled in endoscopes during intracellular internalization [24]. According to our recent findings employing transferrin-conjugated nanoparticles, the duration of drug retention in cancer cells seems to be particularly important in overcoming drug resistance in the resistant cell line [25]. As a consequence, in anticancer therapy, sustained-release formulations may be more effective than alternate drug delivery techniques. The information gathered about the various aspects of drug resistance can be utilized in the development of an effective pharmaceutical delivery system.

11.4.2 Organellar and vesicular barriers

When the nanoparticles are absorbed, they must travel to their intracellular destination to deposit their payload. Endosomal vesicles carry macromolecules to a variety of intracellular locations, including the Golgi, lysosome, ER, mitochondria, and nucleus. Endocytosis occurs through a multitude of pathways and mechanisms, including clathrin-dependent endocytosis, caveolin-dependent endocytosis, micropinocytosis, phagocytosis, and many more. Different NPs are ingested by diverse endocytosis processes. The bulk of these pathways, however, may result in NP trafficking to a non-target organelle or the lysosome, where it is destroyed. This is especially important for delivering highly labile medications, such as genes and peptides [26–31].

11.4.3 A novel strategy for therapeutic target identification

A prevalent misunderstanding seems to be that big data solely refers to enormous data available in multiple volumes. However, in complement to quantity, the actual problems, as well as the potential in the case of big data, stem from various ranges of digital data available. Due to the handling of multiple data sources available, the process of drug development can be made quite challenging but interesting [32]. The big data strategy is a game-changer in the area of drug development. For instance, the data gathered from sources like clinical trials and therapeutic operations serves as a driving force for the identification of new targets that can be explored and the launch of a new drug research campaign directed at undesirable medical conditions, to fulfil the criteria (Figure 11.2). Drug discovery is becoming faster and faster, evolving again and again, with discoveries based on constant knowledge of past successes and failures, in collaboration with big data [33].

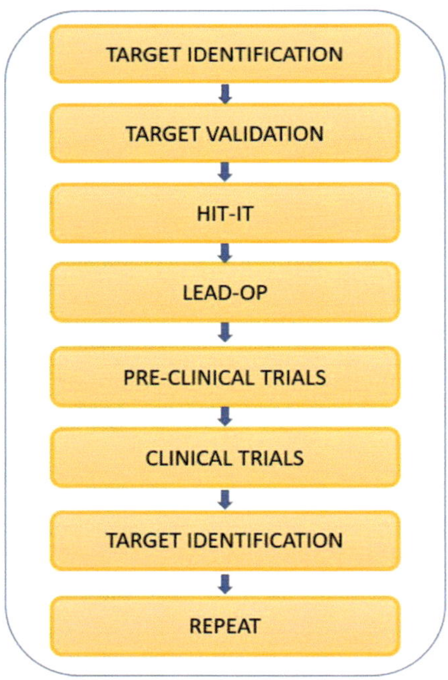

Figure 11.2 Process of driving new drug discovery.

11.4.4 Data integration on drug targets

As previously noted, optimally understanding these judgments necessitates the integration of data from several disciplines. However, combining knowledge from different fields is difficult, and there are currently few multidisciplinary resources available [34]. SAR is one of the world's greatest public domains for discovering a cure for tumor development. It combines clinical, genetic, medicinal chemistry, pharmacology, and three-dimensional protein structure data and presents it in integrated reports to help medication developers. SAR, for example, may be used to help in drug development, evidence-based diagnosis, and target prioritization [35]. Open Targets is a kind of public–private domain that contains information about the multiple targets available for the treatment of human illness. The use of such tools gives a right to the person to explore the possible testable ideas available. Other specialized resources, such as the effective integration of interdisciplinary data on the human intestinal mycobiome and the process of metabolism associated with it, are difficult to acquire and maintain, due to the limited availability of resources [36].

11.5 AI Approaches in Drug Discovery

Artificial intelligence is quite an amazingly ancient subject. The term "artificial intelligence" as well as many of the technologies, was developed in the early 1950s [37]. The term artificial intelligence indicates a computer program that can recognize the patterns of incoming data and then utilize this information to generate innovative calculations by using fresh data [38]. In practice, statistics may be used to forecast correlational data patterns existing between two parameters. Despite their size, the machine learning parameters refer to simple algorithms which are fully accessible to the human researcher; however, the precise method for prediction is still unclear. Figure 11.3 (a) represents different factors responsible for prediction [39].

The information given in Figure 11.3 (b) represents the sparse datasets that the researcher wishes to obtain and distil. These datasets are quite large and complex, including all the interactions and reaction pathways occurring in chemical compounds. DL approaches have allowed advancements in a variety of fields, including image processing with convolutional neural networks (CNNs), as well as text and voice processing with recurrent neural networks (RNNs) [40]. In contrast to ML, deep learning may identify significant patterns buried in a data space that is multidimensional, having multiple layering of different datasets [41]. However, these approaches deal with complexities like unorganized datasets that may increase the computing cost. These methodologies are included by AI, particularly when the algorithms alter and adapt in reaction to new inputs [42]. The Table 11.1 shows the various example of AI tools for drug delivery.

(a) Represents different methodologies of data analysis, how they vary in the amount and intricacy of parameters, and how the transparency with human understanding; (b) presents the proliferation of drug discovery-related data into the public domain, and its relation with cancer and machine learning articles.

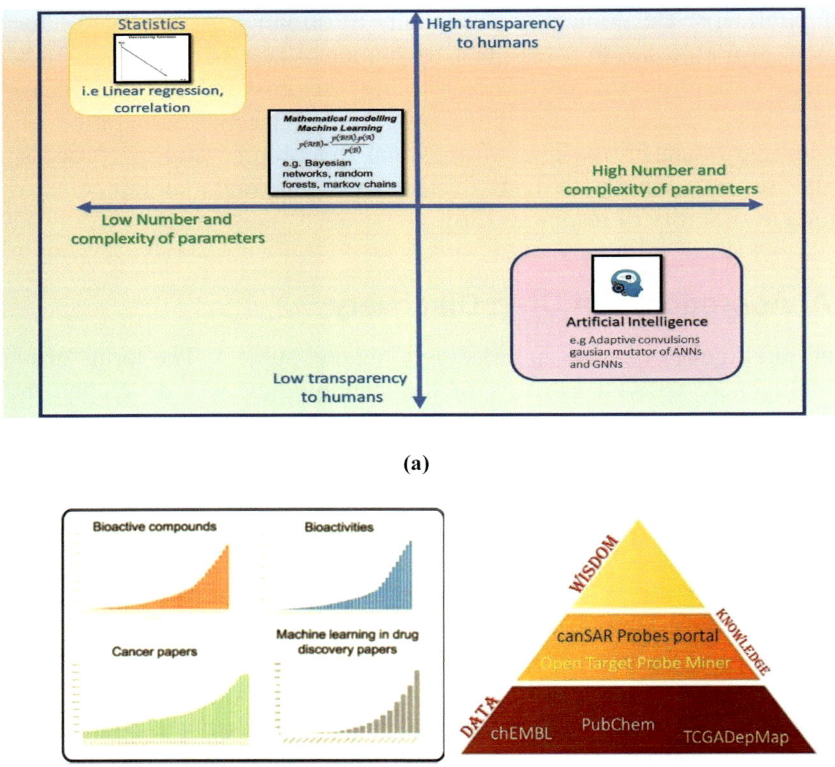

(a) (b) (c)

Figure 11.3 Application of big data analytics in drug discovery.

Table 11.1 Examples of AI tools used in drug discovery.

Tools	Details	Website URL
DeepChem	MLP model that employs a python-based AI system to select a viable candidate in drug discovery	https://github.com/deepchem/dee pchem
DeepTox	The software forecasts the toxicity of over 12,000 medicines	www.bioinf.jku.at/research/Deep Tox
Chemputer	Aids in the reporting of chemical synthesis procedures in a uniform style	https://zenodo.org/record/14817 31
DeltaVina	A scoring algorithm for drug–ligand binding affinity	https://github.com/chengwang88 /deltavina
ORGANIC	A molecular synthesis tool that aids in the creation of molecules with desired characteristics	https://github.com/aspuru-guzikgroup/ORGANIC
Hit Dexter	A machine learning approach is being used to anticipate molecules that may react to biological experiments	http://hitdexter2.zbh.unihamburg.de
PotentialNet	NNs are used to predict ligand-binding affinity	https://pubs.acs.org/doi/full/10.1 021/acscentsci.8b00507
Neural graph fingerprint	It aids in the prediction of the characteristics of new compounds	https://github.com/HIPS/neuralfingerprint
AlphaFold	Predicts 3D structures of proteins	https://deepmind.com/blog/alpha fold
DeepNeuralNet QSAR	Python-based system led by computational tools	https://github.com/Merck/Deep NeuralNet-QSAR

(c) Demonstrates the contrasts between data, knowledge, and wisdom in a pyramid style and the correlation of resources to different heights inside this pyramid.

11.6 Several Approaches Exist for AI to Filter Through Large Amounts of Data to Discover and Understand Patterns

11.6.1 Machine learning

Machine learning employs a pool of algorithms that do not require human involvement, interpretation, or unambiguous instruction for learning. By using big data, several possibilities have been created for machine learning

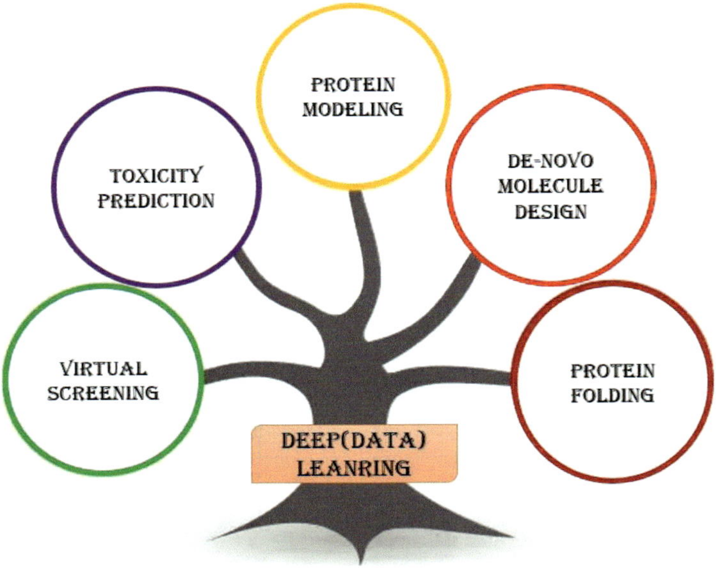

Figure 11.4 AI approaches in deep learning.

approaches in the pharmaceutical business, allowing them to be developed for intriguing models such as medication side-effect prediction. Deep learning neural networks (DLNN) training requires a large amount of data. However, this strategy is particularly effective in the analysis of huge datasets and the design of de novo studies [43].

11.6.2 Regularized regression

GLMNET is an abbreviation used for Elastic-Net Regularized Generalized Linear Models. It is a linear regression model in which the term "loss function penalty" is a linear mixture of L1 (LASSO) and L2 (ridge) penalties. The linear regression model is simple to understand, yet it works well in a variety of circumstances [44].

11.6.3 Variants in the deep learning model

Generative adversarial networks (GANs) are a hybrid of generative and discriminator networks. GANs are used to distinguish between true and false data and are useful in the creation of new molecules and the optimization of novel compounds with desired or required properties. Convolutional neural networks (CNNs) are mostly used for computer vision or picture categorization, but they have also been used to diagnose illnesses such as cancer.

11.6.4 Protein modeling and protein folding methods

Since the introduction of the third generation of predictors, several deep learning algorithms as well as more traditional ML approaches have been used regularly for PSA prediction; some of these are detailed here.

11.6.5 The RF method

The Random Forest regression model consists of a trained bootstrapping sample of the training set with a subset of descriptors whose selection is done haphazardly. The complete model appeared like a forest where the descriptors are investigated as splitting possibilities for each node. Two hyperparameters were regulated in this study:

1) N: The number of trees.
2) The percentage of input variables used at each split (max features).To make predictions, all the outputs are averaged [46].

11.6.6 SVM regression model

The support vector machine is a nonlinear approach that has been effectively used for several drug design challenges. The SVM technique completely renovates descriptors received as input into the high-dimensional space where they fit the linear model. A kernel function is used to perform all the abrupt transitions [45].

11.6.7 Predictive toxicology

Predictive toxicology and model development rely heavily on data, which has traditionally been constrained because of the scarcity of publicly available and high-quality information. Just 20 years ago, even very tiny databases of chemical structures and experimental bioactivities demanded time-consuming and laborious data collection, curation, and compilation. In the past, data scarcity and data quality were the most major issues in predictive toxicity modeling. The data was acquired by measuring physical and chemical properties manually in the laboratory, and therefore slowly [47].

As a consequence, models were developed using the same training sets, and data was only available for a subset of features and endpoints.

Data accessibility has altered and improved as a consequence of developments in the automated workplace and a paradigm shift toward twenty-first-century toxicology, which focuses on adverse effects pathways and mechanisms of action. ToxCast and Tox21, for example, have generated publicly available biologically active data for a large number of chemicals across a wide range of endpoints using high throughput/high content assays.

Furthermore, omic methods for identifying changes in genomes, proteomes, and metabolomes are available [48].

11.7 Implementation of Deep Learning Models in De Novo Drug Design

Implementation of the techniques of DL in the process of drug development is mostly based on notions similar to those employed in conventional new drug research. In the first approach, the ligands are encrypted individually to the tensors, whereas in the second technique, the ligands with the effective properties of the receptor are encrypted to the tensors concurrently. Based on the received input, the model for deep learning is then trained to match the experimental data which includes information like pIC50, pKi, 3D information, etc. using suitable networks (GAN, VAE, etc.). Deep learning models that have been taught can tell the difference between active and inactive medications, along with the conduct of new drug designs to find novel molecules [49].

For the development of novel drug designs, deep learning finds representatives using a domain name database. DeLinker68, for example, may use the three-dimensional structural information to train a molecular generative model which is graph-based. A variational autoencoder (VAE) framework is used to train a generation model using a collection of fragment–molecule pairs. The two segments, with their relative positions as well as their specific confirmation in the receptor pocket, are fed into the model as an input. This iteratively constructs the new linker "bond-by-bond" from the atom pool.

The developed model may generate newer molecules based on the initial substructures and the three-dimensional structural data information. The delinker demonstrates excellent applicability and effectiveness in proteolysis, targeting chimera (PROTAC) design, scaffold-hoping and fragment-linking. The required components may be created by 71 MolAICal54, 55 using a deep learning network trained on FDA-approved pharmaceutical fragments.

The old approach is then used to generate three-dimensional ligands in the receptor pocket based on the DL models [50].

11.7.1 Autoencoder

Unsupervised neural networks are mostly utilized for predicting drug–target interactions and analyzing drug similarity.

11.7.2 Deep belief networks

They have been used in virtual screening, multi-target drug categorization, and classifying tiny compounds as drugs or non-drugs [51].

11.8 Future Prospective

The contemporary healthcare industry is confronting various challenging issues, like escalating costs of treatment and medicines. This area must be looked up to by society to implement desired improvements. With the incorporation of AI into pharmaceutical product production, individualized drugs with the appropriate dosage, release characteristics, and other needed elements may be made based on each patient's need. The incorporation of technologies based on AI will not only shorten the duration of time consumed in launching a product to the market but will also help to improve the overall quality and safety of the product. AI provides aids to better utilize the resources available without boosting their cost, and hence the value of automation can be increased. Apart from this, some issues related to the implementation of AI persist, which are required to be solved, in the hope that AI will become an indispensable tool in the pharmaceutical business soon.

11.9 Conclusion

As mentioned in this article, the revolution of big data has arrived. We now have the capability of generating and storing massive volumes of different data, integrating it, and developing learning algorithms to assist in important decision-making phases along the drug development continuum. Integration of big data knowledge by the experts may benefit the decision-making process of drug development, regardless of the dangers of hype. This module helps us to effectively handle the complex nature of de novo research while also balancing novelty and safety more efficiently. The boon that arises due to the usage of big data by multiple sectors has attracted various biotechnology companies as well, to incorporate AI in several domains of drug discovery, for instance, Benevolent, Atom Smart, and Exscientia. There is indeed a significant opportunity to capitalize, mostly on the enhanced data and analytics available in the business sector. The recent collapse of IBM Watson, on the other hand, serves as a sobering story about the dangers of overlooking the problem's severity. Furthermore, it is critical to continue generating sustainable open-source content. Aside from the apparent benefits of big data and AI for drug discovery, there are still certain industries that are underserved. The amount of chemical and biological data available to the public has increased considerably during the last decade.

The ultimate goal is to employ cloud computing and artificial intelligence to enhance the decision-making process of drug discovery. The application of big data in the process of drug discovery has resulted in the generation of compounds with more drug-like properties. It involves approaches such as selection of the desired target and de novo drug design. Aside from making big data and AI methodology available, it is vital to give the training to ensure that the upcoming generation of drug pioneers is better equipped to employ the powerful techniques and resources.

Acknowledgment

I would like to thank all co-authors for helping in writing, editing, and drafting the content and diagrams of the chapter.

Funding

None.

Conflict of Interest

None.

References

[1] Sung, H., Ferlay, J., Siegel, R. L., Laversanne, M., Soerjomataram, I., Jemal, A., & Bray, F. Global cancer statistics estimates of incidence and mortality worldwide for 36 cancers in 185 countries. CA: a cancer journal for clinicians, 71(3), 209-249.

[2] Pucci, C., Martinelli, C., &Ciofani, G. Innovative approaches for cancer treatment, Current perspectives and new challenges. Ecancermedicalscience, (2019), 13.

[3] Zhu, H, Big data and artificial intelligence modeling for drug discovery, Annual review of pharmacology and toxicology, (2020), 60, 573-589.

[4] Qian, T, Zhu, S., &Hoshida, Y, Use of big data in drug development for precision medicine: an update, Expert review of precision medicine and drug development, (2019), 4(3), 189-200.

[5] Kim, R. S, Goossens, N., &Hoshida, Y, Use of big data in drug development for precision medicine. Expert review of precision medicine and drug development (2016), 1(3), 245-253

[6] Song, C., Kong, Y., Huang, L., Luo, H., & Zhu, X. Big data-driven precision medicine: Starting the custom-made era of iatrology. Biomedicine & Pharmacotherapy, (2020). 129, 110445.

[7] Chan, K. C. Big data analytics for drug discovery. In 2013 IEEE International Conference on Bioinformatics and Biomedicine (BIBM) (2013, December). (pp. 1-1).

[8] Zhang, Z. Big data and clinical research: perspective from a clinician. Journal of thoracic disease, (2014), 6(12), 1659.

[9] Hulsen, T., Jamuar, S. S., Moody, A. R., Karnes, J. H., Varga, O., Hedensted, S. & McKinney, E. F. From big data to precision medicine. Frontiers in medicine, (2019). 6, 34.

[10] Harpaz, R., DuMochel, W., & Shah, N. H. Big data and adverse drug reaction detection. Clinical Pharmacology & Therapeutics, (2016), 99(3), 268-270.

[11] Rejeb, a., Rejeb, k., & Keogh, j. G. Potential of big data for marketing: a literature

[12] Krishna, R., & Mayer, L. D. Multidrug resistance (MDR) in cancer: mechanisms, reversal using modulators of MDR and the role of MDR modulators in influencing the pharmacokinetics of anticancer drugs. European Journal of Pharmaceutical Sciences, (2000), 11(4), 265-283.

[13] Brown, R., & Links, M. Clinical relevance of the molecular mechanisms of resistance to anti-cancer drugs, Expert reviews in molecular medicine, (1999), 1(15), 1-21.

[14] Bennis, S., Chapey, C., Robert, J., &Couvreur, P. Enhanced cytotoxicity of doxorubicin encapsulated in polyisohexylcyanoacrylate nanospheres against multidrug-resistant tumour cells in culture. European Journal of Cancer, (1994). 30(1), 89-93.

[15] Faneyte, I. F., Kristel, P. M., van de Vijver, M. J. Determiningm MDR1/P-glycoprotein Expression in Breast Cancer. Int J Cancer (2001), 93, 114-122.
[16] Molinari, A., Calcabrini, A., Meschini, S., Stringaro, A., Crateri, P., Toccacieli, L., &Arancia, G. Subcellular detection and localization of the drug transporter Pglycoprotein in cultured tumor cells. Current Protein and Peptide Science, (2002), 3(6), 653-670
[17] De Verdiere, A. C., Dubernet, C., Nemati, F., Soma, E., Appel, M., Ferte, J.,& Couvreur, P. Reversion of multidrug resistance with polyalkyl-cyanoacrylate nanoparticles: towards a mechanism of action. British journal of cancer, (1997), 76(2), 198-205.
[18] Maeda, H., Seymour, L. W., & Miyamoto, Y., Conjugates of anticancer agents and polymers: advantages of macromolecular therapeutics in vivo. Bioconjugate chemistry, (1992), 3(5), 351-362.
[19] Kakizawa, Y., & Kataoka, K. Block copolymer micelles for delivery of gene and related compounds. Advanced drug delivery reviews, (2002), 54(2), 203-222.
[20] Kataoka, K., Harada, A., & Nagasaki, Y. Block copolymer micelles for drug delivery: design, characterization and biological significance. Advanced drug delivery reviews, (2012), 64, 37-48.
[21] Calcabrini, A., Meschini, S., Stringaro, A., Cianfriglia, M., Arancia, G., & Molinari, A. Detection of P-glycoprotein in the nuclear envelope of multidrug resistant cells. The Histochemical Journal, (2000), 32(10), 599-606.
[22] Fu, L. W., Zhang, Y. M., Liang, Y. J., Yang, X. P., & Pan, Q. C. The multidrug resistance of tumour cells was reversed by tetrandrine in vitro and in xenografts derived from human breast adenocarcinoma MCF-7/adr cells. European Journal of Cancer, (2002), 38(3), 418-426.
[23] Arancia, G., Molinari, A., Calcabrini, A., Meschini, S., &Cianfriglia, M. Intracellular Pglycoprotein in multidrug resistant tumor cells. Italian journal of anatomy and embryology, (2001), 106, 59-68.
[24] Minko, T., Paranjpe, P. V., Qiu, B., Lalloo, A., Won, R., Stein, S., &Sinko, P. J. Enhancing the anticancer efficacy of camptothecin using biotinylated poly (ethyleneglycol) conjugates in sensitive and multidrug-resistant human ovarian carcinoma cells. Cancer chemotherapy and pharmacology, (2002), 50(2), 143-150.
[25] Sahoo, S. K., &Labhasetwar, V. Enhanced antiproliferative activity of transferrin-conjugated paclitaxel-loaded nanoparticles is mediated via

sustained intracellular drug retention. Molecular pharmaceutics, (2005), 2(5), 373-383.

[26] Netti PA, Berk DA, Swartz MA, Grodzinsky AJ, Jain RK. Role of extracellular matrix assembly in interstitial transport in solid tumors. Cancer Res 2000; 60:2497-503.

[27] Graff BA, Vangberg L, Rofstad EK. Quantitative assessment of uptake and distribution of iron oxide particles (NC100150) in human melanoma xenografts by contrast-enhanced MRI. MagnReson Med. 2004 Apr;51(4):727-35

[28] Pun SH, Tack F, Bellocq NC, Cheng J, Grubbs BH, Jensen GS, Davis ME, Brewster M, Janicot M, Janssens B, et al. Targeted delivery of RNA-cleaving DNA enzyme (DNAzyme) to tumor tissue by transferrin-modified, cyclodextrin-based particles. Cancer

[29] BiolTher 2004; 3:641-50.

[30] Goodman TT, Olive PL, Pun SH. Increased nanoparticle penetration in collagenase-treated multicellular spheroids. Int J Nanomedicine 2007; 2:265-74; PMID:17722554

[31] Kuhn SJ, Finch SK, Hallahan DE, Giorgio TD. Proteolytic surface functionalization enhances in vitro magnetic nanoparticle mobility through extracellular matrix. Nano Lett 2006; 6:306-12; PMID:16464055.

[32] Neeves KB, Sawyer AJ, Foley CP, Saltzman WM, Olbricht WL. Dilation and degradation of the brain extracellular matrix enhances penetration of infused polymer nanoparticles. Brain Res 2007; 1180:121- 32; PMID:17920047.

[33] Workman, P., Antolin, A. A., & Al-Lazikani, B. Transforming cancer drug discovery with Big Data and AI. Expert opinion on drug discovery, (2019), 14(11), 1089-1095.

[34] Sellwood MA, Ahmed M, Segler MH, et al. Artificial intelligence and drug discovery, Futur Med Chem. 2018;10:2025–2028.

[35] Coker EA, Mitsopoulos C, Tym JE, et al. canSAR: update to the cancer translational research and drug discovery knowledgebase. Nucleic Acids Res. 2018, 47, D917–D922.

[36] Koscielny G, An P, Carvalho-Silva D, et al. Open Targets: A platform for therapeutic target identification and Validation. Nucleic Acids Res. 2017, 45, D985–D994.

[37] Paul, D., Sanap, G., Shenoy, S., Kalyane, D., Kalia, K., &Tekade, R. K. Artificial intelligence in drug discovery and development. Drug Discovery Today, (2020).

[38] Kim, H., Kim, E., Lee, I., Bae, B., Park, M., & Nam, H. Artificial Intelligence in Drug Discovery: A Comprehensive Review of Data-driven and Machine Learning Approaches. Biotechnology and Bioprocess Engineering, (2020), 25(6), 895-930.
[39] Chan, H. S. et al. (2019) Advancing drug discovery via artificial intelligence. Trends Pharmacol. Sci. 40(8), 592–604
[40] Rantanen, J. and Khinast, J. (2015) The future of pharmaceutical manufacturing sciences. J. Pharm. Sci. 104, 3612–3638
[41] Jaĺmsaĺ-Jounela, S.-L. (2007) Future trends in process automation. Annu. Rev. Control 31, 211–220
[42] Davenport, T. H. and Ronanki, R. (2018) Artificial intelligence for the real world. Harvard Bus. Rev. 96, 108–116
[43] Noronha A, Modamio J, Jarosz Y, et al. The virtual metabolic human database: integrating human and gut microbiome metabolism with nutrition and disease. Nucleic Acids Res. 2019, 47, D614–D624
[44] Friedman, J., Hastie, T., &Tibshirani, R. (2010). Regularization paths for generalized linear models via coordinate descent. Journal of statistical software, 33(1), 1.
[45] Doucet, J.-P.; Barbault, F.; Xia, H.; Panaye, A.; Fan, B. Nonlinear SVM Approaches to QSPR/QSAR Studies and Drug Design. *Curr. Comput.-Aided Drug Des.*2007, *3*, 263– 289.
[46] Gao, W., Mahajan, S. P., Sulam, J., & Gray, J. J. (2020). Deep learning in protein structural modeling and design. *Patterns*, 100142.
[47] Pakhrin, S. C., Shrestha, B., Adhikari, B., & Kc, D. B. (2021). Deep Learning-Based Advances in Protein Structure Prediction. *International Journal of Molecular Sciences*, *22*(11), 5553.
[48] Zhang, J., Norinder, U., &Svensson, F. (2021). Deep Learning-Based Conformal Prediction of Toxicity. *Journal of Chemical Information and Modeling*.
[49] Bo-Bai, Q., Liu, S., Tian, Y., Xu, T., Banegas-Luna, A. J., Pérez-Sánchez, H., ... & Yao, X. (2021). Application advances of deep learning methods for de novo drug design and molecular dynamics simulation. Wiley Interdisciplinary Reviews: Computational Molecular Science, e1581.
[50] Mouchlis, V. D., Afantitis, A., Serra, A., Fratello, M., Papadiamantis, A. G., Aidinis, V &Melagraki, G. (2021). Advances in de novo drug design: from conventional to machine learning methods. *International journal of molecular sciences*, *22*(4), 1676.

12

Risk Assessment in the Field of Oncology using Big Data

Akanksha Pandey[1], Rishabha Malviya[1*], and Sunita Dahiya[2]

[1]Department of Pharmacy, School of Medical and Allied Science, Galgotias University, Greater Noida, India
[2]Department of Pharmaceutical Sciences, School of Pharmacy, University of Pureto Rico, Medical Sciences Campus, USA
*Correspondence Author: Department of Pharmacy, School of Medical and Allied Sciences, Galgotias University, Greater Noida, India, EMail: rishabhamalviya19@gmail.com

Abstract

Big data refers to massive volumes of information that are unmanageable by traditional systems and web-based applications. It surpasses traditional storage, processing, and analysis systems. The concept of "big data" incorporates not only just quantity but also efficiency and variety. In 2019, 1200 children are expected to die from cancer in the United States. In 2016, there were 1.7 hundred thousand advanced cases of adolescent malignancy. The FDA expects that first-in-child medicine authorization will take at least 6.5 years or half a decade. The need to speed up the throughput of pediatric oncology clinical studies is growing. RCTs have the potential to change a wider range of registries, allowing people to enroll in medical studies. Even though the number of patients enrolled in clinical trials has increased, there are still barriers to clinical trial enrollment. The standardization of data collection across trials and disease types could make medicinal cancer analysis easier, whereas the underlying causes and treatment duration correlation for radiation-induced cancer are uncertain, enormous breakthroughs in IGRT would help physicians better comprehend the ideas and process.

Keywords: Big Data, Oncology, Cancer Prediction, Risk Assessment.

12.1 Introduction

Data has served as the key to improved organisation and new advances. The lot of data we have got, the higher we will be ready to organize ourselves to attain the most effective results. As a result, information assortment is a necessary element of any organization. We will conjointly use this data to forecast current trends in sure parameters further as future events. As we tend to become a lot of responsive to this, we have got begun to supply and collect a lot of information on nearly everything by implementing technological advancements in this direction. We tend to are currently in a very scenario wherever we tend with information from each fact of our lives, together with social activities, science, work, health, and so on. In certain aspects, the current situation is similar to an informational flood. Technology developments have aided the United States in producing colleagues in nursing expanding the amount of knowledge to the point where it is now unmanageable utilising currently available technologies. As a result, the term "big information" was coined to explain data that is massive and unmanageable. We would like to develop new methods to organize this information and derive expectant data in order to meet our current and future social needs. Attention is one such special social would including attention organizations, like each different business, generate an amazing quantity of knowledge, that presents each opportunity and challenge. The fundamentals of massive information, together with its management, analysis, and prospects, are mentioned during this review, with attention to the healthcare sector.

In medication, massive knowledge is moving from packaging to reality. AI algorithms will currently notice respiratory disease through chest X-rays [1] as well as diabetic retinopathy through fundscope screening [2, 3] using effectiveness boosting and generally surpassing practitioner diagnosing talents. Rapid innovations in procedure capability and deep learning possess the LED is the most well-known developments in diagnostic procedures: AI applications will currently notice respiratory disease through chest X-rays [1] as well as diabetic macular edema through fundscope screening [2, 3], with effectiveness enhancing and generally surpassing practitioner diagnosing talents. Prognostic analytics has been notably like-minded in creating a sense of the terabytes of knowledge generated by EHRs. Readmissions from heart disease [4], chronic preventive respiratory organ sickness [5], and infant infection have all been shown to be expected and generally prevented victimization printed prognostic analytic algorithms [6].

Oncology, with its abundance of knowledge, looks to be naturally appropriate oracular analytics considering the necessity for higher projections of duration, critical care utilization, harmful impacts, including cellular and epigenetic uncertainty, and oracular analytics applications in medical unit restricted. We tend to argue that current oracular analytic interventions in medicine may shut down necessary gaps in risk stratification strategies. Oracular analytics' growing use mostly in the understanding of pathologies, the creation of drugs, as well as the supervision of public well-being, pave the strategy for future tools to enter clinical practice. Furthermore, doctors, engineers, and politicians ought to remove the analysis, technological, or administrative constraints that prevent forestall analytics from being utilized in medicine, therefore, grasping this potential.

On Gregorian calendar month 1, 2016, there have been quite 15. Five million trauma victims within the United States, for an anticipated rise of quite 20,000,000 Gregorian calendar month 1, 2026 [7]. For them, the employment of electronic health records is on the increase, to higher harness, the data acquired from cancer patients, in addition to optimise therapeutic, the US Moonshot Foundation for Tumors promotes a complete genomic study involving tumors as well as test links, driving individualized medication forward. Two cancer information is firmly within the realm of huge information in therapy (RT), thanks to ubiquitous pictures accounting for a common fraction of total international storage consumption. RT depends mainly on medical imaging [8]. Its use in radiation, stated IGRT, includes neoplasm identification, staging, survival rate, care planning coming up with, radiological dosing, including post-treatment nursing. As a result of the operation generating an outsized quantity of digital information, IGRT is often renamed information-guided RT [9]. Massive setup errors, ranges of internal mobility-related variations affecting neoplasm location and environment, and form area unit presumably to be detected throughout RT utilizing regular scanning, that has become a demand should obtain simplest neoplasm exposure, up to native management and healthy tissue scotch, and therefore up the quality of life [10, 11]. Advances in IGRT have benefited cancer patients for many years. They are seeming to possess longer survival times, and also the range of victims of malignancy is growing.

The delayed effects of RT were getting a serious supply of worry. Second malignant neoplasms (SMNs) are one of the foremost vital and critical complications for cancer survivors, particularly people who are younger and have an extended expectancy. Various epidemiologic Prospective research shows incontestable endocrine as well as mammary gland tumor hazards associated with radiotherapy, leukemia, and different malignancies [12, 13].

Although, the specific methodology of radiation-induced malignancy, as well as the dose-response connection, are unknown. The danger of cancer related to RT is still a topic of debate and a point of contention in clinical radiation oncology that does have an effect on Patient management and moral choice regarding therapy. Several cancer hazards have been the subject of controversy for at least a decade [14–16]. Furthermore, recent risk-based research demonstrates that many errors in radiation oncology are caused by systemic faults in procedure and workflow [17]. These blunders may aggravate the situation, even more, the danger, in any case, cancer risk research has advanced tremendously. We learned more about clinical radiation oncology.

12.1.1 What is big data?

There does not appear to be a widely agreed-defined statement for big data. However, it appears that at least three of its distinguishing characteristics – the three Vs – are commonly understood: volume, variety, and velocity.

Volume is an important feature of big data. Enormous volumes of knowledge of information tax the capability and ability of ancient data devices for recording, monitoring, as well as recovering like knowledge warehouses. Massive knowledge necessitates adaptable and simply extensile alternatives for data storage as well as administration.

Another quality is variable. Today's healthcare information comes in a very style of formats, together with structured and free-text information nonheritable by EHRs, diagnostic imaging, and information showing live via social platforms as well as operating systems. As a result, a large amount of this information is not used to improve health or healthcare, for example, less than August 15, 1945, of health information in EHRs is also placed in organised information fields, allowing that information to be evaluated by utilising ancient archiving and evaluation methods. Big-data techniques change the economic connecting and analysis of disparately unionized information to answer specific functional, commercial, or analysis queries [18].

The final attribute is velocity. The majority of the current health information technology platforms are incapable of processing and analyzing enormous amounts of continually refreshed, variously formatted data in real-time. As we will see below, the big data architecture allows for more flexible and rapid data management than has previously been possible.

More identically ready information is accessible for analysis higher at any other time. For instance, the CMS is creating medicare claims information on the market for analysis by investigators and how the cheap term "eligible businesses" is used throughout the program to safeguard as well as enable to facilitate healthcare (PPACA). Examining this information in conjunction with claims information from industrial insurers will give respectable edges. However, we would not classify these activities as capitalizing on the facility of massive information. Solely by combining CMS information with otherwise formatted information (e.g., data from EHRs) and fleetly analyzing it may the three Vs of massive information be complete and large data's full potential complete.

12.1.2 The potential benefits of big data

Big information has the potential to return up to around $300 billion in price in the healthcare sector per annum, the simple fraction of which could be produced through cutting medical spending [19]. Brobdingnagian information has been formally tried for its medical and financial value, in numerous cases. First, the delivery of made-to-order medication (individualized diagnoses and coverings supported by a victim's elaborated asset allocation) has been unassailable for the care of victims having carcinoma or completely various ailments [20, 21]. Furthermore, the use of medical decision-making has already been facilitated by automated methods of X-rays, computed tomography (CT) detection images, and MRI, as well as the quarrying of research records to customize therapies based on patient asset allocations. Third, the habit of relying on user-generated information has been unassailable practice android technologies to control diagnostic and treatment selections equally as tutorial messages to encourage fascinating patient behaviors. Twelve as an associate degree example, the VHA has established type of devices healthcare systems mobile healthcare comes which focuses on individual patients and clinicians through the quick assortment and analysis of patient-generated information [22]. Fourth, Brobdingnagian expertise in public health assessments has renowned trends that may be incomprehensible if shorter samples of consistently organized information had been evaluated alternatively. One example is the social scientist program, cooperation seen between the Veterans Health Administration as well as Facebook. It uses an amount of your time forecasting operating system to analyze voluntary, opt-in information from veterans' communal platforms profiles and cell phones for violence risk hindrance [23].

Finally, big data techniques for interference and identification of fraud, like those utilized by CMS, have supplanted older manual documentation strategies. In 2011, alone, these techniques saved nearly $4 billion in prices [24].

12.1.3 Determining what constitutes "big data"

Big data, as the name implies, refers to giant amounts of knowledge that square measure unmanageable with an ancient software package or web-based platforms. It outperforms ancient storage, processing, and analytical capabilities. Although there square measure many definitions for giant knowledge, Stephen A. Douglas Laney's is the hottest and most widely accepted. Lucy Craft Laney detected that (big) knowledge was increasing in three dimensions: volume, velocity, and selection (also called the three Vs) [25]. The "big" part of huge knowledge refers to the sheer quantity of knowledge it contains. The term "big data" encompasses not solely volume but additionally speed and selection. Selection refers to the varied varieties of organized and unorganized knowledge that every business or service would accumulate, like transaction-level knowledge, recordings, clips, or word documents, as well as log files. Speed refers to the speed or rate at that knowledge is collected and created accessible for additional analysis. These three Vs have come back to outline huge knowledge as an entire. Though some have side further Vs to the present definition [26], the foremost widely accepted fourth V is "veracity."

In recent years, the term "big data" has gained a great deal of traction around the world. Almost every field of analysis, whether or not in trade or domain, generates and analyses giant amounts of knowledge for a spread of functions. The foremost tough task in handling this large pile of knowledge, which may be organized or unorganized, is managing it. We want technologically sophisticated programs, as well as programming packages that may be use quickly and also expense-effective slightly elevated machine capability over other activities as a result of huge knowledge, which is unmanageable with the ancient software package. To form a sense of this large quantity of knowledge, computer science (AI) algorithmic techniques and innovative fusion techniques would be needed. Indeed, mistreatment machine learning (ML) strategies, for instance, multilayer perceptron with different AI approaches to attain machine-controlled decision-making would be an enormous accomplishment. Big data, on the opposite hand, are often hazy when not the proper software package and hardware. Higher techniques for handling this "endless sea" of knowledge, additionally as good net applications for economic analysis and unjust insights, are needed (the information/the knowledge) and insights derived from huge data will build essential ethical foundation elements plus strategy development (similar attention, stability, and perhaps even conveyance) additional acknowledge, communicative, or even economical [27] with warehousing as well as examining equipment on available. What is more, an easy mental image of huge knowledge is going to be essential considering social group development.

Pediatric cancer is rare, with associate expected 11,000 new cases known in kids aged from birth to 14 years in the United States in 2019 [28]. Although the rate for these people has minimized by 65% from 1970 to 2016, tumors be the constant and one of the biggest reasons behind the mortality rate of kids, with more or less 1200 kids foreseen to die from cancer within the United States in 2019 [28]. If the teenagers' area unit is enclosed, the entire range of recent cancer diagnoses within the United States every year (from birth to the age of 19 years) is over 14,500 kids and young adults [28, 29]. In the United States, however, there have been 1.7 hundred thousand of advanced instances of adolescent malignancy in 2016 [30], whereas every cancer prediction, as well as the demise of any toddler or teenager, could be a sorrowful misfortune for a family member, tumor in children continues to exist and associate uncommon sickness with an occasional average yearly occurrence. Although the foremost frequent varieties of tumors in youngsters, such as acute myeloid leukemia, CNS tumors, and lymphomas, offer a considerable problem in terms of gathering adequate knowledge to support clinical findings [28].

12.2 Biomedical Research using Big Data

A biological system, like a personality's cell, exhibits a posh interaction of molecular and physical events. A medicine or pharmacological investigation sometimes holds information on a somewhat compact and/or less complicated part to grasp the interdependencies of assorted parts and events of such a posh system. As a result, generating a broad map of the biological development of interest necessitates multiple simplified experiments. This suggests that the additional information we have, the higher we will be the ability to perceive biological processes. Modern techniques have advanced at a fast rate as a result of this idea. Contemplate the quantity of information generated since economical techniques such as NGS plus ordering GWAS were combined to decrypt identically biology. NGS-based information provides info at antecedently inaccessible depths, transporting the experimental situation to an entirely new level. It is improved sensitivity that we can observe or record biological events coupled with specific diseases in real-time. The "-omics" era began with the realization that giant quantities of information will offer the North American nation a major amount of knowledge that is typically unidentified or hidden in smaller experimental ways. Rather than finding out one "gene," scientists will currently study the complete "genome" of Associate in Nursing organism in "genomics" studies in a very affordable quantity of your time. Similarly, rather than finding out the expression or "transcription" of one cistron, we can currently use "transcriptomics" studies to see the expression of all the genes or the complete "transcriptome" of the associate in the nursing organism. Each of those experiments generates a major quantity of information with a bigger depth than ever before. However, this level of detail and backbone might not be adequate to make a case for a selected mechanism or event. As a result, it is normal to search for yourself analyzing an oversized quantity of information gathered from multiple experiments to realize new insights. This truth is insured by a gentle increase in the variety of publications on massive information in care. Therefore, an examination of massive information via medical and care systems may be extraordinarily helpful in developing new care methods. The foremost recent technological advancements in information standards have already been elevated as a result of data gathering, collecting, as well as assessment for an individualized medication movement within the close to future. Omics studies' massive information the use of next-generation sequencing (NGS) has significantly streamlined the overall scheduling method and reduced the value of generating whole ordering sequence information. Complete ordering sequencing has been born in value from millions to many thousand bucks [31]. As a result of NGS technology, the amount of medical information generated by genomic and transcriptomic studies has been redoubled. In keeping with one estimate, between a hundred million and a couple of billion human genomes can be sequenced by 2025 [32]. Integrating differential gene expression data information using protein expression as well as metagenomics analysis information will considerably improve our understanding of a patient's distinctive profile – an idea referred to as "individual, individualized, or exactitude healthcare." Systematic and integrative omics information analysis combined with care analytics will aid in the development of higher treatment methods for exactitude and individualized medication. Together with EMRs, drugstore dose information, financial information, and trials driven by genetics such as genome sequencing, organic phenomenon, and NGS-based investigations are the most common sources of enormous information in medical care. To supply higher treatments and patient care, care needs a robust integration of medical informations from numerous sources. These prospects square measure therefore exciting that business organizations square measure already exploiting human ordering information to help suppliers in creating individualized medical selections, despite the very fact that genomic information from patients would have several variables to account for. This might be a game-changer within the field of medication and health in the future.

12.3 The Big Data "Omics"

For bioinformaticians, massive information from "omics" studies presents a replacement form of challenge. To analyze such advanced information from biological systems, a strong algorithms area unit is required. The ultimate goal is to convert this massive amount of information into a useful mental object. Travel bioinformatics is the use of bioinformatics approaches to remodel medicine and genomic information into prognosticative and preventive health information, which is on the leading edge of data-driven tending. In tending, varied forms of quantitative information, such as experimental data, pharmacological information, and genetic patterns, will be merged, and then it would not be used to create new conceptual information that may aid in the development of precise medicines [33]. That is also often why technological advances are necessary to assist in analyzing digital wealth. Indeed, multimillion-dollar comes like the "big information analysis and development initiative" are launched with the goal of up the standard of massive information tools and techniques for higher organization, economic access, and intelligent analysis of massive information. The process of "omics" information derived through enormous individual ordering operations as well as alternative storylines of communities comes is predicted to produce various advantages. Researchers can have access to a colossal quantity of data in sequences of people that come like one thousand genomes. Similarly, the reference of the desoxyribonucleic acid parts (ENCODE) project, which was supported by the human ordering project, aimed to work out all practical parts within the human ordering exploitation bioinformatics approaches. We have compiled a listing of a number of the foremost widely used bioinformatics-based tools for omics information massive information analytics:

i. SparkSeq is an associate degree economical and cloud-ready platform for interactive genomic information analysis with ester exactitude that's supported by the Apache Spark framework and Hadoop library.

ii. SAMQA detects errors in large-scale genomic information and ensures its quality. This tool was created to spot and detect issues like SAM configuration mistakes or missing data for the Tumor Research Organization at the Ministry Of Health (NIHC) ordering global program.

iii. ART will simulate browse error and browse length profiles for information generated by SOLiD and Illumina are both examples of extreme throughput right-order systems.

iv. DistMap may be a decentralized brief navigation toolkit supported by the Hadoop cluster which is intended to hide a broader variety of assembling purposes. One in all its implementations, the BWA clerk, for instance, will method five hundred million browse pairs in regarding vi hours, which is roughly thirteen times quicker than the standard mono clerk.

v. SeqWare may be a question engine supported by the Apache HBase information system that integrates ordering browsers and tools to produce access to large-scale whole-genome datasets.

vi. Waterspout may be a paradigm of parallel processing that has been utilized in ordering experimentation with localization to extend the measurability of reading massive amounts of sequencing information.

vii. Hydra processes massive amide and spectra databases for genetics datasets exploiting the Hadoop distributed computing framework. On a Hadoop cluster, this tool is capable of grading twenty-seven billion peptides in but an hour.

viii. BlueSNP may be an R package for GWAS examines that uses the Hadoop platform. It focuses on applied mathematics readouts to seek out important links around collections of nucleotide sequence relationships. The instrument's potency is calculable to be ready to associate degree Alyse 1000 phenotypes on 106 SNPs in 104 people in 0.5 an hour.

ix. Myrna, a cloud-based pipeline that features browse alignments, information normalization, and applied mathematics modeling, provides info on organic phenomenon level variations.

The number of disease-specific datasets generated by omics platforms has exploded in recent years. The Array Express Documentation for Operational Genetics is a knowledge base, for example, that contains data from over one million functional assays and 30,000 experiments. The growing amount of data necessitates the development of more efficient bioinformatics-based software to analyze and interpret the data. This has resulted in the development of specialized tools for analyzing such large amounts of data. The following are several of those major prominent business frameworks enabling insights into massive quantities of data.

12.4 Commercial Healthcare Data Analytics Platforms

Various companies have utilized AI to assess revealed findings, matter knowledge, and image knowledge to get relevant leads to order to beat huge knowledge difficulties and perform sander analytics. IBM Corporate entity is merely one of the most significant and well-known players in this sector, providing business analytics services. IBM Corporation is an associate degree AI platform that enables hospitals, providers, and researchers to share and analyze health knowledge. Similarly, iron Health offers technology-driven services in attention analytics, with specific stress on cancer analysis. Alternative major companies, like Oracle Corporation and Google opposition, square measure targeted the event of cloud-based stockpiling and formats for decentralized computing resources. Astonishingly, numerous organizations and start-ups have developed in recent years to produce statistics as well as applications for medical services.

12.4.1 Ayasdi

Ayasdi is a single equivalent massive trafficker that focuses on ML-based approaches to present a foundation supporting emerging technologies moreover as an associate methodology for applications that have been evaluated and tested company quantifiability. It offers a range of care analytics applications, like understanding and managing clinical variance and remodeling clinical care prices. It may analyze and manage however hospitals area unit unions, conversations between doctors, risk-based treatment choices created by doctors, and also the care they supply to patients. It additionally provides associates with any request seeking population health evaluation as well as maintenance, a proactively approach on the far side of typical risk analysis methodologies. It employs machine learning intelligence to forecast prospective risk projections, determines risk factors, and supply solutions for the most effective doable results.

Table 12.1 Lists of the number of vendors within the attention business. Many of the above business choices square measure.

S. no.	Organization	Details	Web address
1.	IBM	Serves customers for the exchange of medicinal as well as well-being concerns between hospitals, investigators, as well as providers to facilitate advancements in research.	https://www.ibm.com/in-en/watson-health
2.	MedeAnalytics	Includes outcomes monitoring services, medical services as well as programs, and medical monitoring as well as, in addition to a long-standing history of patient data management.	https://medeanalytics.com/
3.	Health Fidelity	Serves as a risk-managing alternative for healthcare organizations' procedures, as well as a tool for enhancement and correction.	https://healthfidelity.com/
4.	Roam Analytics	Forums related to delving through massive amounts of aimless medical details to extract useful information.	https://roamanalytics.com/
5.	Flatiron Health	Apps are available for upgrading as well as organizing carcinoma statistics to facilitate the treatment of cancer.	https://flatiron.com/
6.	Enlitic	Offers deep learning for healthcare diagnostics by utilizing explanatory variables representing medical studies on a global basis.	https://www.enlit ic.com/
7.	Digital Reasoning Systems	It offers intellectual virtual machine capabilities as well as computational intelligence options for unstructured data arranging or even organization.	https://digitalreasoning.com/

Table 12.2 (Continued.)

S. no.	Organization	Details	Web address
8.	Ayasdi	Provides an artificial intelligence-enabled foundation supporting analyzing diagnostic variability, global medical care, chances assessment, as well as another aspect of health informatics.	https://www.ayasdi.com/
9.	Linguamatics	It has an information retrieval tool for sifting through unstructured healthcare data in search of critical information.	https://www.linguamatics.com/
10	Apixio	Provides a platform for cognitive technologies to analyze clinical evidence and pdf medical histories to generate in-depth facts.	https://www.apixi o.com/
11.	Roam Analytics	Infrastructure is required for human speech Manipulation in current medical services.	https://roamanalytics.com/
12.	Lumiata	Analytical, as well as risk mitigation services, are provided services to ensure effective consequences in healthcare.	https://www.lumiata.com
13.	Optum Health	Offers medical algorithms, helps modernize the infrastructure of modern health systems and develops innovative and effective services for the medical sector.	https://www.optum.com/

12.4.2 Linguamatics

It is a natural language processing (NLP) algorithm that is built on something like an information retrieval collaborative technique (I2E). I2E can extract and analyze a wide range of data. This technique produces results ten times faster than existing tools and does not require professional expertise for data analysis. This method can extract genetic correlations and facts from unstructured data. To produce clean and filtered outputs, traditional ML demands content that has been carefully chosen for source. However, when NLP is integrated into EHR or healthcare records in general, it helps in extracting organized information from unorganized input information.

12.4.3 IBM Watson

This is one of the several among IBM's distinctive innovations, that focus on huge knowledge analytics in each skilled field. This platform makes significant use of machine learning and computing (AI) algorithms to extract the foremost data from the littlest quantity of knowledge. IBM Watson follows a strict routine of integrating a large variety of attention disciplines to administer helpful and structured knowledge. IBM Watson and Pfizer have created a productive relationship to hurry the event of novel immune-oncology combos in a shot to explore novel therapeutic targets specifically within the cancer malady model. The researchers will analyze giant genomic knowledge sets by combining Watson's deep learning modules with AI technology. IBM Watson has been accustomed forecast distinct types of cancer-supported organic phenomenon profiles derived from a range of big knowledge sets, indicating the presence of many druggable targets. IBM Watson is additionally used in drug discovery programs to supply a whole summary of the molecular landscape during a specific malady model by combining curated literature and manufacturing network maps. The attention space divides analytics into four classes to check a large variety of medical data: explanatory, examination, prediction, and corrective analyses. The diagnostic analysis explains the explanations and factors behind the incidence of bound events, like selecting a treatment choice for a patient-supported clump and call trees. The term "descriptive analytics" refers to the process of characterizing and responding to present healthcare incidents, whereas "diagnostic analysis" refers to the process of explaining the causes and factors that contribute to the occurrence of bound occurrences, such as selecting a treatment option for a patient supported clump as well as call tree branches. By determinative patterns and possibilities, prophetic analytics focuses on the flexibility to anticipate future outcomes. These ways of square measuring principally supported machine learning techniques and square measuring helps determine the potential for a patient to develop issues. Prescriptive analytics is the method of analyzing knowledge to advocate a course of action for creating the simplest selections potential. As an example, a patient's call to avoid a particular medication supported ascertained adverse effects and anticipated issues. Integration of massive knowledge into attention analytics is an important part of the performance of gift medical systems;

notwithstanding, complicated solutions should be devised. Integration of huge knowledge technology to boost outcomes needs associated design of best practices for varied analytics within the attention domain. However, their square measure varied obstacles to beat when implementing such ways [34].

12.5 Big Data in the Field of Pediatric Cancer

BD can be described as "data resources that are so huge, speed, as well as wide range that their transition into something of worth necessitates specific technology using computational techniques" [35, 36]. Big data in oncology might include living organisms, medical, and organizational information regarding tumor sufferers including both organized and unorganized representations, among other things. Such records could be used to find answers to issues concerning genomes, outcomes, and therapy efficacy [37]. Because of the enormous population of sufferers as well as the existence of massive established linked one's datasets, big data has been widely used in adult oncology research [38]. Furthermore, the application of BD in pediatric cancer is even now in its infancy. The National Cancer Institute (NCI) has taken the lead in promoting community use of big data throughout planned therapy for oncology subdisciplines, together with including the establishment of the Children's tumors Data Project, a $50 million action plan announced in 2019 [39] has a strong emphasis upon information transfer.

12.5.1 Data

The National Cancer Institute (NCI) has assembled a group of knowledge efforts toward the CRDC system [40]. The ecological system's purpose is to construct knowledge commons nodes and sources wherever the knowledge framework for data processing and storage, as well as facilities, and resources, including implementations, will be co-located [41]. The therapeutically applicable analysis to come up with functional therapies (TARGET), the genetics knowledge commons (PDC), the screening knowledge commons, the genomic knowledge commons (GDC), the advanced canine knowledge reliable, and also the individual's tumor connections of layouts, at the midst of further several, are all a part of the system. The GDC is presently the only knowledge supply within the CRDC that covers medical specialty cancer statistics. The TARGET pool may be a cluster of researchers junction rectifiers by participants of the younger's age medical specialty cluster (COG), medical examinations, and investigation organization dedicated alone to medical specialty and teenager's malignance analysis [42]. Investigators from TARGET join forces with the COG to achieve control of healthcare data as well as biospecimens from throughout the connections and links, with the target of making genetic knowledge that will change molecular discoveries and translation into successful medicines. ALL, AML, excretory organs, metastatic tumors, and osteogenic sarcoma are all drawn in TARGET [43, 44]. The GDC organizes, standardizes, and makes knowledge from large-scale NCI studies like TARGET and its adult counterpart, the Cancer ordering Atlas [45, 46] accessible. Genomic and clinical knowledge will be unbroken and analyzed at intervals in the GDC, permitting researchers to match findings across trials. One of the first GDC goals has been to harmonize NCI's cancer genetics knowledge, which has each process the genomic knowledge with normal pipelines and building (a knowledge/ an information) model for biospecimens as well as medical evidence with uniform terminologies and descriptions. The public inquiry, healthcare professional, and conclusion of the Centres For Cancer Disease Control (SEER) Program is a written account that compiles statistics on cancer mortality or even well-being knowledge through 19 topographical regions around us, representing approximately 34th of the population [47]. SEER contains knowledge on folks of all ages and has been widely utilized in childhood cancer analysis. The children's medical specialty cluster (COG) is the world's biggest medical specialty analysis pool, conducting late-phase and early-phase clinical studies unitedly with the National Cancer Institute. Over 150 educational medical centers within United States, Canada, and Australia are a part of the COG network. The bulk of kids within us are treated on or by a COG study. COG has several giant patient registries, together with the childhood cancer analysis network and project: every child, that account for over 19th of medical specialty cases within us beneath the age of 15 years [48]. Additionally, in 2017, the COG and also the NCI launched molecular analysis for medical care alternative (Pediatric MATCH), a prospective preciseness drugs trial to build a genetic written account to look at early-phase medical specialty in medical specialty solid tumors [49]. The University of Chicago's pediatric Cancer Knowledge Commons (PCDC) collaborates with stakeholders to make knowledge commons for medical specialty cancer knowledge [50]. The PCDC workgroup generates and votes on accord knowledge dictionaries for medical specialty cancer mistreatment a reiterative consensus-building technique. For metastatic tumors, rhabdosarcoma, reproductive cell cancers, and acute myelogenous cancer of the blood, the PCDC has generated stable versions of international information dictionaries. The PCDC team works with information homeowners to ascertain information sharing and use agreements, allowing information harmonization and migration into the commons for later sharing and analysis. The PCDC has information on over 240 youngsters as of January 2020, and it is quickly increasing. Information within the PCDC is joined in real-time to alternative sources, like genetic information in TARGET and tissue convenience within the Nationwide Children's Biopathology Center, by victimization associate degree uniform COG range provided to every child. The PCDC is growing, with new diseases such as acute white blood corpuscle cancer of the blood, acute myelogenous cancer of the blood, Hodgkin cancer, osteogenic sarcoma, Ewing malignant neoplastic disease, and reproductive cell cancers being added all the time. The accord information dictionaries are utilized to fuel the ensuing generation of clinical studies, additionally permitting the harmonization of information from completed trials. The PCDC team collaborates closely with the National Cancer Institute to stay the NCI synonym finder [51] and caDSR [52] repositories up to now, permitting clinical trials to decide on a listing of balloted components for future information assortment forms. The PCDC was designed to figure with the Cancer analysis information Commons system. The PCDC can interface and interoperate with the opposite information commons nodes within the CRDC additional just by investing in the Gen3 technical design, that was created for the Genomic Information Commons. The Treehouse Childhood Cancer Initiative could be a written account of organic phenomenon information from over 1100 medicine growth samples that is reliable to researchers for comparative genomic analysis and therefore the development the of latest treatments [53]. Nine hospital and syndicate partners have contributed of growth organic phenomenon information further as patient-privacy-protected clinical data like age, gender, and malady sort. This information is downloaded by anyone. SJCARES, an enormous cloud-based medicine cancer written account engineered by St. Jude Children's Analysis Hospital, was specifically designed for low- and middle-income nations to push inter-national cooperative analysis [54]. The PeCan information Portal, sponsored by St. Jude Children's analysis Hospital and its collaborators, provides "interactive visualizations of medicine cancer mutations across many initiatives at St. Jude Children's analysis Hospital and its collaborators" [55]. PeCan currently includes data on 4877 patients with twenty-three diagnoses and over 8,800 mutations [55]. St. Jude Cloud, which contains high-throughput genetics information from St. Jude patients during a searchable interface with advanced analysis capabilities [56, 57], was simply introduced. The Gabriella Miller children's initial medicine analysis Program could be cloud-based medicine genetics written account that aims to research the genetic causes of childhood cancer and structural birth defects, and further advance individualized medication for the detection, treatment, and management of those diseases in youngsters [58]. The youngster's initial information Resource aims to research the genetic origins and linkages between medical cancer and structural birth abnormalities, which sets it except alternative information commons. The Centers for unwellness management and hindrance (CDC) sponsors the National Program of Cancer Registries, which supports state-based cancer registries within the and represents roughly 97 p.c. of the population [59]. The NPCR launched a medicine and Young Adult Early Case Capture program in 2014, to register medicine cancer cases at intervals of 30 days of identification [60].

12.5.2 Research supported a medicine cancer register

Between the years 2018 and 2020, a literature analysis was conducted victimization PubMed information and therefore the search phrases "pediatric cancer" and "registry" to ascertain whether childhood cancer registry area units are presently being employed in medicine medical specialty analysis [61]. The research yielded 751 publications, that the researchers assessed and

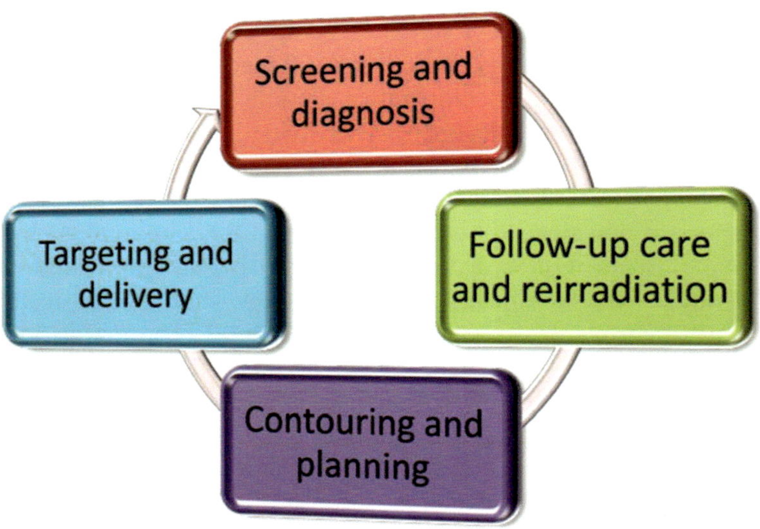

Figure 12.1 From imaging techniques to therapy, a sophisticated RT methodology exists.

located to be similar to modern research in 214 of them. The studies included research that used some co-register for the advancement of medicine or for AYA patients with malignant tumors. Studies that solely reportable analysis on benign tumors or that solely utilized single-injection techniques were eliminated. Every study was coded in line with its primary analysis domain, which was present. Figure 12.1 summarizes the general leads in terms of the number of articles in every domain.

12.5.3 Epidemiologic descriptive analysis

The epidemiological study that utilized registered knowledge retrospectively to outline medicine tumors regarding negative chance categorization, forecasting, health consequences, and medical services analysis was the most important domain. Several studies, as well as those conducted by analysis teams based mostly outside of the United States, used SEER information to analyze a good variety of outcomes. Deng et colleagues determined that numerous prognostic variables, as well as age, radiation use, and gross total surgical operation, were connected with magnified survival in kids pine blastomas victimization SEER [62]. The Finnish Cancer register, the Canadian Cancer register, the France national register of Strong Malignancies in Early life, and therefore the Italian metastatic tumor register were among the registries that conducted the epidemiological analysis. The International Society of medicine medical specialty, Cooperative Weichteilsarkom Studiengruppe, and the European Rhabdoid register, in addition, because of the massive translocation of marrow analysis teams CIBMTR, EBMT, ABMTRR, JSHCT, and therefore the Asia-Pacific Blood and Marrow Transplantation cluster, that represents eighteen countries, were additionally portrayed. Many studies on Aden myosarcoma and metastatic tumors were revealed by the EUROCOURSE cluster, a channel comprising country-wide confirmed cases across twelve nations in Southern and Eastern Europe, which illustrate the facility for integrating specific victims' knowledge through an outsized region, in addition, to illustrating regional disparities in cancer consequences [63]. Several of those consortium-sponsored studies think about the flexibility to pool knowledge into a large knowledge commons, which might additionally profit smaller registries. As an example, between 1994 and 2012, the Cancer register of a recent geographical region, a French colony within the ocean, reportable 162 total incidences of medicine cancer [64]. Combining these cases into an even bigger worldwide register would enable the island to be compared to alternative countries and therefore the register to require half in larger medicine cancer trials. Thai leukemia working party, which is a component of the Thai Society of hematology, was known as a prospective empirical register. This register compared the survival of AYA and grownup people suffering from severe white corpuscle cancer (ALL) and located that people from the UN agency who received somatic cell transplantation had a higher prognosis [65]. Apart from massive therapeutic interventions, consortia like SIOP and COG, the flexibility to perform prospective epidemiological analysis may be a highlight of registries and knowledge commons. Many analyses have utilized huge registries to review uncommon duct malignancies, like solid pseudopapillary exocrine gland tumors with the Italian pediatric Rare neoplasm register [66] or Merkel cell cancer with the SEER [67]. Enormous datasets will improve analysis quality for rare tumors; a single-center register of medicine brain tumors in Kerala, India, was unable to realize applied math significance because of a tiny low sample size and specifically concerned the institution of an Indian neoplasm register [68].

12.6 The Study of Genomics

Several studies, encompassing personalized medicine, hereditary propensity, and genetic susceptibility impacts assessment, employed pediatric cancer registries to conduct genomic research. The PanCareLIFE registry, which includes pediatric cancer victims receiving treatment using cisplatin, carboplatin, or cranial irradiation from across Europe, used genotyping to look for gene variations linked to ototoxicity [69] and infertility [70]. Kim et colleagues used the TARGET database works in collaboration with the World Pleuropulmonary Blastoma/DICER1 Collection to hunt for hereditary DICER1 alterations across varying sorts of pediatric oncology [71].

12.7 Data Sharing Faces Technical Obstacles and Barriers

Despite the massive variety of heterogeneous registries discovered within the literature analysis, important consortia square measure attempting to harmonize intra-registry information. One in all the analyses, for instance, used the metropolis agreement method to standardize the numerical units of busulfan reportage in vegetative cell transplant registries [72]. There have been no reports, however, of investigations geared toward harmonizing information among registries. Rather, the review discovered many registries that square measure are already a part of larger worldwide registries that square measure geographically or disease-specific. Whereas it is been argued that regional registries square measure redundant while national registries square measure cheaper [73], improved harmonization and also the creation of knowledge commons would enable regional registries, like the city Province Cancer register, to participate in international clinical trials. The goal must always be to amass standardized information from cancer patients so connect various information sources. Within the absence of standardized information, they will be consonant to one customary once assortment, as is presently usual procedure. The assembling of clinical, biospecimen, and genetic information for analysis is one example of pediatric cancer. COG clinical information will be consonant to (a common/ a typical) standard and created and offered within the pediatric Cancer Information Commons, standardized genomic information will be found within the Gabriella Miller children initial information Resource Center or the Genomic Information Commons, and biospecimen accessibility will be found within the Biopathology Center (BPC) at Nationwide Children's Hospital. The BPC's given Universal Specimen symbol connects these three information sources. The user will then run a question for a definite patient cohort to see that patients have genomic information and biospecimens accessible. This cannot be attained while not comprehensive information standardization and harmonization processes. Collecting, combining, and exploiting huge information for pediatric cancer analysis has varied issues, several of which square measure common to different areas of medication. Speedy reportage for children's cancer, for instance, is difficult and erring, preventing results from being disseminated quickly [74] because the treehouse project cluster points out, the mistreatment of genetic information is sophisticated and involves challenges like information location, characterization, quality analysis, use approval, and compliance [75]. As a result of specialized implementations and a general lack of reference to information standards, the mistreatment of information from electronic health records (EHR) is especially troublesome [76]. Clinical information assortment is troublesome because of a scarcity of standardization and, particularly, the widespread use of free-text notes. Distinctive children's cohorts for cancer studies necessitates the creation of tailored criteria [77]. Variations in EHR deployments, additionally with widespread customization of the installations, still stymie economical information exchange for analysis [78]. The German MIRACUM cluster has begun to deal with information quality challenges and proposes a framework for a unified, standardized, and consonant EHR for clinical analysis. Similarly, the PEDSnet national network aspires to produce a platform for pediatric analysis discovery, however, it must be seen whether or not information element standardization can extend on the far side of a little variety of clinical parameters. Information-sharing agreements still be a significant impediment to effective and timely analysis. Instead of the legal team being a bottleneck for information use agreement execution, a recent survey of educational researchers found that procedural inefficiencies, incomplete info, a scarcity of incentives and familiarity with educational practices, and school quality were the foremost common problems [79].

12.7.1 Which data should be taken into account, and also how they should be managed?

Lambin et al. have detailed the features that should be considered to be taken into account and also included in a forecasting model [80]. They include:

- Clinical characteristics (patients' performance status, grade, and stage of disease) the tumor, the results of blood tests, and patient questionnaires).
- Treatment characteristics include: intended spatial and temporal dose distribution, as well as concomitant chemotherapy. Data could have been extracted for this purpose to be evaluated directly from the record-and-verify software.
- Scanning characteristics include tumor size and volume, as well as metabolic uptake (read more). Radiomics is a research field that encompasses the entire world.
- Molecular characteristics: intrinsic radiosensitivity.
- Data collection and administration.
- Modern radioactive therapy presents a comprehensive computerized image of the treatment.
- We keep track of the radiation regimens that have been administered to each patient.
- We know exactly where photons go in the body for each patient and therapy session, and according to the predefined statement, we have got it in the technical medium for every particular suffered individual.
- Onboard imaging considers day-to-day fluctuations as well, as a consequence, we are aware of the status of the medication being administered.
- Such methods are capable of providing information on the worldly and spatial dispersion of treatment.

- Information is collected eventually for each victim in each department's documentation plus monitoring program [81].

12.7.2 Data collection and administration

Modern radiation medical specialty provides a transparent processed image of the treatment. We tend to keep track of the radiation regimens that have been administered to every patient. We all know specifically wherever photons come in the body for every patient and medical aid session, as well as the defined statement we have it for every individual through technological medium [82]. A board imaging considers day-to-day fluctuation similarly, therefore we all know wherever the dose is run. These systems will offer info on the temporal and special dispersion of treatments. Knowledge is collected prospectively (for every) patient in each department's record-and-verify program. This extremely digital character permits the quantification and analysis of the healthcare delivery method. The standard of information collected is considerably superior thereto of that most alternative professions of drugs [83]. This knowledge is often extracted at varied levels to be integrated into clinical knowledge warehouses (CDWs) at hospitals. The information contains careful info using scatter plots of dosage quantity, therapy portions, the interval among fractions, total duration of therapy, as well as dosing proportion, plus pictures generated by aboard devices. One more technique that ought to be avoided is removing solely the information that is regarded helpful before desegregating it into the CDW [84]. This approach would well scale back the richness of data and may be avoided. In addition to antecedental delineate knowledge, follow-up is important in radiation medical specialty and drugs normally to site toxicity. During this context, online and mobile device inputs, similar to signals from portable sensors, ought to be promoted. Individuals would therefore be capable of giving further careful, period info on adverse events that occurred throughout and when treatment, while not having to attend their next meeting with radiation oncologists. Many analysis therein already exists in this domain incontestable the worth of consequences as stated by patients in rising review [85, 86]. The amount of information that has got is being accumulated or even regulated regularly increases. We will currently predict that knowledge for one patient would quantity to seven GB, together with raw genomic knowledge, which might account for around the 17th of it. Any organization has important challenges in terms of health knowledge security and accessibility. They must be simply and quickly offered from any place, while not jeopardizing their safety. Remote knowledge access necessitates that the planning considers rigorous security necessities, like robust cardholder recognition and mechanisms that offer most things detectable machining processes. Login methods for relevant tending professionals necessitate an ascendible process at an outsized value, however, they must not be forgotten [87]. Medical history linking and knowledge anonymization are oft-needed processes to provide knowledge for study. They often necessitate the involvement of a reliable third party to handle these procedures. In general, to supply tending knowledge for analysis, the information should be touched from the care zone, wherever it is controlled by the trusty relationship between medical man and patient, to the none-care zone, wherever it is controlled by special knowledge governance bodies, anonymized, and created offered for analysis. Translational analysis platforms are samples of existing solutions to help the storage and access of care. These platforms will mix large amounts of clinical knowledge with omics knowledge [88]. Despite technical developments, some writers worry that growth in knowledge volume is also outpacing hospitals' ability to satisfy the demand for knowledge storage [89]. One choice would be to manage this knowledge in the same manner as most hospitals manage previous medical files, that is, by shifting the oldest and largest files to secondary storage. To preserve speedy and straightforward access to digital knowledge, we might migrate the foremost ample knowledge to a secondary storage-optimized platform distinct from the question platform. Ontologies are wont to extract high-quality knowledge. Standardization of EHR classes and terminologies, treatment protocols, and genetic annotations improves the standard and a likeness of information want to build models. The variety of those characteristics makes it nearly not possible to assemble and mix quality knowledge. Associate in Nursing metaphysics, or a set of shared ideas, may be an important element of any knowledge assortment system and prediction model. Currently, there are about 440 medicine ontologies. SNOMED [90], the NCI wordbook [91], office AE [92], and therefore the UMLS metathesaurus [93] are the foremost unremarkably utilized. These ontologies lack varied radiation medical specialty language, prompting the event of the Radiation medical specialty metaphysics (ROO) [94], which reused alternative ontologies however enclosed artificial language ideas like ROI, focused identities (GTV, CTV, and PTV), and dose–volume histograms (DVH). The usage of ordinary ontologies across the board can modify automatic multicentric knowledge extraction and integration. The quality of the information set and therefore the careful choice of options are important. Once attainable, freelance verification by a second keeper or knowledge checker ought to be used. More confirmation by a competent skill is additionally extraordinarily beneficial, it implies that cooperation among medical man, and therefore human knowledge is necessary to execute this research.

12.7.3 The data deluge

People operating for varied organizations everywhere on the planet generate large amounts of information daily. The term "digital universe" refers to the huge amounts of information that are developed, reproduced, and ultimately ingested during a single year. In 2005, the International Information Corporation (IDC) calculable the scale of the digital universe to be around 130 exabytes (EB). In 2017, the digital universe grew to around 17,000 EB (16 zettabytes) (ZB). By 2020, in line with IDC, the digital universe can have big to 40,000 EB. To urge a way of scale, we would have to be compelled to assign all and sundry concerning 5200 gigabytes (GB) of information. This exemplifies the outstanding rate of enlargement of the digital universe. Google and Facebook, for instance, are aggregating and storing large amounts of information. Google, for instance, might store a spread of knowledge, like the user's location, promotion choices, the directory of programs utilized, the record of web surfing, connections, browser history, and email messages, as well as other pertinent information about the person, betting on our preferences. Facebook, likewise, stores and analyses over 30 petabytes (PB) of user-generated information. "Big data" refers to the massive number of information. The IT business has success used huge information over the last decade to get important info that will generate important revenue.

These observations became therefore distinguished that they need to spawn a brand modern research domain called "statistical data." Information science covers a good variety of topics, together with information administration as well as research, so as may elicit more thoughts and improve a system's practicality or services (e.g., care as well as a transportation network).

Furthermore, the provision of a number of the foremost artistic and purposeful ways to visualize huge information once the analysis has created it easier to grasp the operation of any advanced system. As a growing phase of the population becomes awake to and concerned about the generation of massive information, it is necessary to outline what huge information is. As a result, we are going to plan to give details on the impact of massive information on the world care sector's transformation also as its impact on our daily lives during this review.

12.8 Cancer in Kids in Poor Countries

Numerous papers using minor databases in underdeveloped countries were found to be encouraging, prompting different factions of business for such establishment of greater databases to push medicine cancer analysis. In low- and lower-middle nations, there are clear variations in the assortment and use of huge information on pediatric cancer. These studies underscore the impact of medicine malignancy in low as well as intermediate nations, where prevalence estimates are larger (384,000 new cases annually vs. 45,000 new infections annually) while 5-year survival rates are significantly lower (30% vs. 80%) [95]. To boot, tries are being created to estimate the global burden and form of unknown patients [96]. Cancer registries do not embrace medical cancer cases in Nations with a moderate budget [97]. In underdeveloped nations, there are multiple high-impact potentials to boost the designation and treatment of kids with cancer by exploiting information. For instance, the AN African Cancer written account Network [90] is being coordinated by the International Network for Cancer Treatment and analysis of cancer written account program. Information gathering and "a correct estimate of the worldwide burden of kids cancer may be an important 1st step" [4] are needed to create substantial progress in huge data-driven analysis. In developing countries, huge experimental written account protocols are already in use, like a SIOP adenomyosarcoma protocol employed in southern Asian nations in 2001 [98]. In some developing countries, the supply of enrolling patients during a written account is also a significant impediment.

12.8.1 Screening and diagnosis

Cancers are being diagnosed sooner as a result of advancements in screening and testing technology. Over a previous couple of decades, cancer survival rates have improved. The precise role of screening has been questioned in twelve cases for carcinoma (chest radiation therapy or reduced computerised axial scanning [LDCT]), carcinoma (mammography), body part cancer (computed imaging [CT] colonography, CTC), and alternative malignancies [99–102]. LDCT and diagnostic techniques, severally, are shown in many trials to chop death from carcinoma by two hundredths and carcinoma death by two hundredths [103, 104]. Body part cancer mortality rates are steadily declining over many decades, in line with medical specialty studies.

However, because of the possible hazards, including radiation risks, these routine screening, and diagnostic tests have generated concerns. An LDCT or a CTC's mean effective dose is 1.6–2.1 mSv or 7.0–8.0 mSv, respectively. Mammography screening doses range from 2.0 to 5.0 millisieverts (mSv) [105]. A diagnostic CT's total radiation dosage can range anywhere from 2.2 to 14.0 mSv [106]. Because the dose is so low, these dangers are difficult to quantify. The majority of the quantitative data comes from studies of atomic bomb survivors and cohorts of radiation workers, although it is fraught with ambiguity. The conceptual threat seems to be as described in the following:

- doses less than 20 mSv are associated with a low oncological risk (1< in 1000 patients),
- doses 20–100 mSv are associated with moderate risk (1> in 1000 patients), and
- and doses greater than 100 mSv are associated with conclusive proof of radioactive material melanoma (greater than one in 1000 patients) [107].

Most medical review parties think stated that employing a linear dose–response model is feasible above 100 mSv; however, because of large uncertainties, consensus on dose–response is problematic below 100 mSv.

An LNT approach is nowadays, the most often utilizes model, in which information on the dosing regimen relationship to large doses can essentially continuously be projected backward toward 0 without the use of the limitation of dosage [108]. Many professional advisory groups advocate the LNT concepts for establishing radiation protection rules. The LSS of nuclear weapon victims demonstrates that the dose–response relationship for all solid tumors spanning the 0–2 Gy range, is comparable mostly with the LNT concept [109]. The LNT model, on the other hand, exclusively considers molecular damage and ignores defensive, organismal biological responses. An increasing quantity of experimental and epidemiological research contradicts its use for evaluating or judging the oncological risk. Furthermore, LNT can result in significant stress-related casualties [110]. The use of LNT for estimating carcinogenic hazards generated by low doses below 100 mSv, according to the French Academy of Sciences, is unnecessary and should be discouraged [111].

Apart from the LNT model, there are a number of alternative models that are at odds. A sub-linear model, like the linear-quadratic model supported by the "dual radiation action" hypothesis, which reflects the influence of indefinite quantity and rate, maybe would not make a case for current LSS leukemia knowledge. A hormesis model predicts a protective role below a threshold dose for 37 cancer and other illnesses fatalities. It is a part of associate adjustive response triggered by low-dose radiation, as proven by many animal studies showing increased lifespan in varied class concepts and various human research reports retrospectively showing lower malignancy overall incidence or a minuscule extra hazard ratio [27, 31, 32]. Beneath a threshold dose, like medical specialty suppression, the barrier theory suggests that the danger remains minor. This model is additionally congruent with leukemia and cancer knowledge, implying that cancer thresholds cannot be on top of 60 mSv and noncancer unhealthiness thresholds cannot be on top of 0.9 Sv. 33. Minimal sensitivities gradually decrease with an increase in treatment and finally vanish at dosages >0.5 Gy, consistent with such a proximal hypothesis that fits well enough with existing cancer mortality facts characterized by hypersensitive or witness results [34].

Furthermore, it is crucial to think about the risks and blessings of screening and diagnostic tests. This might indicate that the person ought to build a lot of well-read calls regarding undergoing these operations. The calculable risk of carcinoma connected with diagnostic technique is 86 cancers and 11 deaths/100, ladies screened annually from 40 to 45 years older and biyearly thereafter; the benefit-to-risk quantitative relation for lives saved is 4.5:1 and 9.5:1 for life-years saved [118]. The benefit-to-risk quantitative relation of carcinoma screening with CT is influenced by a variety of things, together with screening effectuality, smoking habits, the tested subject's sex, CT technology, and therefore the patient's age at the time of screening. It will reach a quantitative relation of roughly 10:1 sure enough populations and screening effectuality, and it rises with age [119].

At the degree of screening and diagnostic radiation, the radiation-related cancer risk is not established, however, there are unit risks related to not finishing an AN examination, like missing a designation and/or beginning medical aid too late to enhance the medical outcome. there is presently sturdy proof to support the worth of screening and designation, and it is crucial to hold out screening and diagnostic in a very method that maximizes advantages whereas avoiding hazards. Recently, a replacement rising field referred to as radiomics has shown promise in not solely providing a quantitative thanks to assessing tumor constitution but conjointly in automatic cancer detection, staging determinations, and treatment response prediction by applying an outsized range of quantitative options from screening or designation pictures [120]. CT, MRI, and PET can become additional standardized, duplicatable, and quantitative as a result of this new field. Its potential has been established in a very kind of malignancies, providing inexpensive adaptation and secernment of treatment [121]. In light-weight of this, radiomic alterations, which might be assessed before apparent changes on routine diagnostic imaging, could also be an improved predictor of cancer risk. Radio genetics, which mixes radionics and genomic information, might offer the best level of personalized risk classification.

12.8.2 Distribution as well as identifying

For ages, computed tomography visuals have become the accepted norm technique of corroboration, offering selectivity, destruction, and displacements of tumor cells and/or OARs in two-dimensional (2D), three-dimensional (3D), or 4D to attain the beam-and target alignment, potency identification, and ability to adapt, that is highly dependent on both visual acuity and image processing occurrence.

Delta radionics, a linear form of mammography, may now be predicated on regular imaging to serve as a benchmark for monitoring methods and enhancement, as well as active observation. The osteoblast dosages are also dependent on the scanned region, screening settings, and methodologies defined by various suppliers, in addition to screening regularity [122]. In general, the dosage from 2D imaging is localized epidermal layer, whereas the dosage from 3D/4D scanning is fairly distributed evenly all across the scanned body. Out-of-field dosages via screening can be proportional to dosages by scattered plus absorption radiant energy care-seeking treatment if screening factors are not improved. The testing method, including such orthogonally pictures or CBCT, as well as the procurement feature, also including the CBCT head phase, is also used to film the thorax, stomach, or pelvis to prevent irradiating peripheral regions all are the important considerations, especially for pediatric patients [123].

Other than the first quarter, laser focusing, including imaging techniques, increases the dosage to an even greater extent of curative radioactivity. Clinical studies are not widely ready to evaluate if the combined effect of aiming and delivering is beneficial. The observed reduction in negative impacts of IGRT could be partly attributable to the suitable mechanism of healthy tissues to a minimum level of exposure through the scanning process before the production of various therapeutic concentrations, in concept, resulting in improved precision.

Carcinoma cells' radiation sensitivity produced by the screening dose was recently confirmed by chance [124]. The delayed period during screening and therapy, as per the generalized sequential model, impacts local tumor growth-restricted during adding screening dosage incorporate further into the prescribed dose [125]. Optical imaging techniques, for example, are critical in IGRT since they do not expose the subject to extra radioactivity during the RT administration. Sensors can be enabled in combination with X-ray devices that support information about the underlying anatomical features including such skeletons to enable continuous monitoring and treatment and identify such alterations inside the fatty tissue where another malignancy growth is present [126].

Because adult (7.1%) and pediatric (18.8%) cancer victims have greater rates of pharmaceutical faults, several process quality assurance methods have been designed and implemented to identify defects and ascertain whether such tailored therapies consist of provided properly [127]. Inappropriate actions, operational faults, mechanical problems, starting incidents, accident precursors, close calls, and some other accidents are all examples of these types of errors [128] The majority of mistakes are detected even during setup/treatment and follow-up stages of the therapy. Independent health centers should conduct a risk evaluation of their remarkable practice, categorizing and acquiring knowledge from tragedies, and providing suitable screening frequency bands to maximize scientist time performance and patient therapies effectiveness, upgrading existing procedures as well as proposing new frameworks to maximize scientist time performance and patient therapies effectiveness. However, methods that require a lot of manual tasks are more likely to make mistakes. To further reduce social fault, a variety of apps are being developed that use cutting-edge web tools, information extraction, and advanced analytics to automate and analyze the therapeutic workflows and IGRT method.

12.9 Gaps in Oncology Risk Stratification Strategies

Identification of risk factors in malignancy is hampered by an absence of relevant prognostic information, the requirement for long manual information entry, an absence of comprehensive information, and, in some cases, associate degree overreliance on practicing intuition. Take into account an example of a person who suffers from a pathologic process, debilitating illness, and prognosis. Although between patients' treatment of advanced solid tumors, prospective information suggests that clinicians area unit dangerous at predicting prognosis [129]. Failures in distinguishing people who are in severe danger of death may result in excessively aggressive life-ending attention or the utilization of inessential acute attention by cancer patients [130]. Though prognostic aids for a few cancers exist in a medical specialty, they are seldom used as a result they are doing not apply to the bulk of cancers [131, 132], might not embrace genetic information, and fail to spot the bulk of patients WHO can die among a year [133] and necessitate long manual information entry [134]. Moreover, interclinician variability and bias will affect the assessment of prognostic variables like performance standing [135]. Even fewer printed information area units are accessible for determining the danger of alternative necessary outcomes in cancer patients, like medical aid or adverse effects.

The push for a lot of patient-centered care could be an actuation behind up-risk stratification models in a medical specialty. Moreover, shifting compensation models, like various payment models like bundled payments, can encourage the proper look after the proper patient instead of entirely incentivizing the degree of services provided [136]. Oncologists' area units are progressively being asked to tailor treatment plans supported an actual danger posed by the person of sure outcomes. This necessitates the gathering and analysis of information on communities, service sessions, as well as certain health manifestations. As a part of the medical specialty Care Model, the CMH Services has compiled comprehensive information sets of Medicare beneficiaries and has been operating with EHR vendors to boost information assortment and data wants [137–139]. But, despite the provision of progressively made information group action clinical and use factors, strong prophetical tools area unit needed to see the future risk of acute use or alternative negative outcomes. In areas like admission risk hindrance within the general patient setting, call aids supported by prophetical Statistics has been found to boost value-based real worthwhile public health decision [140, 141]. Similar tools area unit desperately required in medical specialty to boost practician judgment as well as primary care methods.

12.10 Predictive Analytics of Tumors: Presently Employed Case Studies

Modeling observational tools use algorithms derived from historical patient information to foreseeable medical consequences for people or assemblages

[142]. There are many use cases the area unit is probably generalizable because the quantity of EHR, radiology, genomic, and alternative information has adults in a medical specialty.

12.10.1 Management of public health

Orienting therapies toward unsound individuals to minimize adverse effects is a very important facet of population health management. Patients receiving therapy in the United Nations agency area unit in danger of mortality or acute care use will be known as victimization prophetic algorithms. This prediction can be accustomed influence practician behavior across the cancer spectrum, like the following therapy [143], following large intestine tumor operation [144, 145], or throughout medical release coming up with [146]. Mediating on behalf of such vulnerable individuals could facilitate cutting back resource overuse. Indeed, establishments like Penn medication and New Century Health use prophetic algorithms to spot cancer patients in United Nations agency area units at high risk of close-at-hand hospitalization/admission to get care, so use proactive phone calls or visits to focus on care management solutions [147, 148] People at Google [149] have reported victimization the quick tending ability Resources format to hurry up the effortful method of extracting information from EHRs, although EHR information is notoriously tough to use. The team used deep learning ways developed from over forty-six billion information points within the quick tending ability Resources format to accurately predict a range of medical events, as well as in-hospital fatality, inpatient hospitalization, duration of stay, and even release assessments are all measured.

12.10.2 Radiomics

Predictive analytics models area units commencing to be utilized in a medical specialty, as proven by the growing field of radiomics. Radiomics may be a sort of texture analysis that studies growth characteristics victimization quantitative information from scans [157]. These characteristics will facilitate solid growth detection, characterization, and observance [150]. Computer-assisted find ion will be accustomed detect cancerous respiratory organ CT [151] lumps " or " MRI prostatic lesions [152], while it might even be accustomed to automatizing growth staging [153], maybe most intriguing, AI- computational methods focus on (lung/ carcinoma) CTs can predict vital outcomes like change standing and therefore the risk of distant metastases

[154, 155]. Dynamic magnetic resonance imaging will be accustomed find initial reactions to therapy as well as to tell professionals of a patient's growth performance before the customary prediction of the relevant area unit used, as a result of radiologic information systems will inform choices regarding supply [156].

12.11 Big Data in Healthcare Presents Several Challenges

Methods for large knowledge administration as well as evaluation are perpetually modified significantly for period knowledge monitoring, collection, compilation, statistics (including cubic centimeters plus data modeling), and visual image technologies which will aid in the integration of EMRs into care. As an example, within us, the adoption rate of state-tested and approved EHR applications within the care business is about to be done [168]. However, the provision of many government-certified EHR programs, every with its clinical nomenclature, technical necessities, and purposeful capabilities, has hampered knowledge interchange and sharing. However, we can with confidence state that the care business has entered a "post-EMR" implementation section. The first goal now is to derive unjust insights from the huge amounts of information created by EMRs. During this section, we will undergo a number of these problems in additional detail [177].

12.11.1 Storage

One of the key problems is storing huge amounts of knowledge, although several corporations are comfy with knowledge storage on their premises. it is a variety of benefits, as well as management regarding confidentiality, accessibility, as well as uptime. AN on-site server network, on the opposite hand, is often expensive to scale and sophisticated to work. It appears that using web backups that are based on the present system may be a better option than most attention firms have gone with given the declining costs and increased responsibility. Organizations should choose cloud partners World Health Organization recognizes the critical nature of medical regulation as well as privacy challenges. What is more, cloud storage has reduced initial prices, quicker catastrophe recovery, and easier growth. Organizations can even take a hybrid approach to knowledge storage programs, which can be the foremost adjustable and sensible resolution for suppliers with completely different knowledge access and storage necessities.

12.11.2 Cleaning

After the acquisition, the information should be clean or clean to make sure accuracy, correctness, consistency, relevancy, and purity. To attain high levels of correctness and integrity, this cleansing method is often performed manually or mechanically mistreatment logic rules. Machine learning algorithms are utilized in a lot of advanced and precise technologies to save lots of time and cash and to stay dangerous knowledge from derailing huge knowledge comes.

12.11.3 Format unification

Patients generate an outsized quantity of knowledge that is tough to collect with typical EHR formats as a result of its advanced and tough-to-handle. It is quite robust to manage huge knowledge, particularly once it involves attention to suppliers World Health Organization lacks a particular knowledge organization. There was a necessity to systematize all clinically relevant info for the needs of medical statistics, and reimbursements, including invoicing. To capture essential clinical ideas, medical cryptography systems like CPT and ICD scripted pairs have been created. However, such scripted sets possess their unique set of constraints.

12.11.4 Approximation

Several investigations have found that the subject's knowledge reportage into EMRs or EHRs is not altogether correct nevertheless [178–181], possibly as a result of the EHR's limited functionality, advanced workflows, and a misunderstanding of why huge knowledge is thus essential to capture well. All of those components would possibly contribute to huge knowledge quality challenges throughout its lifecycle. Although reports recommend disparities in these things, EHRs promise to enhance knowledge quality and communication in attention workflows. The employment of self-report surveys from patients for their symptoms might improve documentation quality.

12.11.5 Pre-processing of pictures

Various physical factors are reportable in studies to guide significantly changed knowledge reliability but also mischaracterizations of present healthcare information [182]. Medical pictures are oftentimes hampered by technical impediments like many varieties of noise and artifacts. Improper treatment of medical pictures can even lead to image meddling, like the delineation of anatomical structures like veins that are not in step with real-world circumstances.

Some of the measures that may be wont to profit from the aim embody noise reduction, artifact removal, neutering of the distinction of noninheritable pictures, and image quality correction when mishandling.

12.12 Case Studies for Future Applications

12.12.1 Support for clinicians' decisions

Oncology clinicians can progressively use prophetical analytics instruments that can be used to impact everyday elements of treating patients as prophetical analytics tools reach a performance threshold. In prospective studies, prophetical algorithms are shown to scale back the required duration to reply for the sufferers of infection and to confirm quicker planned therapy for stroke patients. Forecasting negative outcomes associated with therapy, the probable period of therapeutic efficacy, the likelihood of recurrence, as well as general expectations at the point of service are all possible applications of statistics to assist physicians in making better decisions. Forty-two time period EHR-based algorithms are developed as a proof of construct to estimate medicine patients' risk of short mortality before beginning therapy [183, 184, 186].

These algorithms are on paper applicable to any EHR and are supported by structured and unstructured EHR information. Although the long-run applications of those algorithms are unknown, oncologists might realize correct mortality predictions are very helpful for care.

12.12.2 Stratification of genomic risk

As the hereditary screening of so next offspring's malignancies scanning becomes more prevalent among cancer patients, strong methods capable of predicting factors based on the number of genomes analyzed are required. The outcome of the prohibitive expenses of instruments such as next offspring's genotyping as a blind screening strategy for an entire society, prophetical tools supported by patient history and clinical characteristics are often accustomed target genetic testing to specific people. It is demonstrated that machine learning techniques implemented to tailor upcoming genotyping screens correctly classify true variations among artifacts; this (can be) a doubtless helpful prophetical tool as a result of variants of uncertain relevance can result in important indecision between doctors and patients in terms of interpretation [187, 188]. Moreover, genomic risk stratification is often accustomed to verify that patients can like carcinoma screening. Compared to the present carcinoma screening paradigm supported age, a bunch within the UK discovered that giving carcinoma diagnostic techniques to ladies with a more genomic chance of carcinoma reduced overmedication and amelioration cost-effectiveness [189]. Predicting risk can still evolve because the field of genetic science advances, and it will possibly be most helpful during a multimodal context that features different clinical information points.

12.13 The Next Breakthrough in Exactitude Medicine Is Prophetical Analytics

Oncology looks to benefit from advancements in procedure methods in medical chances assessment of cancer victims, even as breakthroughs in cellular and epigenetic identification of tumors have made life risk assessment significantly more accurate. Advanced algorithms that predict the likelihood of using it, expenses, and clinical results can nearly definitely play a much bigger role in how medical patients are treated in the future. Combining medical, genomic, as well as biomolecule insights may shape the future of high-accuracy chances assessment in medicine, enabling genuine exactitude medicine [185].

12.13.1 Perspectives for the future

Many different forms of pediatric cancer registries were found in the literature review, ranging from small single-institutional efforts to large global collaborations. The variety of database papers conducted in the last 2 years demonstrates an increasing commitment to pediatric malignancies in using large datasets to address clinical and experimental scientific concerns. The environment of big data in pediatric malignancies, however, is significantly separated, with both tiny and major registries amassing data various forms of information, poor inter-registry harmonization, also the absence of a source of information standardization throughout the sector, as evidenced by the paper studies. In addition, the literature study reveals discrepancies in pediatric malignancy databases for impoverished countries including resource-constrained nations, as well as early initiatives to engage such victims more fully in therapeutic trials. There are numerous chances to assist in resolving these issues and improving cancer care for children.

12.13.2 Clinical trials and the development of new therapies

The FDA estimates a minimum pause for about 6.5 or half the decade among first-in-human as well as first-in-child authorization of medications for oncologic treatments for children [190]. In youngsters, off-label medication accounts for over 90% of all prescriptions [191]. As a result, there is an increasing need to speed up the throughput of pediatric oncology clinical trials, such as real-time registration of pediatric cancer victims who are included in databases.

However, the review of the literature indicated that the overwhelming majority of pediatric cancer written account analyses for the last many decades have perpetually been including some observational research and sometimes even rarer planned randomized controlled trials investigating innovative medications or treatment methods. Whereas tries are created in the past to copy RCTs in adults' exploitation of data-based information, there is only a minor number of those studies in medicine, as well as its method seems to be inaccurately established [192]. As a result, innovative procedure approaches, as well as randomized controlled trials, stay the gold customary for evaluating new therapies; but, traditionally, solely the biggest international medicine registries, like COG, SIOP, and CWS, have generated this sort of analysis. RCTs could alter a broader variety of registries, to enroll in medical studies, particularly tiny or geographically isolated databases. Although the amount of medicine patients registered in clinical trials has raised over time [193], there are still impediments to clinical test enrollment among folks [194] and investigators [195]. Trust, appropriate programming, an easy presentation of the chances and rewards, and motivation for kids to participate are all essential components influencing participation during a clinical test for folks [194]. The shortage of awareness of obtainable examine, the danger of studies, the space to an area that would conduct the examines, and also the time needed to debate the trial with folks are all challenges for suppliers [194, 195]. The EU medicine regulation [196], the pediatric analysis equity act (PREA) within us, the simplest prescription drugs for kids act [197], and also the recently declared medicine Investigation arrange [198] all act as incentives for the event of medicines with medical applications. Made clinical trials in kids, on the opposite hand, necessitate the acquisition of reliable, high-quality analysis information. The standardization of information gathering across studies and malady classes can create medicine cancer analysis considerably easier.

12.14 Conclusion

This article discusses the carcinoma-associated risks with IGRT procedures such as prediction and treatment, segmentation and scheduling, focusing on execution, as well as follow-up treatment and re-irradiation. Whereas the underlying mechanisms and duration of treatment correlation for radiation-induced carcinoma are unknown, immense new advancements in IGRT would then assist physicians in effectively understanding the concepts and process involved, influencing individualized RT approaches and methods for carcinoma risk reduction, resulting in safe and secure RT delivery and improved diagnosis and treatment benefits.

The examination of certain information can yield additional insights into operational, technological, therapeutic, as well as many other aspects of healthcare advancements. A better understanding, diagnosis, and therapy of numerous diseases have resulted from the aggregated sea of information from healthcare organizations and biomedical researchers. Various reputable consulting organizations and healthcare corporations have correctly predicted that the big data healthcare sector will develop at an accelerating rate. The exponential increase of medical data from diverse sectors has compelled professionals to devise novel ways for analyzing and interpreting the massive amount of information in a short amount of time. The next great goal could be to create a defined structure of the living structure using biological information and "omics" approaches. Within the last few years, the development and integration of big data have resulted in significant breakthroughs in the medical field, spanning from the monitoring of medical information to drug discovery initiatives. Big data analytics is predicted to progress toward a predictive system in the following year. This would imply predicting future results in a person's health based on present or historical information collected. Even this may also be possible that designed data collected through any specific location will result in the development of a particular community's well-being data. When integrated with other types of data, BD will improve medical services by enabling the possibilities of an outbreak and the initial diagnosis of illness terms.

Acknowledgment

The authors are highly thankful to the Department of Pharmacy of Galgotias University, Greater Noida for providing all necessary support for the completion of the work.

Funding

There is no source of funding

Conflict of Interest

There is no conflict of interest in the publication of this chapter's content.

References

[1] Rajpurkar P, Irvin J, Ball RL, et al. Deep learning for chest radiograph diagnosis: a retrospective comparison of the CheXNeXt algorithm to practicing radiologists. PLoS Med. 2018;15: e1002686.

[2] Gulshan V, Peng L, Coram M, et al. Development and validation of a deep learning algorithm for detection of diabetic retinopathy in retinal fundus photographs. JAMA. 2016; 316:2402-2410.

[3] Ting DSW, Cheung CY-L, Lim G, et al. Development and validation of a deep learning system for diabetic retinopathy and related eye diseases using retinal images from multiethnic populations with diabetes. JAMA. 2017; 318:2211-2223.

[4] Amarasingham R, Patel PC, Toto K, et al. Allocating scarce resources in real-time to reduce heart failure readmissions: a prospective, controlled study. BMJ Qual Saf. 2013; 22:998-1005.

[5] Shams I, Ajorlou S, Yang K. A predictive analytics approach to reducing 30-day avoidable readmissions among patients with heart failure, acute myocardial infarction, pneumonia, or COPD. Health Care Manage Sci. 2015; 18:19-34.

[6] Escobar GJ, Puopolo KM, Wi S, et al. Stratification of risk of early-onset sepsis in newborns $\geq$ 34 weeks' gestation. Pediatrics. 2014; 133:30-36.

[7] Miller KD, Siegel RL, Lin CC, et al. Cancer treatment and survivorship statistics, 2016. CA Cancer J Clin. 2016;66(4):271–289.

[8] Editorials. Prevention is as good as a cure. Nature. 2016; 539:467.

[9] Verellen D, De Ridder M, Linthout N, Tournel K, Soete G, Storme G. Innovations in image-guided radiotherapy. Nat Rev Cancer. 2007;7(12):949–960.

[10] Dawson LA, Sharpe MB. Image-guided radiotherapy: rationale, benefits, and limitations. Lancet Oncol. 2006;7(10):848–858.

[11] Dawson LA, Jaffray DA. Advances in image-guided radiation therapy. J Clin Oncol. 2007;25(8):938–946.

[12] Swerdlow AJ, Higgins CD, Smith P, et al. Second cancer risk after chemotherapy for Hodgkin's lymphoma: a collaborative British cohort study. J Clin Oncol. 2011;29(31):4096–4104.

[13] Berrington de Gonzalez A, Curtis RE, Kry SF, et al. Proportion of second cancers attributable to radiotherapy treatment in adults: a cohort study in the US SEER cancer registries. Lancet Oncol. 2011;12(4):353–360.

[14] Brenner DJ, Hall EJ. Computed tomography-an increasing source of radiation exposure. N Engl J Med. 2007; 357:2277–2284.

[15] Brenner DJ, Doll R, Goodhead DT, et al. Cancer risks attributable to low doses of ionizing radiation: assessing what we really know. Proc Natl Acad Sci U S A. 2003;100(24):13761–13766.

[16] 10. Moul JW. Radiotherapy: secondary malignancies after prostate cancer treatment. Nat Rev Clin Oncol. 2010; 7:249–250.

[17] Huq MS, Fraass BA, Dunscombe PB, et al. The report of Task Group 100 of the AAPM: Application of risk analysis methods to radiation therapy quality management. Med Phys. 2016; 43(7):4209.

[18] Demystifying big data: A Practical Guide To Transforming The Business of Government https://bigdatawg.nist.gov/_uploadfiles/M0068_v1_3903747095.pdf(accessed on 27 December 2021)

[19] Kayyali B, Knott D, Van Kuiken S. The big-data revolution in US health care: Accelerating value and innovation. Mc Kinsey & Company. 2013 Apr;2(8):1-3.

[20] McAfee A, Brynjolfsson E, Davenport TH, Patil DJ, Barton D. Big data: the management revolution. Harvard business review. 2012 Oct 1;90(10):60-8.

[21] Roski J, Bo-Linn GW, Andrews TA. Creating value in health care through big data: opportunities and policy implications. Health affairs. 2014 Jul 1;33(7):1115-22.

[22] Hodach R, Chase A, Fortini R, Delaney C, Hodach R. Population health management: A roadmap for provider-based automation in a

new era of healthcare. Institute for Health Technology Transformation. Retrieved from. 2012.

[23] Roski J, Bo-Linn GW, Andrews TA. Creating value in health care through big data: opportunities and policy implications. Health affairs. 2014 Jul 1;3(7):1115-22.

[24] Bertolucci J. Big data project analyzes veterans' suicide risk. Retrieved December 31, 2014.

[25] Laney D. 3D data management: controlling data volume, velocity, and variety, Application delivery strategies. Stamford: META Group Inc; 2001.

[26] Mauro AD, Greco M, Grimaldi M. A formal definition of big data based on its essential features. Libr Rev. 2016;65(3):122–35.

[27] Gubbi J, et al. Internet of Things (IoT): a vision, architectural elements, and future directions. Future Gener Comput Syst. 2013;29(7):1645–60.

[28] Childhood Cancers. National Cancer Institute 2015. https://www.cancer.gov/types/childhood-cancers(Accessed31October2019).

[29] Pediatric and Young Adult Early Case Capture | CDC 2019. https://www.cdc.gov/cancer/npcr/early-case-capture.htm(Accessed31October2019).

[30] Surveillance, epidemiology, and end results program n.d. https://seer.cancer.gov/(Accessed4March2017).

[31] Service, R. F. The race for the $1000 genome. Science. 2006;311(5767):1544–6.

[32] Stephens ZD, Lee SY, Faghri F, Campbell RH, Zhai C, Efron MJ, Iyer R, Schatz MC, Sinha S, Robinson GE. Big data: astronomical or genomical?. PLoS biology. 2015 Jul 7;13(7):e1002195..

[33] Li, L., Cheng, W. Y., Glicksberg, B. S., Gottesman, O., Tamler, R., Chen, R., Bottinger, E. P. and Dudley, J. T., 2015. Identification of type 2 diabetes subgroups through topological analysis of patient similarity. Science translational medicine, 7(311), pp.311ra174-311ra174.

[34] Dash, Sabyasachi; Shakyawar, Sushil Kumar; Sharma, Mohit; Kaushik, Sandeep (2019). Big data in healthcare: management, analysis and future prospects. Journal of Big Data, 6(1), 54–. doi:10.1186/s40537-019-0217-0

[35] De Mauro A, Greco M, Grimaldi M. A formal definition of Big Data based on its essential features. Library Review. 2016 Apr 4.

[36] ACCP Journals. American College of Clinical Pharmacology n.d. https://accp1.onlinelibrary.wiley.com/doi/full/10.1002/jcph.1141 (Accessed20January 2020).

[37] Chambers DA, Amir E, Saleh RR, Rodin D, Keating NL, Osterman TJ, Chen JL. The impact of big data research on practice, policy, and cancer care. American Society of Clinical Oncology Educational Book. 2019 May 17;39: e167-75.

[38] Andrade J, Cox SM, Volchenboum SL. Large-scale data sharing initiatives in genomic oncology. Adv Mol Pathol 2018; 1:135–48 https://doi.org/.doi:10.1016/j.yamp.2018.06.009.

[39] The childhood cancer data initiative: sharing for progress. National Cancer In- stitute 2019. https://www.cancer.gov/news-events/cancer-currents-blog/2019/lowy-childhood-cancer-data-initiative (Accessed 20 January 2020).

[40] Major A, Cox SM, Volchenboum SL. Using big data in pediatric oncology: Current applications and future directions. In Seminars in Oncology 2020 Feb 1 (Vol. 47, No. 1, pp. 56-64). WB Saunders.

[41] Grossman RL. Progress towards cancer data ecosystems. Cancer journal (Sudbury, Mass.). 2018 May;24(3):122.

[42] Therapeutically Applicable Research to Generate Effective Treatments. NCI Office of Cancer Genomics. 2013. https://ocg.cancer.gov/programs/target (Ac- cessed 20 January 2020).

[43] Ma X, Liu Y, Liu Y, et al. Pan-cancer genome and transcriptome analyses of 1,699 paediatric leukaemias and solid tumours. Nature 2018; 555:371–6.

[44] TARGET Data Matrix. Office of Cancer Genomics n.d. https://ocg.cancer.gov/programs/target/data-matrix (Accessed21January 2020).

[45] Newly launched Genomic Data Commons to facilitate data and clinical infor- mation sharing. National Institutes of Health (NIH); 2016. https://www.nih.gov/news-events/news-releases/newly-launched-genomic-data-commons-facilitate-data-clinical-information-sharing. (Accessed January 21, 2020).

[46] Home | NCI Genomic Data Commons n.d. https://gdc.cancer.gov/ (Accessed 21 January 2020).

[47] About the SEER Program - SEER. SEER n.d. https://seer.cancer.gov/about/overview.html(Accessed20January2020).

[48] O'Leary M, Krailo M, Anderson JR, Reaman GH, Group Children's Oncology. Progress in childhood cancer: 50 years of research collaboration, a report from the Children's Oncology Group. Semin Oncol 2008; 35:484–93 .

[49] NCI-COG Pediatric MATCH. National Cancer Institute 2015. https://www.cancer.gov/about-cancer/treatment/clinical-trials/nci-supported/pediatric-match (Accessed 20 January 2020).

[50] Pediatric Cancer Data Commons –Connect. Share. Cure n.d. http://commons.cri.uchicago.edu(Accessed21January2020).

[51] NCI Thesaurus n.d. https://ncithesaurus.nci.nih.gov/ncitbrowser/ (Accessed21 January 2020).

[52] CDE Browser 5.3.5 n.d. https://cdebrowser.nci.nih.gov/cdebrowserClient/cdeBrowser.html (Accessed21January2020).

[53] Bjork I , Peralez J , Haussler D , Spunt SL , Vaske OM . Data sharing for clinical utility. Cold Spring Harb Mol Case Stud 2019:5 https://doi.org/10.1101/mcs.a004689.

[54] SJCARES Registry n.d. https://www.stjude.org/global/sjcares/registry.html (Ac- cessed 20 January 2020).

[55] St. Jude PeCan Data Portal n.d. https://pecan.stjude.org(Accessed 27 March 2017).

[56] Home. St Jude Cloud n.d. https://www.stjude.cloud/ (Accessed 21 January 2020).

[57] Home. St Jude Cloud n.d. https://www.stjude.cloud/ (Accessed 21 January 2020).

[58] Seven Bridges announces the launch of largest pediatric data resource as a member of the Gabriella Miller Kids First Data Resource Cen- ter. Seven Bridges 2018. https://www.sevenbridges.com/press/releases/gabriella-miller-kids-first-data-resource-center/(Accessed4November2019).

[59] About NPCR | Cancer | CDC 2019. https://www.cdc.gov/cancer/npcr/about.htm(Accessed 20 January 2020).

[60] Pediatric and Young Adult Early Case Capture | CDC 2019. https://www.cdc.gov/cancer/npcr/early-case-capture.htm(Accessed20January2020).

[61] pubmeddev. Home-PubMed-NCBI n.d. www.ncbi.nlm.nih.gov/pubmed (Accessed 21 January 2020). https://pubmed.ncbi.nlm.nih.gov/

[62] Deng X, Yang Z, Zhang X, et al. Prognosis of Pediatric Patients with Pineoblas- toma: A SEER Analysis 1990-2013. World Neurosurg 2018;118: e871–9.

[63] Doganis , Pnagopoulou P , Tragiannidis A , et al. Survival and mortality rates of Wilms tumour in Southern and Eastern European countries: Socioeco- nomic differentials compared with the United States of America. Eur J Cancer 2018; 101:38–46.

[64] Hartmann E, Missotte I, Dalla-Pozza L. Cancer Incidence Among Children in New Caledonia, 1994 to 2012. J Pediatr Hematol Oncol 2018; 40:515–21.
[65] Limvorapitak W, Owattanapanich W, Utchariyaprasit E, Niparuck P, Puavilai T, Tantiworawit A, Rattanathammethee T, Saengboon S, Sriswasdi C, Julamanee J, Saelue P. Better survivals in adolescent and young adults, compared to adults with acute lymphoblastic leukemia– A multicenter prospective registry in Thai population. Leukemia Research. 2019 Dec 1; 87:106235.
[66] Crocoli A, Grimaldi C, Virgone C, De Pasquale MD, Cecchetto G, Cesaro S, Bisogno G, Cecinati V, Narciso A, Alberti D, Ferrari A. Outcome after surgery for solid pseudopapillary pancreatic tumors in children: Report from the TREP project—Italian Rare Tumors Study Group. Pediatric blood & cancer. 2019 Mar;66(3):e27519.
[67] Paulson KG, Nghiem P. One in a hundred million: Merkel cell carcinoma in pediatric and young adult patients is rare but more likely to present at advanced stages based on US registry data. Journal of the American Academy of Dermatology. 2019 Jun 1;80(6):1758-60.
[68] Govindan A, Parambil RM, Alapatt JP. Pediatric intracranial tumors over a 5-year period in a tertiary care center of North Kerala, India: a retrospective analysis. Asian journal of neurosurgery. 2018 Oct;13(4):1112.
[69] Clemens, Eva; Broer, Linda; (2019). Genetic variation of cisplatin-induced ototoxicity in non-cranial-irradiated pediatric patients using a candidate gene approach: The International PanCareLIFE Study. The Pharmacogenomics Journal, (), –. doi:10.1038/s41397-019-0113-1
[70] Van der Kooi AL, Clemens E, Broer L, Zolk O, Byrne J, Campbell H, van den Berg M, Berger C, Calaminus G, Dirksen U, Winther JF. Genetic variation in gonadal impairment in female survivors of childhood cancer: a PanCareLIFE study protocol. BMC cancer. 2018 Dec;18(1):1-7.
[71] Kim J, Schultz KAP, Hill DA, Stewart DR. The prevalence of germline DICER1 pathogenic variation in cancer populations. Mol Genet Genomic Med. 2019 Mar;7(3): e555. doi: 10.1002/mgg3.555
[72] McCune JS, Quinones CM, Ritchie J, Carpenter PA, van Maarseveen E, Yeh RF, Anasetti C, Boelens JJ, Hamerschlak N, Hassan M, Kang HJ. Harmonization of busulfan plasma exposure unit (BPEU): a community-initiated consensus statement. Biology of Blood and Marrow Transplantation. 2019 Sep 1;25(9):1890-7.

[73] Steliarova-Foucher E, Stiller C, Colombet M, Kaatsch P, Zanetti R, Peris-Bonet R. Registration of childhood cancer: Moving towards pan-European coverage? European Journal of Cancer. 2015 Jun 1;51(9):1064-79.

[74] Puckett M, Neri A, Rohan E, Clerkin C, Underwood JM, Ryerson AB, Stewart SL. Evaluating early case capture of pediatric cancers in seven central cancer registries in the United States, 2013. Public Health Reports. 2016 Jan;131(1):126-36.

[75] Learned K, Durbin A, Currie R, Kephart ET, Beale HC, Sanders LM, Pfeil J, Goldstein TC, Salama SR, Haussler D, Vaske OM. Barriers to accessing public cancer genomic data. Scientific data. 2019 Jun 20;6(1):1-7.

[76] Meehan RA, Mon DT, Kelly KM, Rocca M, Dickinson G, Ritter J, Johnson CM. Increasing EHR system usability through standards: Conformance criteria in the HL7 EHR-system functional model. Journal of biomedical informatics. 2016 Oct 1; 63:169-73.

[77] Phillips CA, Razzaghi H, Aglio T, McNeil MJ, Salvesen-Quinn M, Sopfe J, Wilkes JJ, Forrest CB, Bailey LC. Development and evaluation of a computable phenotype to identify pediatric patients with leukemia and lymphoma treated with chemotherapy using electronic health record data. Pediatric blood & cancer. 2019 Sep;66(9): e27876.

[78] Bowles KH, Potashnik S, Ratcliffe SJ, Rosenberg MM, Shih MN, Topaz MM, Holmes JH, Naylor MD. Conducting research using the electronic health record across multi-hospital systems: semantic harmonization implications for administrators. The Journal of nursing administration. 2013 Jun;43(6):355.

[79] Mello MM, Triantis G, Stanton R, Blumenkranz E, Studdert DM. Waiting for data: Barriers to executing data use agreements. Science 2020; 367:150–2.

[80] P. Lambin, R. G. P. M. van Stiphout, M. H. W. Starmans, E. Rios-Velazquez, G. Nalbantov, H. J. W. L. Aerts, et al., Predicting outcomes in radiation oncology– multifactorial decision support systems, Nat. Rev. Clin. Oncol. 10 (2013) 27–40, doi:10.1038/nrclinonc.2012.196.

[81] J.-E. Bibault, I. Fumagalli, C. Ferté, C. Chargari, J.-C. Soria, E. Deutsch, Personalized radiation therapy and biomarker-driven treatment strategies: a systematic review, Cancer Metastasis Rev. 32 (2013) 479–492, doi:10.1007/s10555-013- 9419-7.

[82] Q.-T. Le, D. Courter, Clinical biomarkers for hypoxia targeting, Cancer Metastasis Rev. 27 (2008) 351–362, doi:10.1007/s10555-008-9144-9.

[83] P. Okunieff, Y. Chen, D. J. Maguire, A. K. Huser, Molecular markers of radiationrelated normal tissue toxicity, Cancer Metastasis Rev. 27 (2008) 363–374, doi:10.1007/s10555-008-9138-7.

[84] J. Kang, R. Schwartz, J. Flickinger, S. Beriwal, Machine learning approaches for predicting radiation therapy outcomes: a clinician's perspective, Int. J. Radiat. Oncol. Biol. Phys. 93 (2015) 1127–1135, doi:10.1016/j.ijrobp.2015.07.2286.

[85] F. Denis, S. Yossi, A.-L. Septans, A. Charron, E. Voog, O. Dupuis, et al., Improving survival in patients treated for a lung cancer using self-evaluated symptoms reported through a web application, Am. J. Clin. Oncol. (2015) doi:10.1097/ COC.0000000000000189.

[86] A. D. Falchook, G. Tracton, L. Stravers, M. E. Fleming, A. C. Snavely, J. F. Noe, et al., Use of mobile device technology to continuously collect patient-reported symptoms during radiotherapy for head and neck cancer: a prospective feasibility study, Adv, Radiat. Oncol. (2016) doi:10.1016/j.adro.2016.02.001.

[87] M. Li, S. Yu, K. Ren, W. Lou, Securing personal health records in cloud computing: patient-centric and fine-grained data access control in multi-owner settings, in: S. Jajodia, J. Zhou (Eds.), Security and Privacy in Communication Networks, Springer Berlin, Heidelberg, 2010, pp. 89–106. (Accessed 21.05.16).

[88] V. Canuel, B. Rance, P. Avillach, P. Degoulet, A. Burgun, Translational research platforms integrating clinical and omics data: a review of publicly available solutions, Brief. Bioinform. 16 (2015) 280–290, doi:10.1093/bib/bbu006.

[89] V. Huser, J. J. Cimino, Impending challenges for the use of Big Data, Int. J. Radiat. Oncol. Biol. Phys. (2015) doi:10.1016/j.ijrobp.2015.10.060.

[90] Systematized nomenclature of medicine – clinical terms – summary | NCBO BioPortal. n.d. (accessed 07.03.16).

[91] https://bioportal.bioontology.org/ontologies/NCIT

[92] https://bioportal.bioontology.org/ontologies/CTCAE

[93] National Library of Medicine https://www.nlm.nih.gov/research/umls/knowledge_sources/metathesaurus/index.html(Last accessed on 25 december 2021)

[94] Radiation oncology ontology, 2018. https://bioportal.bioontology.org/ontologies/ROO(Last accessed on 26 December, 2021)

[95] Lam CG, Howard SC, Bouffet E, Pritchard-Jones K. Science and health for all children with cancer. Science. 2019 Mar 15;363(6432):1182-1186. doi: 10.1126/science.aaw4892.

[96] Rodriguez-Galindo C, Friedrich P, Alcasabas P, Antillon F, Banavali S, Castillo L, Israels T, Jeha S, Harif M, Sullivan MJ, Quah TC. Toward the cure of all children with cancer through collaborative efforts: pediatric oncology as a global challenge. Journal of Clinical Oncology. 2015 Sep 20;33(27):3065.

[97] https://www.inctr.org/programs/pediatric-oncology/index.html

[98] John R, Kurian JJ, Sen S, Gupta MK, Jehangir S, Mathew LG, Mathai J. Clinical outcomes of children with Wilms tumor treated on a SIOP WT 2001 protocol in a tertiary care hospital in south India. Journal of Pediatric Urology. 2018 Dec 1;14(6):547-e1.

[99] National Lung Screening Trial Research Team. Results of initial low-dose computed tomographic screening for lung cancer. New England Journal of Medicine. 2013 May 23;368(21):1980-91.

[100] Evans WP. Breast cancer screening: successes and challenges. CA Cancer J Clin. 2012; 62:5–9.

[101] Shaukat A, Mongin SJ, Geisser MS, et al. Long-term mortality after screening for colorectal cancer. N Engl J Med. 2013;369(12): 1106–1114.

[102] Smith RA, Andrews K, Brooks D, et al. Cancer screening in the United States, 2016: A review of current American Cancer Society guidelines and current issues in cancer screening. CA Cancer J Clin. 2016;66(2):96–114.

[103] Humphrey LL, Deffebach M, Pappas M, et al. Screening for lung cancer with low-dose computed tomography: a systematic review to update the US Preventive services task force recommendation. Ann Intern Med. 2013;159(6):411–420.

[104] Myers ER, Moorman P, Gierisch JM, et al. Benefits and harms of breast cancer screening: a systematic review. JAMA. 2015;314(15): 1615–1634.

[105] Mullenders L, Atkinson M, Paretzke H, Sabatier L, Bouffler S. Assessing cancer risks of low-dose radiation. Nat Rev Cancer. 2009;9(8):596–604.
[106] Asha S, Curtis KA, Grant N, et al. Comparison of radiation exposure of trauma patients from diagnostic radiology procedures before and after the introduction of a panscan protocol. Emerg Med Australas. 2012;24(1):43–51.
[107] Kritsaneepaiboon S, Jutiyon A, Krisanachinda A. Cumulative radiation exposure and estimated lifetime cancer risk in multiple-injury adult patients undergoing repeated or multiple CTs. Eur J Trauma Emerg Surg. 2018;44(1):19–27.
[108] Calabrese EJ. Origin of the linearity no threshold (LNT) dose-response concept. Arch Toxicol. 2013;87(9):1621–1633.
[109] Ozasa K, Shimizu Y, Suyama A, et al. Studies of the mortality of atomic bomb survivors, Report 14, 1950-2003: an overview of cancer and noncancer diseases. Radiat Res. 2012;177(3):229–243.
[110] Doss M. Adoption of linear no-threshold model violated basic scientific principles and was harmful. Archives of toxicology. 2014 Mar;88(3):849-52.87:2063-2081) and the letter from Ralph J Cicerone (Arch Toxicol (2014) 88:171-172). Arch Toxicol. 2014; 88:849–852.
[111] Tubiana, M. (2005). Dose–effect relationship and estimation of the carcinogenic effects of low doses of ionizing radiation: The joint report of the Académie des Sciences (Paris) and of the Académie Nationale de Médecine. International Journal of Radiation Oncology*Biology*Physics, 63(2), 317–319. doi:10.1016/j.ijrobp.2005.06.013
[112] Calabrese EJ, O'Connor MK. Estimating risk of low radiation doses–a critical review of the BEIR VII report and its use of the linear no-threshold (LNT) hypothesis. Radiation research. 2014 Nov;182(5):463-74.
[113] Preston DL, Pierce DA, Shimizu Y, et al. Effect of recent changes in atomic bomb survivor dosimetry on cancer mortality risk estimates. Radiat Res. 2014; 162:377–389.
[114] Doss M. Linear no-threshold model vs. radiation hormesis. Dose Response. 2013; 11:480–497.
[115] Siegel JA, Pennington CW, Sacks B, Welsh JS. The birth of the illegitimate linear no-threshold model: an invalid paradigm for estimating risk following low-dose radiation exposure. Am J Clin Oncol. 2018;41(2):173–177.

[116] Doss M, Little MP, Orton CG. Point/counterpoint: low-dose radiation is beneficial, not harmful. Med Phys. 2014; 41:070601.
[117] Jacob P, Meckbach R, Kaiser JC, Sokolnikov M. Possible expressions of radiation-induced genomic instability, bystander effects or low-dose hypersensitivity in cancer epidemiology. Mutat Res. 2010;687(1–2):34–39.
[118] Warner E. Clinical practice. breast-cancer screening. N Engl J Med. 2011; 365:1025–1032.
[119] Mascalchi M, Belli G, Zappa M, et al. Risk-benefit analysis of X-ray exposure associated with lung cancer screening in the Italung-CT trial. AJR Am J Roentgenol. 2006;187(2):421–429.
[120] Yip SS, Aerts HJ. Applications and limitations of radiomics. Phys Med Biol. 2016;61(13): R150–R166.
[121] Aerts HJ, Velazquez ER, Leijenaar RT, et al. Decoding tumour phenotype by noninvasive imaging using a quantitative radiomics approach. Nat Commun. 2014; 5:4006.
[122] Schneider U, Hälg R, Besserer J. Concept for quantifying the dose from image guided radiotherapy. Radiat Oncol. 2015; 10:188.
[123] Ding GX, Munro P. Radiation exposure to patients from image guidance procedures and techniques to reduce the imaging dose. Radiotherapy and Oncology. 2013 Jul 1;108(1):91-98.
[124] Yang W, Wang L, Read P, Larner J, Sheng K. Increased tumor radio resistance by imaging doses from volumetric image guided radiation therapy. Med Phys. 2009; 36:2808.
[125] Flynn RT. Loss of radiobiological effect of imaging dose in image guided radiotherapy due to prolonged imaging-to-treatment times. Med Phys. 2010;37(6):2761–2769.
[126] Pallotta S, Vanzi E, Simontacchi G, et al. Surface imaging, portal imaging, and skin marker set-up vs. CBCT for radiotherapy of the thorax and pelvis. Strahlenther Onkol. 2015;191(9):726–733.
[127] Walsh KE, Dodd KS, Seetharaman K, et al. Medication errors among adults and children with cancer in the outpatient setting. J Clin Oncol. 2009;27(6):891–896.
[128] Portaluri M, Fucilli FI, Gianicolo EA, et al. Collection and evaluation of incidents in a radiotherapy department: a reactive risk analysis. Strahlenther Onkol. 2010;186(12):693–699.
[129] Christakis, N. A., Smith, J. L., Parkes, C. M. and Lamont, E. B., 2000. Extent and determinants of error in doctors' prognoses in terminally ill patients: prospective cohort study Commentary: Why do doctors

overestimate? Commentary: Prognoses should be based on proved indices not intuition. Bmj, 320(7233), pp.469-473.
[130] Sborov K, Giaretta S, Koong A, et al. Impact of accuracy of survival predictions on quality of end-of-life care among patients with metastatic cancer who receive radiation therapy. J Oncol Pract. 2019;15: e262-e270.
[131] Fong Y, Evans J, Brook D, et al. The Nottingham prognostic index: five- and ten-year data for all-cause survival within a screened population. Ann R Coll Surg Engl. 2015; 97:137-139.
[132] Alexander M, Wolfe R, Ball D, et al. Lung cancer prognostic index: a risk score to predict overall survival after the diagnosis of non-small-cell lung cancer. Br J Cancer. 2017; 117:744-751.
[133] Lakin JR, Robinson MG, Bernacki RE, et al. Estimating 1-year mortality for high-risk primary care patients using the "surprise" question. JAMA Intern Med. 2016; 176:1863-1865.
[134] Morita T, Tsunoda J, Inoue S, et al. The Palliative Prognostic Index: a scoring system for survival prediction of terminally ill cancer patients. Support Care Cancer.1999;7:128-133.
[135] Chow R, Chiu N, Bruera E, et al. Inter-rater reliability in performance status assessment among health care professionals: a systematic review. Ann Palliat Med.2016; 5:83-92.
[136] Burwell SM. Setting value-based payment goals: HHS efforts to improve U.S. health care. N Engl J Med. 2015; 372:897-899.
[137] Center for Medicare & Medicaid Innovation. Oncology care model. https://innovation.cms.gov/initiatives/oncology-care/.AccessedOctober17,2018.
[138] Kline R, Adelson K, Kirshner JJ, et al. The Oncology Care Model: perspectives from the Centers for Medicare & Medicaid Services and participating oncology practices in academia and the community. Am Soc Clin Oncol Educ Book. 2017; 37:460-466.
[139] Kline RM, Bazell C, Smith E, et al. Centers for Medicare and Medicaid Services: using an episode-based paymentmodel to improve oncology care. J Oncol Pract. 2015; 11:114-116.
[140] Ostrovsky A, O'Connor L, Marshall O, et al. Predicting 30- to 120-day readmission risk among Medicare fee-for-service patients using non-medical workers and mobile technology. Perspect Health Inf Manag. 2016; 13:1e.
[141] Conn J. Predictive analytics tools help hospitals reduce preventable readmissions. Mod Healthc. 2014; 44:16-17.

[142] Elfiky AA, Pany MJ, Parikh RB, et al. Development and application of a machine learning approach to assess short-term mortality risk among patients with cancer starting chemotherapy. JAMA Netw Open. 2018;1: e180926.
[143] Brooks GA, Kansagra AJ, Rao SR, et al. A clinical prediction model to assess risk for chemotherapy-related hospitalization in patients initiating palliative chemotherapy. JAMA Oncol. 2015; 1:441-447.
[144] Yeo H, Mao J, Abelson JS, et al. Development of a nonparametric predictive model for readmission risk in elderly adults after colon and rectal cancer surgery. J Am Geriatr Soc. 2016;64: e125-e130.
[145] Fieber JH, Sharoky CE, Collier KT, et al. A preoperative prediction model for risk of multiple admissions after colon cancer surgery. J Surg Res. 2018; 231:380-386.
[146] Manning AM, Casper KA, Peter KS, et al. Can predictive modeling identify head and neck oncology patients at risk for readmission? Otolaryngol Head Neck Surg. 2018; 159:669-674.
[147] Vogel J, Evans TL, Braun J, et al. Development of a trigger tool for identifying emergency department visits in patients with lung cancer. Int J Radiat Oncol Biol Phys. 2017;99: S117.
[148] Furlow B. Predictive analytics reduces chemotherapy-associated hospitalizations. Managed Healthcare Executive. https://www.managedhealthcareexecutive.com/mhe-articles/predictive-analytics-reduces-chemotherapy-associated-hospitalizations.AccessedMarch13,2019.
[149] Rajkomar A, Oren E, Chen K, et al. Scalable and accurate deep learning with electronic health records. npj Digital Med. 2018;1.
[150] Bi WL, Hosny A, Schabath MB, et al. Artificial intelligence in cancer imaging: clinical challenges and applications. CA Cancer J Clin. 2019; 69:127-157.
[151] Chan H-P, Hadjiiski L, Zhou C, et al. Computer-aided diagnosis of lung cancer and pulmonary embolism in computed tomography-a review. Acad Radiol. 2008;
[152] Wang S, Burtt K, Turkbey B, et al. Computer aided-diagnosis of prostate cancer on multiparametric MRI: a technical review of current research. BioMed Res Int. 2014; 2014:789561.
[153] Song SE, Seo BK, Cho KR, et al. Computer-aided detection (CAD) system for breast MRI in assessment of local tumor extent, nodal status, and multifocality of invasive breast cancers: preliminary study. Cancer Imaging. 2015; 15:1.

[154] Aerts HJWL, Velazquez ER, Leijenaar RTH, et al. Decoding tumour phenotype by noninvasive imaging using a quantitative radiomics approach [published correction appears in Nat Commun. 2014; 5:4644

[155] Coroller TP, Grossmann P, Hou Y, et al. CT-based radiomic signature predicts distant metastasis in lung adenocarcinoma. Radiother Oncol. 2015; 114:345-350.

[156] Sorace AG, Wu C, Barnes SL, et al. Repeatability, reproducibility, and accuracy of quantitative MRI of the breast in the community radiology setting. J Magn Reson Imaging. 2018; 48:695-707.

[157] Wong AJ, Kanwar A, Mohamed AS, et al. Radiomics in head and neck cancer: from exploration to application. Transl Cancer Res. 2016; 5:371-382. 15:535-555.

[158] Yu K-H, Zhang C, Berry GJ, et al. Predicting non-small cell lung cancer prognosis by fully automated microscopic pathology image features. Nat Commun. 2016; 7:12474.

[159] Sooriakumaran P, Lovell DP, Henderson A, et al. Gleason scoring varies among pathologists and this affects clinical risk in patients with prostate cancer. Clin Oncol (R Coll Radiol). 2005; 17:655-658.

[160] Ehteshami Bejnordi B, Veta M, Johannes van Diest P, et al; CAMELYON16 Consortium. Diagnostic assessment of deep learning algorithms for detection of lymph node metastases in women with breast cancer. JAMA. 2017; 318:2199-2210.

[161] Or-Bach, Z., 2017, October. A 1,000-x improvement in computer systems by bridging the processor-memory gap. In 2017 IEEE SOI-3D-Subthreshold Microelectronics Technology Unified Conference (S3S) (pp. 1-4). IEEE.

[162] Mahapatra NR, Venkatrao B. The processor-memory bottleneck: problems and solutions. XRDS. 1999;5(3es):2.

[163] Voronin AA, Panchenko VY, Zheltikov AM. Supercomputations and big-data analysis in strong-field ultrafast optical physics: filamentation of high-peak-power ultrashort laser pulses. Laser Phys Lett. 2016;13(6):065403.

[164] Dollas, A. Big data processing with FPGA supercomputers: opportunities and challenges. In: 2014 IEEE computer society annual symposium on VLSI; 2014.

[165] Saffman M. Quantum computing with atomic qubits and Rydberg interactions: progress and challenges. J Phys B: At Mol Opt Phys. 2016;49(20):202001.

[166] Nielsen MA, Chuang IL. Quantum computation and quantum information. 10th anniversary ed. Cambridge: Cambridge University Press; 2011. p. 708.

[167] Raychev N. Quantum computing models for algebraic applications. Int J Scientific Eng Res. 2015;6(8):1281–8.

[168] Harrow A. Why now is the right time to study quantum computing. XRDS. 2012;18(3):32–7.

[169] Lloyd S, Garnerone S, Zanardi P. Quantum algorithms for topological and geometric analysis of data. Nat Commun. 2016;7:10138.

[170] Buchanan W, Woodward A. Will quantum computers be the end of public key encryption? J Cyber Secur Technol. 2017;1(1):1–22.

[171] De Domenico M, et al. Structural reducibility of multilayer networks. Nat Commun. 2015;6:6864.

[172] Mott A, et al. Solving a Higgs optimization problem with quantum annealing for machine learning. Nature. 2017;550:375.

[173] Rebentrost P, Mohseni M, Lloyd S. Quantum support vector machine for big data classification. Phys Rev Lett. 2014;113(13):130503.

[174] Gandhi V, et al. Quantum neural network-based EEG filtering for a brain-computer interface. IEEE Trans Neural Netw Learn Syst. 2014;25(2):278–88.

[175] Nazareth DP, Spaans JD. First application of quantum annealing to IMRT beamlet intensity optimization. Phys Med Biol. 2015;60(10):4137–48.

[176] Reardon S. Quantum microscope offers MRI for molecules. Nature. 2017;543(7644):162.

[177] Reisman M. EHRs: the challenge of making electronic data usable and interoperable. Pharmacy and Therapeutics. 2017 Sep;42(9):572.

[178] Valikodath, N. G., Newman-Casey, P. A., Lee, P. P., Musch, D. C., Niziol, L. M. and Woodward, M. A., 2017. Agreement of ocular symptom reporting between patient-reported outcomes and medical records. JAMA ophthalmology, 135(3), pp.225-231.

[179] Fromme EK, Eilers KM, Mori M, Hsieh YC, Beer TM. How accurate is clinician reporting of chemotherapy adverse effects? A comparison with patient-reported symptoms from the Quality-of-Life Questionnaire C30. Journal of Clinical Oncology. 2004 Sep 1;22(17):3485-90.

[180] Beckles GL, Williamson DF, Brown AF, Gregg EW, Karter AJ, Kim C, Dudley RA, Safford MM, Stevens MR, Thompson TJ. Agreement between self-reports and medical records was only fair in a cross-sectional study of performance of annual eye examinations

among adults with diabetes in managed care. Medical care. 2007 Sep 1:876-83.

[181] Echaiz JF, Cass C, Henderson JP, Babcock HM, Marschall J. Low correlation between self-report and medical record documentation of urinary tract infection symptoms. American journal of infection control. 2015 Sep 1;43(9):983-6.

[182] Belle A, et al. Big data analytics in healthcare. Biomed Res Int. 2015; 2015:370194.

[183] Elfiky, A. A., Pany, M. J., Parikh, R. B. and Obermeyer, Z., 2018. Development and application of a machine learning approach to assess short-term mortality risk among patients with cancer starting chemotherapy. JAMA network open, 1(3), pp. e180926-e180926.

[184] Bertsimas D, Dunn J, Pawlowski C, et al. Applied informatics decision support tool for mortality predictions in patients with cancer. JCO Clin Cancer Informatics.2018;2:1-11.

[185] Parikh RB, Kakad M, Bates DW. Integrating predictive analytics into high-value care: the dawn of precision delivery. JAMA. 2016; 315: 651-652.

[186] Burki TK. Predicting lung cancer prognosis using machine learning. Lancet Oncol. 2016;17: e421.

[187] van den Akker J, Mishne G, Zimmer AD, et al. A machine learning model to determine the accuracy of variant calls in capture-based next generation sequencing. BMC Genomics. 2018;19:263.

[188] Welsh JL, Hoskin TL, Day CN, et al. Clinical decision-making in patients with variant of uncertain significance in BRCA1 or BRCA2 genes. Ann Surg Oncol. 2017;24:3067-3072.

[189] Pashayan N, Morris S, Gilbert FJ, et al. Cost-effectiveness and benefit-to-harm ratio of risk-stratified screening for breast cancer: a life-table model. JAMA Oncol.2018;4:1504-1510.

[190] Neel DV, Shulman DS, DuBois SG. Timing of first-in-child trials of FDA-approved oncology drugs. European Journal of Cancer. 2019 May 1; 112:49-56.

[191] Hwang TJ, Tomasi PA, Bourgeois FT. Delays in completion and results reporting of clinical trials under the Paediatric Regulation in the European Union: A cohort study. PLoS medicine. 2018 Mar 1;15(3):e1002520.

[192] Christensen ML, Davis RL. Identifying the 'blip on the radar screen' : leveraging big data in defining drug safety and efficacy in pediatric practice. The Journal of Clinical Pharmacology. 2018 Oct;58:S86-93.

[193] Parsons HM, Penn DC, Li Q, Cress RD, Pollock BH, Malogolowkin MH, Wun T, Keegan TH. Increased clinical trial enrollment among adolescent and young adult cancer patients between 2006 and 2012–2013 in the United States. Pediatric blood & cancer. 2019 Jan;66(1):e27426.

[194] Greenberg RG , Gamel B , Bloom D , et al. Parents' perceived obstacles to pedi- atric clinical trial participation: findings from the clinical trials transformation initiative. Contemp Clin Trials Commun 2018; 9:33–9

[195] Smith B, Benjamin D, Bradley J, et al. Investigator barriers to pediatric clin- ical trial enrollment: Findings and recommendations from the Clinical Trials Transformation Initiative. Pediatrics 2018;142 796–796.

[196] Global pediatric drug development. Curr Ther Res Clin Exp 2019; 90:135–42.

[197] Ward RM, Kauffman R. Future of pediatric therapeutics: reauthorization of BPCA and PREA. Clin Pharmacol Ther 2007; 81:477–9.

[198] Paediatric medicine: Paediatric Investigation Plan - EUPATI. EUPATI 2016. https://www.eupati.eu/clinical-development-and-trials/paediatric-medicine-paediatric-investigation-plan/.

13

Challenges for Big Data in Oncology

Deepika Bairagee[1,2*], Neetesh Kumar Jain[2], Sumeet Dwivedi[2], and Kamal Dua[3,4]

[1]Pacific College of Pharmacy, Pacific University, India
[2]Oriental College of Pharmacy and Research, Oriental University, India
[3]Discipline of Pharmacy, Graduate School of Health, University of Technology Sydney, Australia
[4]Faculty of Health, Australian Research Centre in Complementary and Integrative Medicine, University of Technology Sydney, Australia
*Corresponding Author: EMail: bairagee.deepika@gmail.com

Abstract

During this era of data innovation, big data research is making its way into the biological sciences. Big data's capacity to identify patterns and translate vast amounts of data into usable information for correct medical decision-making is determined by its capability to notice patterns and translate large capacities of data into valuable information for accurate medicine and decision-making. In a few cases, the usage of big data in medical services is now supplying answers for increasing patient consideration and the age of significant value in medical care associations. The fundamentals of big data are presented, with a focus on cancer. There is an enormous number of informative collections in oncology that collect data on malignant development genome, transcriptome, clinical information, neurotic, personal happiness information, and so on. When it comes to physical and biological data exchanges, radiotherapy data is unique. The authors examined developments and discussed current obstacles in using top-bottom and bottom-top approaches to cross-examine large data in radiotherapy in this chapter. They discussed the specifics of big data in radiotherapy and the challenges that bioinformatics tools for data aggregation, sharing, and privacy provided. The problem of big data in cancer is incorporating all of this disparate data into a unified platform that can be examined and translated into comprehensible files. The capability to extract info from all of the obtained data resulted in advancements in cancer patient treatment and resolution.

Keywords: Big Data, Oncology, Radiotherapy, Bioinformatics, Challenges, Cancer Genome, Clinical Data.

13.1 Oncology

Oncology is a medical specialty that deals with the prevention, diagnosis, and treatment of cancer. Oncologists are doctors who specialize in the treatment of cancer. In 1618, the term "oncology" was used in neo-Greek to refer to Galen's treatise De tumribus parameter naturam, which dealt with unusual tumors. Improved preventative measures to reduce exposure to risk factors including cigarettes, smoking, and alcohol intake, improved identification of a few malignancies, and improved treatment have all contributed to increasing cancer endurance. Multidisciplinary cancer conferences bring together clinical oncologists, careful oncologists, radiation oncologists, pathologists, radiologists, and organ-specific oncologists to determine the best treatment for a specific patient, taking into account the patient's physical, social, mental, enthusiastic, and financial circumstances. Oncologists must stay up with the latest advancements in oncology since cancer therapy is always changing. Because a cancer diagnosis can induce anxiety and distress, doctors may utilize several approaches to convey the bad news, including SPIKES.

Factors that are at risk

- **Tobacco:** Tobacco openness is the most common reason for malignant development and death. Tobacco use is strongly linked to an increased risk of lung, laryngeal, mouth, and throat cancers, as well as cancers of the bladder, kidney, liver, stomach, pancreas, colon, rectum, and cervix. Smokeless tobacco (snuff or biting tobacco) has been linked to an increased risk of mouth, throat, and pancreatic cancer.
- **Liquor:** Drinking alcohol increases the risk of developing diseases of the mouth, throat, larynx, liver, and bosom. People who consume alcoholic drinks and smoke cigarettes have a significantly increased risk of cancer.
- **Obesity:** Breast, colon, rectum, endometrial, throat, kidney, pancreas, and gallbladder cancers are more common in obese persons.

- **Age:** For some cancers, old age is a risk factor. The median time it takes for cancer to be diagnosed is 66 years.
- **Malignancy:** Malignancy is induced by changes in particular qualities that affect the way our cells work, which are known as inducing substances. Some of them are the consequence of natural holes causing DNA damage. Compounds like the synthetics present in cigarette smoke, or radiation like powerful sun rays and other cancer-causing elements, might fill these gaps.
- **Irresistible specialists:** Oncoviruses, microorganisms, and parasites are examples of irresistible cancer-causing experts.
- **Immunosuppression:** The immune system helps to defend the body from illnesses, as evidenced by the fact that particular malignant growths arise at a considerably greater incidence in those who are immunosuppressed [1].

Fighting cancer is like searching for the Holy Grail of medicine. Oncologists have a wide variety of challenges in the current period, including the standardization of treatment, maintaining expertise in the interdisciplinary management of a wide variety of cancers, and collecting information across multiple dimensions to define an infection.

Malignant growth appears to be tightening its grip, as reported by TOI, with 1,000,000 new instances accounted for annually. By 2025, some experts predict, the incidence of executioner's disease will have increased by fivefold. It has been estimated that the annual rate of new cases of malignant growth is somewhere between 1000 and 1100 per million people.

Cancer is one of the most difficult diseases because of how quickly it spreads, how it is constantly changing, and how dangerous it can be. The second leading cause of death in the world. There are around 14 million new cases of malignant growth per year, as reported by the World Health Organization. In addition, it is anticipated that this figure would increase by almost 70% during the following two decades. Recent years have seen tremendous progress in the fight against cancer, with the survival rate increasing and the overall fix remaining very subtle.

The question now is, how can oncologists mitigate these challenges? In that spot, a flood of data enters the picture. Studies show that by 2021, the global big data market would be worth an amazing $6.6 billion. As it evolves, it will improve infection analysis and open up new horizons in traditional medical care. It is already being utilized in several regions of the world to better identify and diagnose diseases like cancer in their earliest stages.

Enormous information is enabled with strong capacities and it is generally appropriate for taking care of a lot of information. The job of large information in medication is to fabricate better well-being profiles and better prescient models around individual patients to analyze better and treat an infection. For example- IBM Watson Health has settled a portion of the squeezing difficulties through Big information. By combining the wisdom of many professionals, we can better decipher data, analyze trends, and make informed decisions for people all across the world.

13.2 Big Data

Big data, together with the computer system technologies used to analyze it, has been called one of the most important innovations of the past decade (Figure 13.1). Similar effects to the Internet, cloud computing, and, more recently, block-chains (popularized by cryptocurrencies like the tiny bitcoin) are anticipated. Almost every sector is being affected by the big data phenomena. The first users to make extensive use of them are those in the information energy sector (IBM, Google, Facebook, and Amazon). These massive IT-focused corporations develop and apply algorithms to predict consumer behavior, then put that information to use in targeted advertising based on their customers' specific traits and interests. Innovations in big data have also attracted the attention of health insurance companies and governments, and the field has spread into the life sciences [2].

Although "big data" is used frequently, it is not necessarily referred to the same thing or has the same meaning. People generally have a concept of

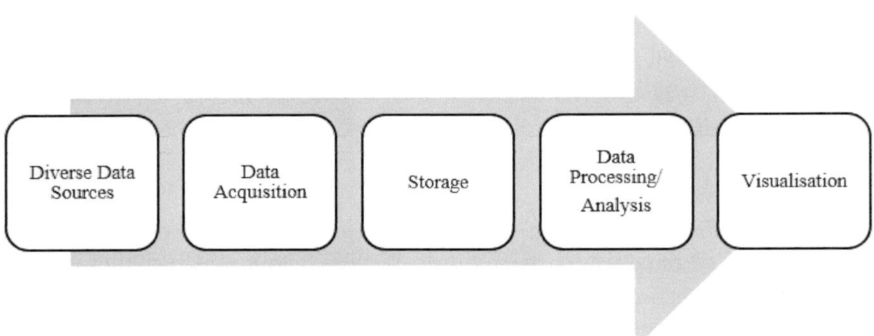

Figure 13.1 Building blocks of big data system.

what is "big" (i.e., "something that would not fit on a page"), but massive data encompasses much more. Big data in medical care is characterized using the 5Vs as shown in Figure 13.2. According to this definition, big data includes:

Volume: Big data is frequently huge in scale, containing a large number of data points/records on a wide range of topics. Treatment information (surgery, systemic treatment, radiation, and their combinations), reaction information, and so forth.

Velocity: There are dual types of velocity in big data: big data is being generated at an increasingly rapid rate, and it must be calculated and digested quickly. The disease is becoming more common over the world, and patients are staying longer. With technology advancements and device tracking, a growing range of information must be processed on time. This is undoubtedly the same.

Variety: An enormous variety of information forms can be found in big data. This diversity presents both opportunities (a wide range of data types improves the excellence and utility of the data) and contests (the variability of the data necessitates standardization) (synoptic reporting).

Variability: It is critical to understand that data fluctuates depending on the location and time of capture. The majority of reported data is contingent on obtaining a (predefined) essential dataset. This necessitates not simple agreement from the information, which can be sparse; it also necessitates unambiguous definitions (e.g., recurrence vs. recurring disease).

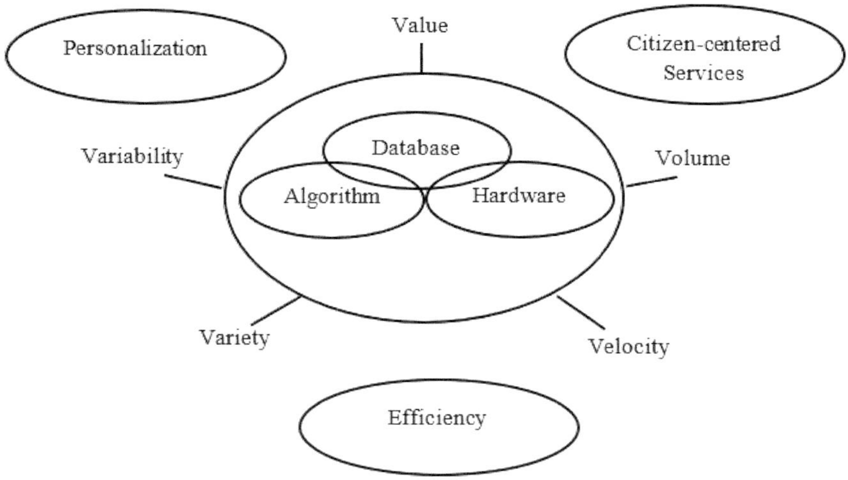

Figure 13.2 Big information in medical care is its description in line with the 5Vs.

Value: Installing associate degree info infrastructure to assemble and comprehend information is merely helpful if it permits the formulation of data-derived conclusions or measurements supported by precise data, leading to demonstrable advantages or impacts in health care.

Data sources are often geographically distributed throughout the globe and placed in ways that make merging the data challenges, which is an even bigger problem than the sheer volume of data itself. Transferring such large amounts of data over the internet raises potential scalability issues, but incorporating it into a work that is routinely used to answer an overall question requires harmonization and standards efforts. We need data since Big Data has so many potential applications. Concerns about travel times, costs, and distribution might be reduced with the help of location tracking, which logistics firms may implement.

Establishing a data infrastructure to collect and analyze data is only beneficial if it facilitates the development of data-derived conclusions or measurements that are grounded on accurate data, ultimately allowing for the demonstration of advantages or impacts in health care. The dominant players in data are most akin to advertising and marketing agencies. Information gathered from social media sites, microblogging platforms, and search engines like Google and Facebook allows for highly targeted advertising campaigns. Drones are used by breeding companies to collect data in the agricultural sector that is then used in the breeding process. With the use of Big Data, hospitals can keep closer tabs on patients undergoing intensive.

In the fitness area, Big Data is probably treasured in growing and revising disorder prevention policies, in addition to measuring remedy efficacy and expecting epidemic breakouts at an early stage. Combining huge databases of genes with environmental records can assist decide if people/companies are liable for getting illnesses and cancer. This may result in extra-targeted interventions geared toward changing environmental and behavioral variables that make contributions to fitness dangers in sure populations. Big data can also be used to assess cutting-edge preventative measures and devise new techniques to enhance them. In a healing setting, huge records can be used to music the results of sure medicines, like as high priced oncolytic, on customers and cyst faculties. This might aid knowledge of drugs and gases, which is critical in calculating the cost-effectiveness of various treatment regimens.

13.3 Utility of Big Data

The destiny capability of huge information (in a biomedical investigation) is but unknown. Massive information will function as an incentive for everyday diagnoses, nature of care/life (counting PROMs and PREMs), and organic studies for the time being (and with inside the destiny). We will display to you some samples of presently-to-be-had apps (Figure 13.3).

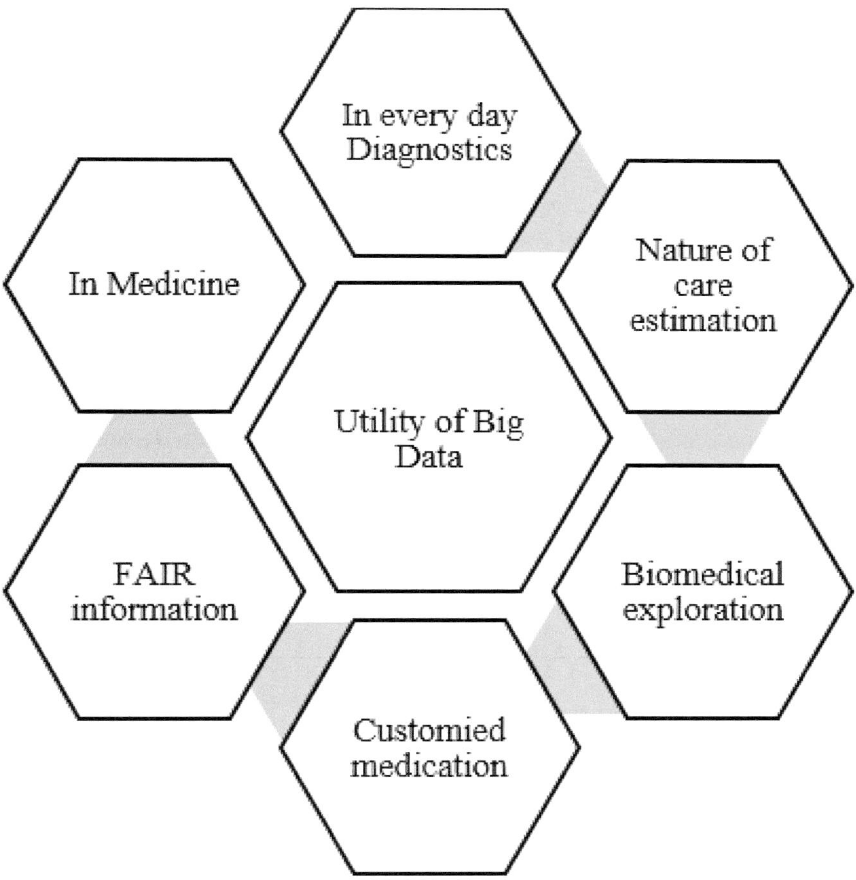

Figure 13.3 Utility of big data.

13.3.1 In every day diagnostics

Huge volumes of data would already be consistent in clinical practice. Pathologists in the Dutch have a model system of close, ongoing access to the cross-border histological follow-up of each patient. Since 1971, the Netherlands has entrusted all of its advanced histopathology records to PALGA. PALGA is one of the world's largest biomedical data collections, with over 72 million records from more than 1.2 million Dutch patients. There are 55 different pathology labs in the country, and this applies to all of them. When a Dutch pathologist signs a histology report, one copy is kept in the local medical clinic data system and the other in the PALGA database. As a consequence, this database includes each patient's continuous neurotic progression, which is evident for each PALGA segment (pathologist or atomic scientist). This has enormous potential in recognizing relevant patient (oncological) history, such as preventing a new treatment in cases of speculative growth of obscure essential; or providing obsessive documentation on past important neurotic components (such as resection edges and positive lymph hub) if pathology was performed in a different lab. This information can also be used to probe for hidden connections between diseases with a low prevalence that, at first glance, seem unrelated.

The electronic documentation of a patient's medical history generates a mountain of scientific data ideal for illustrative purposes in diagnostics. One of the earliest prediction fashions for HNC sufferers getting care at scientific centers in advanced countries is www.oncoloogiq.nl. The atomization of the measurable anticipation manner lets in fashions to be routinely refreshed while new statistics are gained. These statistics will also be utilized to create scientific dynamic gadgets which can offer stepped-forward affected person recommendations and non-parallel affected person-associated final results assessments.

13.3.2 Nature of care estimations

Connecting data based on long-term results with data on understanding qualities and treatment has tremendous promise for improving back quality and care feasibility. A recent French report revealed the location of sub-atomic testing for designated treatment in non-small cell lung cancer (NSCLC) in France, as well as the treatment regimens that resulted as a result of this. This opens the door to direct critique of ideal test-treatment relationships. More prominently, it could be a powerful motivator for failing to fulfill expectations for labs to modify their conventions/work processes to progress their ideal of care. In addition, by combining data from the public malignant growth vault (which contains a clinical-stage, therapy, and result information) with the PALGA dataset, researchers in the Netherlands were able to demonstrate the wide range of clinical considerations in head and neck malignant growth. While working on the nature of care requires transparency in such information, it should be acknowledged that criticizing such information, particularly results in information and when benchmarked, must be done with extreme caution, since laboratories and clinics would dread being named and shamed. Most emergency clinics are ready to engage in such mirror critiques when they are discreetly supplied to the general public and cared for on an individual basis. As a result, the Dutch Institute for Clinical Auditing, for example, has improved calculations for the regular entry of pathology and treatment-related data (www.dica.nl). Higher repeat rates than similar medical clinics, according to mirror data, might be a reason to focus on the foundation chain to detect (and correct) any faults.

13.3.3 Biomedical exploration

In the sphere of research, the principal length of "genome extensive association considers" (GWAS) has been growing closer to a time of "data extensive association examines" (DWAS), with a focal point on big data. Increment of information, both due to the expanded utilization of imaging and sub-atomic examinations and blends with different information, offer an incomparable Walhalla for every information researcher and bioinformatician. Large information fills a neglected need in biomedical exploration. For instance, Our little understanding of the science of illness is a major limitation of today's medical treatment. Accurately gathering large amounts of big data allows for the summarization and coordination of all the crucial multisource components, such as DNA, RNA, protein, and metabolomics information, to better predict how tumors will behave and which patients will help great shape specific treatment. For example, this integrated multi-omics data will give a deeper comprehension of natural conduct and instruments that drive development patterns and metastasis merely as a response to (defined) HNSCC therapy.

13.3.4 Customized medication

According to the viewpoint of turning our present agreement and accessible information into significant experiences that can be utilized to improve treatment results, customized medication is reliant upon large information. The quantity of data available to the biomedical community is continually growing, especially as cutting-edge developments, including sequencing and imaging, generate terabytes of data. The majority of data comes from calculated programmed information assessments like radionics and computerized image inspection, rather than from instantaneous, patient-related records available in ordinary clinical practice. Head and neck malignancies offer a unique set of analytical and healing hurdles due to the fact to their complex existence structures and heterogeneity. Radiomics has the capcaner those obstacles.

Radiomics is a non-invasive and cost-powerful approach to amassing and mining a massive range of scientific imaging highlights. Radiomics is predicated on the idea that specific features visible in an imaging study can be used to identify the phenotypic characteristics of a whole organism. For example, radiomics in accuracy oncology and malignancy for the defining (or customizing) of sufferers' (expected/anticipated) patience and remedy result expectancies to assist withinside the choice of the maximum suitable remedy for sufferers with head and neck disease. Clinical and radiation oncologists can be capable of boom primary remedy and illumination dosages in unique affected person populations because of this.

13.3.5 FAIR information

To guarantee information can be reused in auxiliary examinations, it is fundamental they stick to the FAIR standards. These FAIR information standards have been first distributed in 2014. Since then, the G20 (2016) and G7 (2017) have recognized and embraced the principles, while the EU has placed FAIR data at the heart of the European Open Science Cloud (EOSC). The metadata of an information asset is an important part of its FAIRness, with findability (F) requiring an industrious identifier, availability (A) requiring clearly defined access rules (information security imperatives are included in the definition) and permitting, and interoperability (I) requiring the use of a local area perceived cosmology for portraying the information. Finally, the information's provenance and precision, as well as the fulfillment of meta-information, are critical for the information's reusability (R) [4].

13.3.6 Data sourced elements of big data in medicine

Big Data reasserts exist, and they could take many exceptional forms. In oncology, records received from sufferers are critical. These records, which consist of more than a few records, are mechanically saved in computerized affected person documents for healing purposes. This record consists of medical records approximately sufferers, tumors, treatment, and outcomes, in addition to demo image information which includes gender and age, providing symptoms, own circle of relatives history, comorbidity, radiological records (which include CT, MRI, PET, and US), and stable and fluid tissue-primarily based evaluation which include histopathological diagnosis/features, immunohistochemistry, DNA/RNA sequencing experiments, bloodstream evaluation, and e-health. In addition, records from in vitro studies are normally relevant, for this reason, this is probably a hand useful resource. The computational evaluation of the records is a useful resource this is second in phrases of records. These processed records consist of radionics and virtual photograph evaluation, in addition to hereditary look and alteration evaluations, in addition to oblique and calculated records. Increasingly, those organized records come from tools gaining knowledge of and regularly incorporate massive quantities of dependent records saved on a computer. Clients' personal which include diligent associated final results measurements (PROMs) and affected person associated enjoy measurements (PREMs), are the third supply of big data. These people file all styles of measurements for the use of packages on computer systems and cell devices, both supplied with the aid of using their precise caregivers (eHealth and telemedicine) or on their initiative. Published literary works are the fourth supply of supply (IBM task). Because around 1 million biomedical papers are produced every year, no health practitioner on the planet can examine even a fragment of the records issued every year, plenty on my own all associated textbooks and different net assets.

Nonetheless, there is one thing of huge statistics in oncology this is critical: the depth (volume) of records in keeping with patients. Even though traditional patron cohort sizes are small, oncology frequently creates and maintains thousands, if not millions, of observables. This disparity in the number of records furnished to every patron in opposition to the size is certainly extra said in uncommon cancers and tumor kinds, inclusive of head and neck cancer. Current methodological improvements in system-gaining knowledge of neural networks, on the opposite hand, are extraordinarily beneficial if you could find out conditions to observe. For instance, object detection in photos is successful, but it takes thousands of hours to perfect such systems. As a result, while looking at a wide range of samples, we also require data depth if we wish to use this to improve tailored treatments. This necessitates the need for good information, information sharing, data provenance, and information interchange protocols in oncology, especially in the field of head and neck oncology [5, 6].

Complex datasets that conventional records processing structures cannot store, manage, or examine speedily or cheaply are treated with the aid of using huge records analytics solutions. Big records analytics can convert fitness care with the aid of using supporting clinicians, providers, and policymakers with intervention-making plans and implementation, [4] quicker sickness detection, healing selection support, final results prediction, and multiplied personalized medicine, ensuing in lower-cost, higher-pleasant care with higher outcomes [7].

In 2018, the World Health Organization (WHO) proposed the expedited thirteenth General Program of Work (GPW13), which became authorized and followed with the aid of using its 194 Member States. The GPW13 specializes in measurable influences on people's fitness on the kingdom stage to convert public fitness with three middle features: greater normal fitness coverage, fitness emergency protection, and stepped-forward fitness and well-being. The GPW13 generated forty-six final results intention signs that addressed a huge variety of fitness issues. Big records analytics would possibly assist with fitness coverage decisions, expediting the attainment of the GPW13 number one desires and targets, and directing the European region's roadmap primarily based totally on the European Program of Work (EPW) 2020–2025 [8, 9].

13.4 Big Data in Oncology

Cancer is one of our society's maximum essential fitness issues, and the trouble is most effective probable to worsen as the world's populace grows and ages. According to the State of Health withinside the EU reports, maximum cancers are one of the predominant reasons for untimely mortality withinside the EU. It additionally has a terrible effect on the economic system as it reduces labor marketplace participation and productivity. Most cancer researchers now have powerful new strategies to extract benefits from various reasserts of facts and ways way to advance in Big Data analytics.

These several reasserts consist of a large number of facts on an unmarried patient. Because most cancers is a molecularly unexpected complex sickness with good-sized intra- and intertumoral heterogeneity amongst diverse cancers kinds or maybe sufferers, the gathering of diverse styles of oomicsfacts can offer a particular molecular profile for every patient, helping oncologists in their look for customized remedy strategies [10, 15].

In Comprehensive Cancer Centres, the approach of mixing the reasserts of facts is used (CCCs). The Molecularly Aided Stratification for Tumor Eradication Research (MASTER) venture is being performed through the National Center of Tumor Disease, which is one of every of Germany's 13 CCCs. Using high-overall performance complete exome or complete genome sequencing and RNA sequencing, facts applicable to diagnostic information of more youthful sufferers with advanced-level malignancies are collected, analyzed, and debated in the MASTER trial [11, 15].

Another instance of an achievement story supplied within the assessment is the Personalized Treatment for Recurrent Malignancies in Children (INFORM) registry, which tries to cope with relapses of high-chance tumors in pediatric patients. Data from whole-exome, low-coverage whole-genome, RNA sequencing, and microarray-primarily based DNA methylation profiling are utilized to recognize patient-unique recuperation targets. The INFORM registry commenced as a German-huge initiative and has in the end grown to encompass European and Australian countries. In addition to the said tasks, numerous greater tasks concentrate on the loose use of Big Data in oncology. The EU budget is greater than 90 tasks in this field (tasks with an expenditure of greater than Ă499.999). Cancer Core Europe [11, 15] is an ability umbrella for bringing collectively national activities, such as those stated above, to a European level. Associations are of extraordinarily excessive significance particularly inside the case of pediatric or different uncommon varieties of most cancers, wherein the information accumulated for one affected person is certainly extensive, but the variety of sufferers an unmarried center could have to get admission to is just too low to acquire statistical strength excessive sufficient to attain significant results. One of the principal demanding situations of those collaborations is the get admission to the information in addition to the possibility to examine the massive quantity of information in a green way. Physicians, researchers, and informatics specialists will be able to gain the most benefit from gathered data and professional understanding if they have easy and simple access to personal or companion data. At the German Cancer Research Center, for example, technology has been developed to provide methods for gaining access to and analyzing personal information as well as information from companions.

Furthermore, the Cancer Genome Atlas (TCGA) and the International Cancer Genome Consortium (ICGC), which offer researchers with getting entry to a big wide variety of sequenced sufferers with numerous sorts of cancers, are used to perform worldwide exemplary techniques of facts sharing among partners or the overall public. The availability of that facts, whilst joined with records from different sources, has enabled big meta-analyses and gadgets to get to know algorithms, taking into consideration the identity of the novel most cancers riding pressure genes that belong to specific pathways and are probably possible healing targets.

Several public databases, inclusive of the Catalogue of Somatic Mutations in Cancer, additionally permit get entry to a listing of mutations that have been proven to be implicated withinside the majority of cancers (COSMIC). These common records reassessments, in mixture with the fame quo and pointers furnished via way of means of CCCs throughout Europe, have the ability to boom the wide variety of sufferers who can advantage of molecular profiling and personalized remedy primarily based totally on Big Data analysis. The use of Big Data in healthcare poses sizable ethical and crook worries because of the non-public nature of the facts involved [15].

13.5 Ethical and Criminal Troubles for the Powerful Use of Big Data in Healthcare

Private autonomy is crucial further to the effects of the general public's attraction for openness, belief, and equality withinside the use of Big Data. Data heterogeneity, facts security, analytical methods in records processing, and a loss of the right records garage infrastructure have been diagnosed as key technological and infrastructural demanding situations that could jeopardize a big-data-pushed healthcare system [12, 15].

Skovgaard et al. [13, 15] looked at people's attitudes toward the reuse of fitness data in the European Union. People benefit from utilizing fitness data for purposes other than therapy, according to the research, as long as the data is of equivalent quality. The latest moniker revealed in Science by some of the world's most eminent scientists for the unfettered use of public genetic information has found fertile ground among the general population. There are many factors to consider when providing personal information, including the potential for profit, the importance of privacy, and the frequency with which the information is used [14].

The maximum modern EU Data Protection Regulation (GDPR) tries to strike the stability among affected person privateness and the cap potential to change affected person statistics for healthcare and studies reasons. On January 23, 2017, the Council of Europe's Consultative Committee on Information Security followed Guidelines for the Protection of People with ChalRegard to the Processing of Personal Data in a World of Big Data, the primary file on privateness and Big Data, which gives advocated measures for stopping any cap potential bad effects of Big Data use on human rights and freedoms [15].

13.6 Challenges with Big Data in Oncology

It is well known among researchers and doctors that advancements in malignant growth therapy are dependent on better knowledge of disease science. The advancement of so-called "omics" methods, like genomes, transcriptomics, proteomics, and epigenomics, to mention a few, are expanding our understanding of growth science while simultaneously creating a massive amount of data on the disease. The high throughput advances utilized to investigate the "omics" sciences have ushered in an era of massive data on the disease. Genomic organization, structures, alterations, rehash content, and development are all covered by genomics. Next-generation sequencing (NGS), also known as high-throughput sequencing, is a new technique that makes DNA and RNA sequencing significantly faster and less expensive than before. The NGS FASTQ papers are colossal. It is all about the "information storm." A transcriptome analysis examines all of the RNA data delivered by a genome. Because the transcriptome is controlled, it may be modified under specific conditions, allowing researchers to investigate features that are conveyed differently in distinct populations of cells or under different medications. In the field of cancer, these types of investigations are commonly used. High-throughput techniques, such as continual quantitative PCR (qPCR), microarrays, and RNA-Seq, are used to examine transcriptomic data (NGS sequencing). Proteomics is the study of a specific proteome, including data on protein design and capacity, protein articulation profiling, and their variations and changes, with the goal of better understanding cell functions.

Proteomics research has been used extensively in disease research to identify specific biomarkers linked to determination, visualization, and user expectation. Mass spectrometry is used in high-throughput proteomics research (MS). Epigenomics is the study of epigenetic changes in a cell's hereditary substance, known as the epigenome. Tumorigenesis is critical for epigenetic alterations that control cell activities. The progress of genome portrayal, including the identification of epigenetic modifications, has been made possible by NGS. They include vast information assortments of disease patients, as well as clinical datasets containing the findings, therapies, and patient results, as well as clinical preliminary data. The concept of "big information" is relatively recent, having emerged in the mid-2000s. The three

Vs were coined to describe large data from the start: volume, speed, and variety. A fourth V was added to complete the concept; it refers to the veracity or the unwavering quality of the aggregated data.

The disease research test is to look at all informational collections from growth science and clinical facts regarding patients more easily. One challenge is to develop an intelligible document that unifies the many forms of information into a cohesive stage that allows them to interact with one another. The amount of information now available is enormous, and it is growing at an alarming rate. It is critical to have the ability to bridle at that moment, posing difficult inquiries to separate fresh information from current data. New types of figure examination are becoming increasingly necessary. In actuality, many databases facilitate data sharing, allowing everyone with an interest to add to our collective knowledge of the subject by drawing from the accumulated information. This new era of massive-scale science is exemplified by the following databases. TP53, which codes for the p53 protein, is the most often altered quality in human tumors.

This trait has been studied extensively in oncology for over 30 years. A TP53 dataset containing only TP53 modifications associated with human malignant growths is discarded by the International Agency for Research on Cancer (IARC) in Lyon, France. There is also information about quality capacity, clinicopathologic aspects of malignancies, and patient data in addition to TP53 groupings. The database can be downloaded in its entirety and is available to researchers and doctors. This database is a repository for TP53 aggregated data. The most recent issue, R18, was released in April 2016 and contains information on over 29,000 physical alterations. It is a valuable tool for specialists to better understand the role of the p53 protein in oncology.

In 2006, the National Institutes of Health (NIH) faced a difficult task. The Data Portal for the Cancer Genome Atlas (TCGA) began as a three-year experimental project. Their goal was to compile a massive and comprehensive database of changes that occur in certain cancer types. The goal was to create a public organization that would collect the findings of many exploration organizations working on related projects. The pilot's success prompted the National Institutes of Health to expand the project, which lasted ten years. They had completed the tissue collection. The information is open to the public and unrestricted, allowing scientists all across the world to make and approve key revelations. Without a doubt, this initiative has made a significant contribution to the understanding of cancer genomes, such as lung adenocarcinoma or bosom disease.

The database includes almost 11,000 patient tests and 33 different forms of growth, including rare illnesses. The study of disease genetics is driving the development of personalized medicine. The use of a medicine that expressly targets a genetic transformation is required for the objective treatment. The TCGA provided professionals with comprehensive indexes of essential genetic changes in a variety of important illness types, allowing doctors to better analyze and treat patients. TCGA will close in 2016, although it has prepared for new cancer genomic initiatives based on its methodology. Another important NIH program is the Roadmap Epigenomics Mapping Consortium. The primary purpose is to provide a public repository of human epigenomic data. The collaboration examines DNA methylation, histone modifications, chromatin openness, and short RNA recordings in a range of normal cells and tissues using next-generation sequencing (NGS). The dataset provides a structure or reference for common epigenomes that might be used to measure tissues and organ frameworks that are frequently connected with human illness. In light of epigenetic components of growth, the Roadmap allows researchers to make relevant disclosures and physicians to further develop medication. This data bank has led to significant advances in disease research, such as the discovery of an epigenetic system that protects T cells against specific treatment in T-cell intensive lymphoblastic leukemia.

Patients and clinicians will be able to exchange information on treatments and results thanks to ASCO's big data initiative CancerLinQ. It starts and ends with the patient, creating a never-ending cycle of discovery. Patients' names are anonymized in databases to safeguard their reputations, which is a critical component. Patients and their primary care physicians may both contribute to and benefit from the information gathered in the data collection. Regardless matter whether it is obvious these days, they traveled a long distance to arrive. CancerLinQ was founded in 2010, and after two years, they created their first model to demonstrate the feasibility of the venture gathering only in breast cancer. More than 170,000 clinical records of individuals with bosom cancer had been compiled by 2013. Until now, CancerLinQ would share information from twelve locations, including malignant growth therapy centers. ASCO agreed to a deal with SAP, a product company, to develop a massive data programming platform that will enable.

CancerLinQ is based on SAP's HANA technology. The CancerLinQ dataset can help doctors for a better understanding of therapy and outcomes. A type of malignant growth with a specific hereditary alteration, for example, has been discovered to promote protection from a specific treatment. The common information can keep the doctor from prescribing a comparable drug in another quiet with a similar change. The internet is expected to be the first tool this year, according to ASCO. Consortiums, which bring together a variety of disease groups, are not the only ones that can generate massive datasets. Significant cancer centers, such as the MD Anderson Cancer Center, are also concerned by the massive data. The MD Anderson Cancer Center created the massive information and APOLLO (Adaptive Patient-Oriented Longitudinal Learning and Optimization) stages to cope with this issue. The massive data is an adaptable learning environment created by the Institutional Longitudinal Patient Disease Registry, which securely houses clinical and omics data as well as a set-up of massive data examination that can be barbecued and give end-customers sensible and imperative answers to their clinical or research questions.

The two phases are designed to enable experts to perform science-based medicine, whittle away at the MD Anderson standard's level of patient thought, and assist specialists from around the world to practice with the MD Anderson standard. All of the data used to create the massive data projects comes from patients, and it is now being translated into advantages for them. There are several approaches to dealing with the massive data problem in cancer. The need for cooperative effort can be found in a variety of stages. One of the most difficult problems is the strange speed with which data can be analyzed when there is so much. What is the best way to make sense of data? Having excellent collaboration methods is already a good start. There will be a lot of assumptions in the future. Real work is being done, and we will figure out how much we can put into one of those drives.

13.6.1 Data acquisition

Making incredible hazard-define fashions primarily based totally on the studies of a massive range of sufferers increases expenses and outcomes, however, there is a key stumbling block: a loss of vital well-worth data. The maximum principal hurdle to status up towards hazard-primarily based fashions, in particular in cancer, is the restriction of a few affected person data. Visits to the emergency room and admissions to hospitals are often no longer collected and consolidated into easily accessible huge scholarly files in claims-primarily based medical files. Estimating mortality rates is difficult because, for example, determining an exact date of death usually requires looking through several different data sources. Furthermore, patients' data is nearly never collected at home, where they spend the majority of their time. By uncovering plans that exist in patients at the earliest stages of sickness, novel techniques to routinely gather progressing data on cancer patients may be useful in reducing needless hospitalizations.

Genuine data sources may theoretically consider a never-ending range of EHR-based data. Perceptive computations susceptible to clinical basics, which continually deny relevant sections of the population, could be fast and critical and may be more material than perceptive computations based on authentic data. Real-world datasets, such as those from Flatiron Health and ASCO CancerLinQ, may be able to satisfy this need, but they have major limitations due to manual curation and barriers generated by the UI's changeability with the clinical success record.

13.6.2 Impending validation of algorithms

Several recent FDA certifications of farsighted estimations have been founded on improvements in measurable endpoints, such as location under the chairman brand name twist or positive judicious worth. Nonetheless, few investigations, particularly in cancer, have focused exclusively on the influence of meticulous calculations on significant clinical endpoints, such as persistence or collaborative estimations like the freedom to locate. Several recent FDA certifications of farsighted estimations have been founded on improvements in measurable endpoints, such as location under the becoming chairman brand name twist or positive judicious worth. Nonetheless, few investigations, particularly in cancer, have focused exclusively on the influence of meticulous calculations on significant clinical endpoints, such as persistence or collaborative estimations like the freedom to locate.

13.6.3 Representativeness and mitigating bias

The use of verified evaluation data to grow predictive logical models has the potential of reinforcing existing medical care biases. Calculations primarily based totally on summary medical facts or admission to scientific offerings would possibly effectively classify humans into special categories. Consider the case of a foresight calculation primarily based totally on gathering genetic facts for sure cancer. The informative indexes utilized to put together the computation screen a low variety of sufferers from awesome ethnic minorities. This would possibly cause misguided boom-order genetic variations in minority groups. A loss of facts from under-represented groups, on the opposite hand, would possibly prevent the potential to expect genetic adjustments in under-represented populations, jeopardizing the generalizability of the predictive model. When growing predictive fashions for chance delineation, it will be important to make sure representativeness of all populations of hobby in an education setting and to make sure assessment devices after the prescient equipment is created to make sure that under-addressed gatherings do now no longer revel in specific inclination in prescient output [17].

Large data producers have challenges in making them optimally useful in life sciences, even though they are being used in clinical practice with a huge promise ahead of them. The volume of data continues to grow rapidly as new technology emerges, and the complexities of today's sequencing (particularly, entire exome/genome sequencing) and radiomradionics are likely to prevent data translation. This is especially true when the increase in facts (pace and scope) is accompanied by an increase in facts heterogeneity (changeability), which includes medicines, outcomes, variations in plan focus, logical strategies, and pipeline understanding, all of which make it difficult to make business decisions completely based on the facts. A follow-up look follows the right management of facts, particularly when they come from a couple of sources. What facts are made available, how are they made available, and who owns the facts? Is the affected person nonetheless affected in terms of his or her datasets? Or, if it truly is the case, who is in rate: the expert, the treating doctor, the fact producer, or the individual that tries to make the most of the facts (along with a laptop scientist or scientific bioinformatician)?

13.6.4 Stores and datasets for documenting and sharing biomolecular patient information

Although the reason for (bio) clinical studies is to acquire as a lot of facts as viable from as many sufferers and clinical clinics as feasible, protection considerations, including the General Data Protection Regulation (GDPR), normally restrict get admission. One key take look is to hold the affected person's identifiable facts, including genetic sequences so that they can be used for different motives at the same time as nonetheless protecting the privateness of those who furnished the facts (www.phgfoundation.org). While openness can be favored in a few disciplines, including software program engineering, protection issues ought to trump a preference for overall transparency in this situation. The European Genome Archive (EGA) is a facts series created in particular to keep uncooked sequencing facts. There is a stable activity for facts to get admission to the board of trustees (DAC) for every examination with facts saved in EGA, that is administered via way of means of the exploration power that received the facts and might pick out to provide admission to the facts upon call for via way of means of different scientists. When an expert wishes to appear over dealt with biomolecular facts without being capable of tuning returned character markers, an auxiliary takes a look at what is to be had. There are a pair of alternatives to be had right here that summarize the facts without showing particular marks, or unique shops address this thru fine-graining to get admission to control. Finally, a follow-up takes a look at what will securely join the unique facts belongings, permitting customers to hint at the particular computational remedy achieved at the facts; some drives have already made the primary steps closer to this aim.

Massive data has tremendous promise in life sciences and head and neck cancer. It has the potential to change the way we share clinical and research data. The (near) continuous/spilling of information, along with the massive volume of data, will make trading informational indexes as we know it impossible. Rather than combining all types of datasets into a single large database, it is feasible that large data clients may develop more natural, decentralized virtual organizations, such as those proposed by the Dutch Techcentre for Life Sciences in the individual wellness train (DTL). Datasets are associated as hubs within these organizations and are exposed to customers under preset criteria. Increased availability and (hence) complexity will necessitate better methods for deciphering information, as well as interpreting and relating this information to the individual patient. To be meant the particular "little information" environment of the personal care subordinate patient, this last demands huge information computed from this load of informational collections. Clinical expertise is required for this final step, which includes critically incorporating natural and passionate opinions. For a long time, bedside habits will be far apart from massive amounts of data or AI [3].

13.7 Big Data Challenges in Healthcare

Big Data evaluation is proving to be one of the maximum difficult and demanding situations in current reminiscence for the scientific offerings industry. Suppliers who have simply mastered inputting statistics into their electronic health records (EHR) are being approached to extract huge chunks of statistics from them – and observe what they have found out to complicated tasks that immediately affect their payback rates. The rewards for scientific offerings enterprises that efficaciously combine information-pushed reviews into their scientific and useful cycles can be enormous. Among the numerous benefits of changing information resources into information, experiences are better patients, reduced healthcare expenses, increased permeability into execution, and improved employee and buyer satisfaction rates. The way to a significant medical services examination is a rough one, nonetheless, loaded up with difficulties and issues to address.

By its actual nature, large information is mind-boggling and inconvenient, requiring supplier associations to investigate their ways to deal with gathering, putting away, examining, and introducing their information to staff individuals, colleagues, and patients.

What are a number of the maximum standard demanding situations that corporations enjoy whilst launching a large-scale facts evaluation program, and the way can they conquer those boundaries to obtain their facts-pushed scientific and monetary goals? The big data challenges are given in Table 13.1.

13.7.1 Capture

All information comes from someplace, but it does not necessarily come from a place with strong information management practices, which is problematic for certain medical care providers. Obtaining data that is flawless, comprehensive, true, and well-designed for use in a variety of frameworks is a never-ending struggle for organizations, many of which are losing.

EHR records matched patient-particular records in 23.5% of the information in a single ongoing observation at an ophthalmology center. Patients who indicated that they skilled as a minimum three eye fitness signs had their EHR records contradicted them. Poor EHR usability, convoluted painting procedures, and a muddled knowledge of why big records are so vital to seize nicely may all make contributions to records' exceptional issues.

Suppliers can start to enhance their statistics seize schedules via way of means of focusing on the maximum essential statistics kinds for their projects, enlisting the information of fitness records to board specialists in statistics control and honesty, and growing scientific documentation development programs that mentor clinicians on the way to make certain that statistics is beneficial for downstream examination.

13.7.2 Cleaning

Medical care suppliers are personally acquainted with the significance of tidiness in the center and the working room, however, may not be very as mindful of the fact that scrub their information, as well. Filthy information can rapidly wreck a major information examination project, particularly when

Table 13.1 Challenges of big data in healthcare.

	Category	Description
	Capturing	Catching information that is spotless, finished, exact, and designed effectively for use in various frameworks is a continuous fight for associations, a significant number of which are not on the triumphant side of the contention.
	Cleaning	Medical care suppliers are personally acquainted with the significance of tidiness in the center and the working room, however, may not be very as mindful of the fact that scrub their information, as well.
	Capacity	Although current clinicians hardly ever remember in which their facts are stored, it is far a simple cost, security, and execution subject for the IT department. As the extent of scientific facts grows exponentially, a few companies are nevertheless unprepared to cope with the charges and effects of on-premise server farms.
	Security	Big data presents serious privacy concerns about the personal information obtained from users, yet this is the price the patient must pay in exchange for considerably greater benefits.
	Stewardship	Information on medical services, especially on the clinical side, has a lengthy period of realistic usage. Suppliers may seek to utilize de-recognized datasets for research initiatives, making ongoing stewardship and curation a major challenge. Information can also be reused or reassessed for a variety of reasons, such as quality estimation and execution benchmarking.
Challenges	**Questioning**	Strong metadata and wonderful stewardship practices additionally make it less difficult for organizations to question their facts and locate the solutions they want. The potential to question facts is crucial for deep evaluation, however, hospital therapy establishments have to frequently conquer some of the demanding situations earlier than doing sizeable evaluation in their big facts resources.
	Detailing	Suppliers should create a report that is clear, brief, and open to the interest group after they have confirmed the question cycle.
	Perception	At the place of care, a spotless and drawing-in information representation can make it a lot simpler for a clinician to ingest data and use it suitably.
	Refreshing	Information about medical services is not static, and most components will need to be updated in stages to be current and relevant. These updates may occur regularly for some datasets, analogous to patient essential signs. Other information, such as a person's address or marital status, may only change a few times over their lifetime.
	Sharing	Scarcely any suppliers work in a vacuum, and fewer patients get all of their consideration in a solitary area. This implies that imparting information to outer accomplices is fundamental, particularly as the business moves toward populace well-being on the board and worth-based consideration.

uniting unique information sources that might record clinical or functional components in somewhat various arrangements. Information cleaning – otherwise called purging or scouring – guarantees that datasets are exact, right, steady, significant, and not undermined at all.

While the majority of data cleaning is still done by hand, certain IT suppliers do provide automated scouring machines that use logic rules to evaluate, contrast, and rectify large datasets. As AI approaches to progress, these tools will likely get more polished and accurate, lowering the time and money necessary to assure high levels of accuracy and trustworthiness in medical data warehouses.

13.7.3 Capacity

Although modern clinicians hardly ever consider in which their statistics are stored, the IT branch has essential cost, safety, and execution issues. Some carriers are nevertheless unprepared to cope with the fees and ramifications of on-premise server farms because the quantity of clinical statistics rises dramatically. While many agencies select on-premise statistics garages as it permits them to keep manipulate over safety, get right of entry to, and uptime, on-premise server surroundings can be costly, complicated to administer, and liable to growing facts silos throughout many offices.

The distributed garage is turning into an extremely appealing opportunity as fees lower and reliability increases. According to a 2016 report, approximately 90% of clinical provider organizations use cloud-primarily based fitness IT foundations, which include ability and applications. Although the cloud gives fast catastrophe recovery, fewer front-quit expenses, and less complicated growth, agencies have to be careful whilst selecting companions that admire the relevance of HIPAA and different clinical-associated consistency and safety issues.

Many organizations turn out to be the usage of a hybrid method to manipulate their statistics garage programs, which can be the maximum adaptive and useful technique for providers with converting facts to get the right of entry and ability requirements. However, at the same time as growing a 1/2 of-and-1/2 of the foundation, carriers have to take care to make sure that diverse frameworks can also additionally proportion and proportion facts with exceptional components of the corporation as needed.

13.7.4 Security

Information security is a significant responsibility for medical care organizations, particularly in the aftermath of a sequence of high-profile data breaches, hacking, and ransomware events. Medical data is subject to an almost unlimited amount of weaknesses, from phishing attempts to ransomware to workstations left in a cab. For organizations holding protected health information (PHI), the HIPAA Security Rule provides a broad range of technological precautions, including transmission security, confirmation requirements, and access, integrity, and review powers.

By and by, these protections convert into the presence of mind security methods, for example, utilizing around-date hostile-to-infection programming, setting up firewalls, encoding touchy information, and utilizing multifaceted confirmation. In any case, even the most firmly got server farm can be brought somewhere around the uncertainty of human staff individuals, who will in general focus on accommodation over extensive programming refreshes and muddled requirements on their admittance to information or programming. Medical services associations should oftentimes help their staff individuals to remember the basic idea of information security conventions and reliably survey who approaches high-esteem information resources to keep malevolent gatherings from causing harm.

13.7.5 Stewardship

Information on medical services, especially on the clinical side, has a lengthy period of realistic usage. Suppliers may seek to utilize de-recognized datasets for research initiatives, making ongoing stewardship and curation a major challenge. Information can also be reused or reassessed for a variety of reasons, such as quality estimation and execution benchmarking.

For scientists and information examiners, knowing when the information was created, by whom, and for what purpose – as well as who has lately used the information, why, how, and when – is critical.

Creating total, exact, and cutting-edge metadata is a critical part of an effective information administration plan. Metadata permits experts to precisely duplicate past inquiries, which is indispensable for logical investigations and exact benchmarking, and forestalls the formation of "information dumpsters," or disconnected datasets that are restricted in their value.

Medical services organizations should often repeat to their employees of the fundamental concept of information safety procedures and regularly monitor who approaches high-value information resources to prevent malicious groups from inflicting harm.

13.7.6 Questioning

Strong metadata and exceptional stewardship practices additionally make it less complicated for corporations to question their records and locate the solutions they want. The capability to question records is important for deep evaluation, however, hospital treatment firms have to frequently triumph over some of the demanding situations earlier than doing sizable evaluation in their massive records resources.

Right off the bat, they ought to overcome statistics siloes and interoperability troubles that prevent inquiry gadgets from attending to the association's entire keeping of records. On the occasion that diverse components of a dataset are held in several walled-off frameworks or diverse arrangements, it can now no longer be possible to create a complete illustration of an association's fame or a unique patient's well-being.

Furthermore, whether or not records are housed in an ordinary stockroom or now no longer, there can be a loss of requirements and quality. Without using scientific coding structures like ICD-10, SMOMED-CT, or LOINC, which lessen innovative conceptions to a not unusual place cosmology, it can be tough to guarantee that a question recognizes and gives the right records to the consumer.

Many organizations utilize structured query language (SQL) to dive into great datasets and social records bases, however, it is simplest useful if a consumer can agree with the records' integrity, completeness, and normalization.

13.7.7 Detailing

Suppliers should create a report that is clear, brief, and open to the interest group after they have confirmed the question cycle. The precision and dependability of the data eventually impact the report's exactness and unchanging quality. Helpless data at the start of the encounter will result in questionable reports at the conclusion, which can be troublesome for physicians attempting to treat patients with the data. Suppliers should likewise comprehend the distinction between "investigation" and "detailing."

Reporting is regularly essential for examination – the information should be separated before it very well may be analyzed – yet revealing can likewise remain all alone as a final result.

While a few reviews are designed to spotlight a positive pattern, reap an authentic resolution, or convince the reader to take a particular action, others ought to be supplied in a manner that lets the reader make their very own inferences approximately what the entire variety of facts means. Organizations ought to be extraordinarily clean approximately how they need to utilize their reviews to make sure that facts set heads can deliver the facts they desire.

Because administrative and quality assessment programs frequently seek huge amounts of data to manage value measurements and payment models, the majority of the disclosure in the medical care industry occurs outside of the industry. Certified vaults, announcing devices incorporated into electronic health records, and web-based interfaces supplied by CMS and other parties are among the solutions available to suppliers for meeting these many criteria.

13.7.8 Perception

An easy and drawing-in records illustration of the factor of care could make it plenty less difficult for a doctor to digest facts and practice them appropriately. Shading coding is a famous record belief approach that commonly outcomes in a fast response — for example, red, yellow, and inexperienced are broadly understood to symbolize stop, alert, and go.

Associations should likewise reflect great information show rehearses, for example, outlines that utilization of legitimate extents to delineate differentiating figures, and right marking of data to decrease expected disarray. Tangled flowcharts, squeezed or covering text, and inferior-quality illustrations can baffle and irritate beneficiaries, driving them to overlook or misjudge information. Hotness maps, bar graphs, pie graphs, scatterplots, and histograms are all common examples of information representations, each with its own set of purposes for separating concepts and data.

13.7.9 Refreshing

Information about medical services is not static, and most components will need to be updated in stages to be current and relevant. These updates may occur regularly for some datasets, analogous to patient essential signs. Other information, such as a person's address or marital status, may only change a few times over their lifetime.

Understanding the instability of huge information, or how regularly and how much it changes can be a test for associations that do not reliably screen their information resources. Suppliers should know which datasets require human refresh and which may be automated, how to finish this cycle without disrupting end users, and how to guarantee that updates are made without compromising the dataset's quality or dependability. Associations ought to likewise guarantee that they are not making pointless copy records while endeavoring an update to a solitary component, which might make it hard for clinicians to get to essential data for patient dynamics.

13.7.10 Sharing

Scarcely any suppliers work in a vacuum, and fewer patients get all of their consideration in a solitary area. This implies that imparting information to outer accomplices is fundamental, particularly as the business moves towards populace well-being on the board and worth-based consideration. Interoperability of information is an ongoing challenge for organizations of all sizes and stages of information development. Key disparities in how digital fitness facts are created and carried out can substantially obstruct the potential to alternate records among enterprises, leaving physicians without the records they want to make vital choices, circle around patients, and sell techniques to enhance ordinary performance.

Currently, the company is striving to collaborate on data alternate past specialized and authority boundaries. Emerging devices and procedures like FHIR and public APIs, in addition to collaborative enterprises like CommonWell and Carequality, are making it less complicated for designers to proportion data efficaciously and safely. However, the public's recognition of those techniques has but to attain crucial mass, leaving many enterprises blocked off from the blessings inherent withinside the non-stop sharing of affected person records.

To foster a major information trade biological system that interfaces all individuals from the consideration continuum with reliable, ideal, and significant data, suppliers should beat each challenge on this rundown. Doing as such will set aside time, responsibility, subsidizing, and correspondence – yet achievement will facilitate the weight of that load of concerns [18, 19].

13.8 Big Data Applications in Healthcare

Table 13.2 lists applications that demonstrate how a data-driven approach may be used to enhance processes, improve patient care, and, ultimately, save lives [20].

Table 13.2 Big data applications in healthcare.

S. no.	Big data applications	Description
	Patient predictions for improved staffing	We will examine one of the maximum regular troubles confronted by any shift manager: what number of humans must I have on obligation at any person's time? If you recruit too many individuals, you dangerous incurring extra labor expenditures. Customer providers are degraded when there are too few employees, which can be horrible for sufferers in that commercial enterprise.
	Electronic health records (EHRs)	It is the maximum famous utility of huge records in medicine. Each affected person has a computerized document that includes, among different things, demographics, scientific history, allergies, and laboratory take look at effects. Records are exchanged over steady facts networks, and they are reachable to each public and personal quarter provider. Every document is made of an unmarried editable file, which permits clinicians to make modifications over the years without managing paper or the chance of records duplication.
	Real-time alerting	Real-time notifications are a function of many different records analytics structures in healthcare. In hospitals, clinical decision support (CDS) software programs analyze scientific records in actual time, helping clinicians in making prescriptive judgments.
	Enhancing patient engagement	Many clients – and consequently potential sufferers – are already interested in clever devices that degree their walks, coronary heart rate, napping habits, and different records consistently.
	Using health data for informed strategic planning	Due to elevated insights into humans' motives, the usage of huge records in healthcare allows for strategic planning. Care managers can examine the effects of check-ups amongst folks from numerous demographic corporations to find out what elements deter humans from looking for scientific attention.

Table 13.2 Continued

Big data might just cure cancer	Another interesting instance of huge records in healthcare is the Cancer Moonshot program. Before the cessation of his 2nd period, President Obama installed this approach to accomplish 10 years' well worth of development in the direction of curing most cancers in 1/2 of the time.
Predictive analytics in healthcare	Predictive analytics has been dubbed one of the maximum crucial commercial enterprise intelligence improvements withinside the remaining years, however, its capacity makes use of making bigger a long way past the economic global and into the future. Optum Labs, a US-primarily based studies organization, has accumulated EHRs from extra than 30 million sufferers to set up a database for predictive analytics gear that could decorate healthcare delivery.
Reduce fraud and enhance security	A records breach has been said in ninety-three percent of healthcare organizations, in step with one study. The motive is simple: non-public records are fairly treasured and worthwhile in the illicit market. Any violation might have long way-achieving consequences. With this in mind, many corporations are turning to analytics to assist discover safety worries through detecting modifications in community site visitors or different signs of a cyber-attack.
Telemedicine	Telemedicine has been around for extra than forty years, however with the appearance of online video conferencing, cellphones, wi-fi devices, and wearables, it has simply currently expanded. The period refers to the usage of an era to provide healing offerings over lengthy distances.
Learning and development	To enlarge on our remaining point, your employees' talents, confidence, and competencies at a medical institution or clinic are probably the distinction between existence and death. Doctors and surgeons are, without a doubt, specialists in their disciplines. Most scientific institutions, on the alternative hand, rent various institutions of employees, starting from porters and administrative clerks to cardiologists and mind surgeons.

13.9 Conclusion

The volume, velocity, variety, and authenticity of numerous, often complicated datasets determine the usefulness of big data gathering. Integration of various sources is critical and will benefit biological research, patient treatment, and quality-of-care monitoring. All of the data used in big-data projects comes from patients and has previously proven to be beneficial to them. There are several approaches to solving the big data problem in cancer. We recognize the necessity for collaboration in the numerous existing platforms. When there is so much data, one of the primary issues is how quickly it can be evaluated. Even if there are numerous efforts in place, there is still room for improvement. It is a terrific start to have great ways to collaborate. The future is full of possibilities. There are serious efforts underway, and we will consider how much we can give to one of these programs.

Acknowledgment

The authors are grateful to the administration of Oriental University, Indore, for their assistance.

Conflict of Interest

There is no potential for a conflict of interest.

Funding

There is no funding issued.

References

[1] Janssen, D. F. (2021). Oncology: etymology of the term. *Med. Oncol. 38*, 1-3. DOI: 10.1007/s12032-021-01471-4.
[2] Roman-Belmonte, J. M., De la Corte-Rodriguez, H., & Rodriguez -Merchan, E. C. (2018). How blockchain technology can change medicine. *Postgrad. Med. 130*, 420-427. DOI: 10.1080/00325481.2018.1472996.
[3] Willems, S. M., Abeln, S., Feenstra, K. A., de Bree, R., van der Poel, E. F., de Jong, R. J. B., ... & van den Brekel, M. W. (2019). The

potential use of big data in oncology. *Oral Oncol. 98*, 8-12. DOI: 10.1016/j.oraloncology.2019.09.003.

[4] Agrawal, A., & Choudhary, A. Health services data: big data analytics for deriving predictive healthcare insights. *Health Serv. Eval.* 2019. DIO: 10.1007/978-1-4899-7673-4_2-1.

[5] Shaikh, A. R., Butte, A. J., Schully, S. D., Dalton, W. S., Khoury, M. J., & Hesse, B. W. (2014). Collaborative biomedicine in the age of big data: the case of cancer. *J. Med. Internet Res. 16*, e2496. DOI: 10.2196/jmir.2496.

[6] Ristevski, B., & Chen, M. (2018). Big data analytics in medicine and healthcare. *J. of Integr. Bioinform. 15*. DOI: 10.1515/jib-2017-0030.

[7] Panahiazar, M., Taslimitehrani, V., Jadhav, A., & Pathak, J. (2014, October). Empowering personalized medicine with big data and semantic web technology: promises, challenges, and use cases. In *2014 IEEE Int. Conf. Big Data (Big Data)* (pp. 790-795). IEEE. DOI: 10.1109/BigData.2014.7004307.

[8] World Health Organization. (2018). 13th General program of work (GPW13) WHO impact framework. *Geneva: World Health Organization.*

[9] World Health Organization. (2021). *Compendium of the Roadmap for Health and Well-being in the Western Balkans (2021-2025): European Programme of Work (2020–2025)-United Action for Better Health* (No. WHO/EURO: 2021-3436-43195-60509). World Health Organization. Regional Office for Europe.

[10] Jameson, J. L., & Longo, D. L. (2015). Precision medicine—personalized, problematic, and promising. *Obstet. Gynecol. Surv. 70*, 612-614. DOI: 10.1097/01.ogx.0000472121.21647.38.

[11] Joos, S., Nettelbeck, D. M., Reil-Held, A., Engelmann, K., Moosmann, A., Eggert, A., ... & Baumann, M. (2019). German Cancer Consortium (DKTK)–A national consortium for translational cancer research. *Mol. Oncol. 13*, 535-542. DOI: 10.1002/1878-0261.12430.

[12] Eggermont, A. M., Apolone, G., Baumann, M., Caldas, C., Celis, J. E., de Lorenzo, F., ... & Calvo, F. (2019). Cancer Core Europe: a translational research infrastructure for a European mission on cancer. *Mol. Oncol. 13*, 521-527. DOI: 10.1002/1878-0261.12447.

[13] Ienca, M., Ferretti, A., Hurst, S., Puhan, M., Lovis, C., & Vayena, E. (2018). Considerations for ethics review of big data health research: A scoping review. *PloS One. 13*, e0204937. DOI: 10.1371/journal.pone.0204937.

[14] Skovgaard, L. L., Wadmann, S., & Hoeyer, K. (2019). A review of attitudes towards the reuse of health data among people in the European Union: The primacy of purpose and the common good. *Health policy*, *123*, 564-571. DOI: 10.1016/j.healthpol.2019.03.012.

[15] Pastorino, R., De Vito, C., Migliara, G., Glocker, K., Binenbaum, I., Ricciardi, W., & Boccia, S. (2019). Benefits and challenges of Big Data in healthcare: an overview of the European initiatives. *Eur. J. Public Health*, *29*(Supplement_3), 23-27. DOI: 10.1093/eurpub/ckz168.

[16] Barbosa, C. D. (2016). Challenges with big data in oncology. *J. Orthop. Oncol. 2*, 2. DOI: 10.4172/joo.1000112.

[17] Parikh, R. B., Gdowski, A., Patt, D. A., Hertler, A., Mermel, C., & Bekelman, J. E. (2019). Using big data and predictive analytics to determine patient risk in oncology. *Am. Soc. Clin. Oncol. Educ. Book*, *39*, e53-e58. DOI: 10.1200/EDBK_238891.

[18] Bresnick, J. (10). High-Value Use Cases for Predictive Analytics in Healthcare. HealthITAnalytics.

[19] Sin, K., & Muthu, L. (2015). Application of Big Data in Education Data Mining and Learning Analytics–A Literature Review. *ICTACT journal on soft computing*, *5*.

[20] Durcevic, S. (2020). 18 Examples of Big Data Analytics in Healthcare that can save people. *Business Intelligence*.

Author Biography

Deepika Bairagee

Ms. Deepika Bairagee, B. Pharm, M. Pharm (Quality Assurance), is an Assistant Professor at the Oriental College of Pharmacy and Research, Oriental University in Indore (India). She has 5 years of teaching experience and 2 years of research experience. She has spoken at more than 20 national and international conferences and seminars, presenting over 20 research articles. She has over 20 publications in international and national journals. She is the author of over 18 books. She has over 50 abstracts that have been published at national and international conferences. Young researchers, young achievers, and excellent researchers were among the honors bestowed upon her. Proteomics and Metabolomics are two areas of research that she is currently interested in.

Index

About the Editors

Neeraj Kumar Fuloria is presently working as senior associate professor at the Faculty of Pharmacy, AIMST University, Malaysia. He has extensive experience, with 20 years in academics, research and industry. He completed his B.Pharm in 1998 (Gulbarga University, India), M.Pharm in 2003 (Rajiv Gandhi University Health Sciences, India), MBA in 2004 (Madurai Kamaraj University, India), and Ph.D. in Pharmacy in 2010 (Gautam Buddha Technical University, India). So far, he has supervised 6 Postgraduate and 24 undergraduate research scholars, and is currently supervising 6 Ph.D. scholars. He has published 96 research and review articles, 4 books, 3 MOOC, and 2 patents (Australia). For his research work Dr. Fuloria has received 8 national and international grants. He is also a member of various professional bodies like the Indian Society of Analytical Scientists and the NMR Society of India. Apart from his work in academics and research Dr. Fuloria has received various awards such as Inspirational Scientist Award 2020 (at Trichy, India), Young Achiever Award 2019 (at 10th Indo-Malaysian Conference, India), and an Appreciation Award 2017 (at the Teachers Award Ceremony of Kedah State Malaysia, at AIMST University, Malaysia).

Dr. Rishabha Malviya completed B. Pharmacy from Uttar Pradesh Technical University and M. Pharmacy (Pharmaceutics) from Gautam Buddha Technical University, Lucknow Uttar Pradesh. His PhD (Pharmacy) work was in the area of Novel formulation development techniques. He has 12 years of research experience and presently working as Associate Professor in the Department of Pharmacy, School of Medical and Allied Sciences, Galgotias University since past 8 years. His area of interest includes formulation optimization, nanoformulation, targeted drug delivery, localized drug delivery and characterization of natural polymers as pharmaceutical excipients. He has authored more than 150 research/review papers for national/international journals of repute. He has 58 patents (19 grants, 38 published, 1 filed) and publications in reputed National and International journals with total of 191 cumulative impact factor. He has also received an Outstanding Reviewer award from Elsevier. He has authored/edited/editing 42 books (Wiley, CRC Press/Taylor and Francis, IOP publishing. River Publisher Denmark, Springer Nature, Apple Academic Press/Taylor and Francis, Walter de Gruyter, and OMICS publication) and authored 26 book chapters. His name has included in word's top 2% scientist list for the year 2020 and 2021 by Elsevier BV and Stanford University. He is Reviewer/Editor/Editorial board member of more than 50 national and international journals of repute. He has invited as author for "Atlas of Science" and pharma magazine dealing with industry (B2B) "Ingredient south Asia Magazines".

Swati Verma completed her B.Pharm from KIET (AKTU), Ghaziabad and M. Pharm (pharmaceutical chemistry) from Banasthali Vidyapith, Tonk, Rajasthan. She has joined BBDNIIT as assistant professor and is currently working in Galgotias University, Greater Noida. Her area of interest is computer aided drug design (CADD), peptide chemistry, analytical chemistry, medicinal chemistry, artificial intelligence, neurodegeneration, and gene therapy. She has attended and organized more than 15 national and international seminars/conferences/workshops.

Professor Balamurugan Balusamy has served up to the position of associate professor in his stint of 14 years of experience with VIT University, Vellore. He completed his Bachelors, Masters and Ph.D. degrees at top premier institutions in India. His passion is teaching and he adapts different design thinking principles while delivering his lectures. He has published 30+ books on various technologies and visited 15 plus countries for his technical course. He has published over 150 quality journal, conference and book chapters. He serves on the advisory committee for several startups and forums and does consultancy work for industry on industrial IOT. He has given over 175 talks at various events and symposiums. He is currently working as a professor at Galgotias University and teaches students and does research on Blockchain and IOT.

For Product Safety Concerns and Information please contact our EU representative GPSR@taylorandfrancis.com Taylor & Francis Verlag GmbH, Kaufingerstraße 24, 80331 München, Germany